The Guide to Kansas Birds

AND BIRDING HOTSPOTS

- WATERFOWL
- OTHER SWIMMING BIRDS
- WADING BIRDS
- SHOREBIRDS
- GULLS & TERNS
- UPLAND GAME BIRDS
- DIURNAL RAPTORS
- OWLS & NIGHTJARS
- DOVES & CUCKOOS
- WOODPECKERS & KINGFISHERS
- AERIALISTS
- INSECT CATCHERS
- WARBLERS
- SMALL ACTIVE SONGBIRDS
- THRUSHES & OTHERS
- GROUND DWELLERS & SPARROWS
- BLACKBIRDS, CORVIDS, & ALLIES
- COLORFUL SONGBIRDS

The Guide to Kansas Birds and Birding Hotspots

The Guide to
Kansas Birds
and
Birding Hotspots

Second Edition, Revised and Expanded

PETE JANZEN AND BOB GRESS

Foreword by KENN KAUFMAN

 University Press of Kansas

  Publication made possible, in part, by gifts from Audubon of Kansas
and the Kansas Ornithological Society.

Published by the University Press of Kansas (Lawrence, Kansas 66045), which was organized by the Kansas
Board of Regents and is operated and funded by Emporia State University, Fort Hays State University, Kansas
State University, Pittsburg State University, the University of Kansas, and Wichita State University.

Library of Congress Cataloging-in-Publication Data

Names: Janzen, Pete, author. | Gress, Bob, author.
Title: The guide to Kansas birds and birding hotspots / Pete Janzen and Bob Gress.
Description: Second edition, revised and expanded. | [Lawrence, Kansas] : University Press of Kansas, [2025] |
   Includes index.
Identifiers: LCCN 2024030684 (print) | LCCN 2024030685 (ebook) | ISBN 9780700638598 (paperback) |
   ISBN 9780700638604 (ebook)
Subjects: LCSH: Birds—Kansas—Identification. | Bird watching—Kansas—Guidebooks. | Kansas—
   Guidebooks. | BISAC: NATURE / Birdwatching Guides | NATURE / Animals / Birds
Classification: LCC QL684.K2 G74 2025 (print) | LCC QL684.K2 (ebook) |
   DDC 598.09781—dc23/eng/20241227
LC record available at https://lccn.loc.gov/2024030684.
LC ebook record available at https://lccn.loc.gov/2024030685.

British Library Cataloguing-in-Publication Data is available.

Book design by Karl Janssen

Printed in China

10 9 8 7 6 5 4 3 2 1

The paper used in this publication is acid free and meets the minimum requirements of the American National
Standard for Permanence of Paper for Printed Library Materials Z39.48–1992.

Contents

SMALL ACTIVE SONGBIRDS

THRUSHES & OTHERS

GROUND DWELLERS & SPARROWS

BLACKBIRDS, CORVIDS, & ALLIES

COLORFUL SONGBIRDS

KANSAS BIRDING HOTSPOTS

Foreword to the Second Edition

Kansas appeared to me first as a golden state. It was the end of January, just past my ninth birthday, and my family was moving here from northern Indiana. The world had looked gray for most of the drive, but as we angled south toward Wichita, the sun came out, casting warm evening light over the tawny winter grass of the Flint Hills. The light was as bright as the sunflower on the state flag, bright as the yellow on the meadowlarks along the highway. For a boy already obsessed with birds, this seemed an omen of golden possibilities ahead.

That shining promise was borne out as I explored my new home. I was surrounded by variety and novelty, dazzled by new discoveries in every season. From gorgeous birds like Scissor-tailed Flycatchers and Painted Buntings to elegant migrants like Hudsonian Godwits and Buff-breasted Sandpipers, from smartly patterned Harris's Sparrows to big, ponderous American White Pelicans, I found abundant rewards every time I took to the field.

Within a couple of years, it turned out that another kid at my school, Jeff Cox, was just as avid about birding as I was. The two of us together had the potential to be obnoxious, I know, but the patient adults of the Wichita Audubon Society and Kansas Ornithological Society put up with us anyway. They took us along on field trips, included us in bird counts, and always encouraged us to keep learning.

And there was so much to learn. Kansas is a phenomenally good state for birding. Its central location means that birds typical of both East and West live within the state's borders. As a boy I was delighted to find Eastern and Western Meadowlarks in the same fields, Eastern and Western Kingbirds along the same fence lines. If we travel to the corners of the state, we find even more variety. Northeast of Topeka and Lawrence, we can see essentially the same mix of birds that we might find in the forests of Pennsylvania, while southwest of Dodge City, the birdlife takes on a distinct flavor of Arizona. It's remarkable to see these utterly different bird communities thriving only hours apart.

Migration season ramps up the excitement because Kansas lies at the heart of a great bird migration corridor. Several species spend the winter on the grasslands of southernmost South America and then migrate thousands of miles north to the prairies or tundra of North America, and these long-distance travelers come right across Kansas on their journey. Birds like Swainson's Hawks and Upland Sandpipers fly north from the Argentinian pampas to raise their young here in Kansas, while American Golden-Plovers, Baird's Sandpipers, and numerous others merely pause here before continuing to nesting grounds in the Arctic. The hundreds of thousands of Franklin's Gulls that circle above Kansas prairies every spring have wintered on the coast of Peru, while the graceful Mississippi Kites that depart from our woodlots in fall are headed to the Amazon basin. This is a bird superhighway. At some point, as a kid, I watched the movie *The Wizard of Oz*, and I remember thinking that Dorothy probably wanted to get back to Kansas because the birding was better there.

I lived in Kansas for only about a decade, but I still go back for frequent visits, so I was pleased to see the publication of *The Guide to Kansas Birds and Birding Hotspots* in 2008. That book proved to be incredibly useful and deservedly popular, so now I'm even more excited to welcome this greatly expanded new edition.

Every bird book needs to be updated occasionally, simply because the status of species changes over time. For example, back in 2008, the Neotropic Cormorant and Black-bellied Whistling-Duck were only scarce visitors to Kansas. Their numbers have increased, and now they merit full treatment. But the updates go much further: this edition covers more bird species, brings all the information on each bird up-to-date, includes far more photos, and describes almost four times as many hotspots as the previous version. This strikes me as a whole new book, and it's even better and more essential than before.

Which is just what I expected from these authors. Pete Janzen has led birding trips in Kansas and throughout the United States and has written for regional and national publications. He has made a particular study of Kansas birds. His birding efforts in all 105 counties have given him an exceptional knowledge of the state's birdlife, greatly enhancing the Hotspots section of this guide. Bob Gress, the founder and former director of the Great Plains Nature Center, is also a birder and wildlife photographer. His photos of birds, mammals, and other nature subjects have been published in a wide variety of books and magazines. These two men have national reputations and world-class skills, and you could not ask for better guides to the birdlife of this golden state of Kansas.

Kenn Kaufman

Introduction

The pursuit of birding is growing in popularity as a hobby each year. According to recent surveys, some seventy million people in the United States observe birds at least a few times each year. Kansas offers wonderful birding opportunities. An amazing 482 species of birds have been seen within our borders! Among the fifty states, Kansas ranks in the top fifteen for species of documented birds that have been observed at least once. For an agricultural state in the Great Plains, this is an impressive ranking in comparison to the other high-ranking states, all of which benefit from some combination of ocean coasts, mountain ranges, southwestern deserts, and mild southern climates.

It is difficult to classify birdwatchers, or "birders," as they are usually called. In Kansas, there are thousands of avid birders and tens of thousands more who feed birds in their yards or occasionally venture a short distance away from home to view birds. Some birders are highly competitive and work to build an impressive list of birds they have observed. Others are serious volunteer researchers who are primarily interested in contributing to scientific data gathering and research. Many become immersed in the pursuit of their hobby, traveling widely and often throughout Kansas and beyond each year. For others, one or two birding trips per year are enough. Some are mainly interested in birding in their local area or county, and others confine their interests to the birds they observe in nearby parks or their own yards. Regardless of their individual interests and perspectives, all birders share a love of these special winged creatures. Looking for birds as well as other wildlife, wildflowers, and natural landscapes is a great form of stress relief!

With this growing interest in Kansas birds, we are pleased to present a revised and expanded second edition of *The Guide to Kansas Birds and Birding Hotspots*. Like our original book, published in 2008, this volume is for beginning and intermediate birders. It does not include every species ever seen in Kansas but instead features those species most likely to be encountered as your interest in birding expands its horizons. To assist your explorations, we have greatly expanded the chapter on hotspots to include more places to find birds. We've also increased the number of species in this book and replaced and increased the number of maps, photographs, and other illustrations. We hope this book helps you find, identify, and appreciate our wild avian neighbors. Whatever aspect of birding you find enjoyable and rewarding, we hope you will discover and love the subtly alluring aspects of the Kansas landscape and its unique natural treasures as much as we do.

Using This Guide

Today there are many books available to those wanting to learn more about birds. Among these are several field guides devoted to identifying North American species. Nearly half of the birds shown in these guides do not occur in Kansas. For a person new to birding, these guides may seem overwhelming.

This book is intended to be a less intimidating introduction to the birds of Kansas. It is not a comprehensive guide to all the bird species that can be found in the state. Instead, from the 482 on the Kansas checklist, we have selected 326 species to be included in this book. These are the species that we consider to be most likely to be encountered during the year by those who are developing a serious interest in Kansas birds.

Organization of Species

Typical bird guides are arranged in taxonomic order. This order is based on the classification of bird orders and families, which has been revised many times in the past and is usually revised in some way each year. The full checklist of Kansas birds that appears at the end of this book is arranged in the current taxonomic order of species published by the American Ornithological Society as of 2022. The order of species that we have used in this book follows this taxonomic order only partially. We have frequently departed from it so that similar species can be grouped together even if they are not closely related. The selected species have been divided into eighteen groups. Each group consists of birds that are similar in appearance, habitats, behavior, or a combination of these.

Species Profiles

There are 310 full-page species profiles that follow the format outlined below. A special section on migratory warblers includes an additional sixteen species.

Photographs: A color photograph of an adult of each species is provided. For selected species, additional photographs are included to illustrate male and female plumages, color morphs, immature plumages, and seasonal plumages. Except where otherwise credited, all photographs are by Bob Gress.

Common Name: The common (English) name of each species is given. This and the scientific name that follows are those designated by the seventh edition of the American Ornithologists' Union "Checklist of North American Birds" and its supplements through July 2023. Over time, some of the species names used in this book will undoubtedly be changed, as this is an ongoing process.

Scientific Name: The scientific (Latin) name of each species is given. The first word of each scientific name designates the genus to which the bird belongs, and the second word designates the species name within that genus. The scientific names used in this book are current as of July 2024.

Field Identification: A brief discussion of the major field marks is presented. Where necessary, distinctions between male and female plumages are addressed, as are seasonal changes in plumage. If immature birds are markedly different in appearance and are likely to be seen in Kansas, their field marks are also discussed.

Size: The measurements given are the length from the tip of the bill to the end of the tail and the wingspan when the wings are fully extended. Remember that these are average lengths, based on measurements of many specimens. Individuals observed in the field may vary substantially in size due to age, sex, subspecies, or other factors.

Habitat and Distribution: The species' preferred habitat is briefly discussed. This section also includes information on where in Kansas the species is likely to be found. The number of Kansas counties where the species has been reported is included.

Seasonal Occurrence: The presence of the species throughout the year is outlined, with the months of arrival and departure usually specified for migratory species. These are only general guidelines, and individuals often occur outside the dates given. If this is a species that breeds (nests) in Kansas, the number of counties where nesting has been fully confirmed is included. Ornithologists use a list of criteria to rank breeding activity as possible, probable, or confirmed. The number of counties shown for each species in this book is for fully confirmed breeding records only. Thus, it excludes breeding records that are classified only as possible or probable.

Field Notes: The field notes provide additional information about each species and its status in Kansas.

Birding Basics

Optics and Field Guides

Binoculars are necessary to view most birds well. There are many excellent binocular models on the market today. It is beyond the scope of this book to discuss the pros and cons of all the choices. The price range is also broad, but with careful shopping, you can obtain a good-quality pair of binoculars for a relatively modest price. Some dealers will allow a free trial to see if you like them. All binoculars are labeled with their magnification power and objective lens size, such as 8×40. The first number is the magnification power of the eyepiece, and the second one is the diameter of the objective lens. For birding, many people prefer binoculars with a 7× or 8× magnification, as they are lighter in weight, easier to hold steady, and usually offer a closer focus. Smaller objective lenses allow the binoculars to be small and compact. Larger objective lenses offer a wider field of view and a brighter image. All these factors should be considered before you make your selection.

Many birders eventually obtain another optical instrument called a spotting scope to view birds at greater distances. Scopes are most useful at large lake and wetland areas, allowing you to view areas that cannot otherwise be clearly seen. Spotting scopes are more costly than binoculars and require a tripod. Various "digiscope" adapters are available that attach to spotting-scope eyepieces, allowing good-quality long-range photographs to be taken with a smartphone.

While this book is intended to be a useful introduction to Kansas birds, you will eventually want to purchase one of the comprehensive field guides to birds. These guides include all the species that occur in the mainland United States and Canada and illustrate the multiple plumages of each species. There are a variety of these guides on the market today. For many, the preferred field guide for those who are learning birds for the first time is the *Kaufman Field Guide to Birds of North America*, which uses digitally enhanced photographs and has pointers showing the critical field marks of each species. The *Sibley Guide to Birds* is a comprehensive and advanced field guide, and most serious birders in the United States own a copy. The *National Geographic Field Guide to the Birds of North America* is another excellent and popular guide, which is smaller than the Sibley guide. The companion volume, *National Geographic*

Complete Birds of North America is also on many birders' bookshelves. This invaluable reference book offers a substantial amount of in-depth information that is well beyond that provided by any of the field guides.

If your interest in Kansas birds and birding grows beyond the information that this book provides, you will also want to obtain a copy of *Birds of Kansas*. Published in 2011 by the University Press of Kansas, this is the scientific compilation of all known information for every species of bird that has ever occurred in Kansas. It provides detailed information on migration, distribution, and breeding in far more detail than this book. Some additional books specific to Kansas birding are discussed in the Birding Resources section at the end of this book.

Birding Techniques

A few tips are in order when you are birding in the field. Bird activity is greatest during the first few hours of daylight and occurs to a lesser degree in the late afternoon and evening. Most birders are in the field at those times of the day. Avoid wearing brightly colored or white clothing, as birds perceive more of the light spectrum than humans and generally react negatively to humans wearing these colors. When birding on foot, walk slowly and quietly. Pause frequently to look and listen. The longer you are motionless, the more birds will become used to your presence. When you see a bird, restrain the impulse to hurry toward it. Instead, approach it slowly and avoid sudden movements. When you spot a distant bird, keep your eyes on it as you raise your binoculars. This is hard to do at first but soon becomes second nature.

Weather conditions are also an important consideration, especially during migration. When low-pressure systems dominate, birds sense this and are often more active. Strong low-pressure systems that occur during the spring and fall migrations sometimes cause large numbers of migrating birds to interrupt their flight until the weather has moderated. These hoped-for events are referred to as "fallouts" by birders. When high pressure dominates, stronger winds and clear skies are the norm, and fewer migrants can be expected at these times. In Kansas, the ideal weather conditions for birding are a combination of low wind speed, overcast skies, and low pressure. While awareness of these factors is useful, "perfect" days are few, so do not let less-than-ideal conditions keep you at home.

When you are learning to bird for the first time, identifying the birds you encounter can seem bewildering. Do not grab your field guide and try to use it while observing a bird. Instead, observe the bird carefully and try to remember everything you can about it before it flies away. Then consult your field guide. After you learn to identify the common birds, it will be easier to identify species that are new to you. Enjoy the learning process, and remember that the most skilled birders in the field today learned to identify birds one species at a time.

Most identification involves using a process of logical elimination. Several criteria should be taken into consideration as you study an individual bird.

Size and Shape: Compare the size of the bird to those you are already familiar with, such as Canada Goose, American Crow, Blue Jay, or House Sparrow. The shape of a bird is another significant clue. Look for how slender or plump the bird is and how long its wings and tail are.

Bill and Legs: Observing the leg and bill structure of a bird will allow you to swiftly narrow your identification to a few families of birds. Herons have long legs for wading in water and a large spiked bill for catching fish. Sparrows have short legs and a conical bill for cracking seeds. Sandpipers have long, thin bills that allow them to probe mud for invertebrates and longer legs for wading in shallow water. Hawks have prominent talons and hooked beaks for catching and consuming prey.

Color and Pattern: Look for the presence or absence of plumage patterns such as spotting or streaking. The presence or absence of wing-bars is often a useful clue on songbirds. Head markings are often important. Look for eye-rings, eye-stripes, and any distinctive markings on the crown or nape (back of the neck). Tail patterns often provide clues to identification.

Posture: Note the bird's posture. Herons and shorebirds stand upright on long legs. Woodpeckers cling to tree trunks. Hawks perch on fence posts or utility poles.

Habitat: Habitat is often a major clue to identification. Herons and shorebirds wade in streams and marshes. Nesting Dickcissels sing from perches in the prairies and fields. Most warblers and vireos forage in the canopy of mature woodlands.

Vocalizations: In many cases, you will hear birds before you see them. Seasoned birders identify as many birds by song as by sight. Learning the songs and calls of birds is an acquired skill. There are a variety of websites and commercially available recordings of bird vocalizations, which will assist you in becoming familiar with these songs. However, there is no substitute for spending time in the field and learning these calls from your own experiences.

Sometimes making a sound that birders call "spishing" will draw birds into the open, as it mimics the alarm or scolding calls of other birds. Try saying the word "spish" using only your mouth (no vocal cords). Exaggerate the "s" sound, eliminate the vowel sound, and extend the "sh" sound at the end of the word. You end up with a sound like "sssssssspisssssssssssssshhhhhh." Repeating this several times in a row and varying the speed and tempo will often coax birds into the open. Some birders make wet squeaking sounds instead of "spishing."

Mixed flocks of songbirds will gather to noisily mob predators such as cats, snakes, and owls when they become aware of them. Playing the calls of Screech-Owls sometimes triggers this response, drawing birds into view as they attempt to locate the "predator." This technique is effective but should not be overdone. Another method for attracting birds into view is playing a recording of their songs or calls. This can be an effective technique but should not be overdone, as it can potentially cause birds to leave the area or even abandon their nests in response to what they perceive as competition from a more aggressive member of their species.

As benign as birding may seem, there are ethical standards that you should adhere to. Birding ethics mostly involve using common sense. Avoid disturbance of natural habitats, nesting birds, and rare or threatened species. Respect property rights. Birds should not be disturbed by the actions of birders. If you discover an exceptionally rare bird, try not to disturb it so that others may have an opportunity to see it.

Where to Find Birds

If you are just becoming interested in birding, start in your own yard. You can attract many birds to your yard by providing their basic requirements. This usually involves a combination of landscaping, plantings that offer food and shelter, birdbaths, bird feeders, and birdhouses.

Once you have familiarized yourself with the birds around your home, you may decide to venture elsewhere in your local area to search for new species. When you are considering where to look for birds, remember that birders are not wired to observe the outdoors in the same way that most people are. Birders do not really see landscapes; they see *habitats*. As they do so, they consider what species might be present within those habitats. Try to sample a variety of habitats, and do so at different seasons of the year. Check woodlands along rivers and streams, especially in areas where trees are otherwise scarce. If there are state or county parks located near your home, these usually have areas of native habitat. Native grasslands offer a different set of birds than those found in woodlands or urban situations. Lakes and wetlands should also be investigated. Is there a low-lying area that always floods after significant rains? It probably attracts migrating shorebirds in the spring and fall, herons in the summer, and waterfowl in the winter. The brushy fringes of both woodlands and grasslands (often referred to as "transitional" or "edge" habitats) typically provide the most productive birding in a given area. Weedy patches of sunflowers attract many birds in the fall and winter months. In many cities and towns, the local wastewater treatment ponds attract birds, especially in western Kansas, where other bodies of water are absent. These are not always accessible, nor are they glamorous locations to visit, but they are great places to see birds you otherwise would not find. Cemeteries with numerous pines and cedars are another good place to look for birds, especially in the winter and during migration. These are examples of the kinds of places that birders frequent, but there are many others. Each local area is different and has its own habitat dynamics.

If you become interested in visiting new or unfamiliar locations to see birds, consider one of the many field trips offered by the local Kansas chapters of the National Audubon Society. Nearly all of the local chapters offer a variety of trips throughout the year, led by experienced birders. These trips are free and open to the public. The statewide organization of birders, the Kansas Ornithological Society, hosts meetings twice a year in addition to publishing two periodicals and maintaining an excellent website with a wealth of useful information and links. More information on these organizations is found in the Birding Resources section of this book.

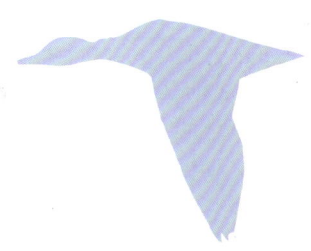

Kansas Geography and Ecosystems

The exceptional diversity of Kansas bird life is due in large part to the state's location in the center of the continent. Species typical of both the eastern and western United States are well represented. Kansas is located far enough to the north that resident species of the Canadian boreal forests appear here, especially during the winter months. Yet it is also far enough south that species more typical of the southern states and the desert Southwest also occur. Its location on the Central Flyway assures that huge numbers of migratory birds occur during the spring and fall migrations. The diversity of bird life in Kansas is well represented throughout the state. Thirty-six of the 105 counties have recorded 300 or more species of birds, including counties located on both the Colorado and Missouri borders. An additional 48 counties have a checklist of 250–299 species of birds.

Habitat diversity also contributes to the quality of birding in Kansas. Although it can seem to the traveler on Interstate 70 to be a barren and uniform landscape, Kansas in fact has a variety of natural habitats. As one travels from the eastern border to the western border of the state, the elevation above sea level increases by over three thousand feet, though this change is largely imperceptible. Because of the "rain shadow" of the Rocky Mountains, average annual rainfall amounts at the Colorado border are less than twenty inches a year, increasing to nearly forty inches along the Missouri border. These changes in elevation and precipitation contribute to the diversity of habitat within the state.

The geography of Kansas is also diverse and plays an important role in shaping its bird communities. One of the most widely accepted ways of delineating Kansas geography is to classify the state by its physiographic regions, which the American Geosciences Institute defines as "the classification of the surface features of Earth on the basis of similarities in geologic structure and the history of geologic changes; descriptions that also include vegetation and/or land use." The basic delineation of Kansas physiographic regions separates the state into eleven distinct regions. Detailed discussion of these regions and their bird populations can be found in the Hotspots portion of this book. They are briefly summarized here.

In northeast Kansas, north of the Kansas River and east of the Big Blue River, is the **Glaciated Region** which marks the southernmost penetration of prehistoric northern glaciers into the state. This region features tall hills and extensive deciduous woodlands. Most of the four eastern tiers of counties south of the Kansas River lie in the **Osage Cuestas** region, which is generally less hilly than the Glaciated Region. Woodlands are widespread in the Osage Cuestas, especially in the counties along the Missouri border and farther westward along riparian corridors. In the western part of this region, grasslands gradually become the dominant plant community. In the southeastern corner of the state are the **Cherokee Lowlands**, which cover about one thousand square miles, mostly in Cherokee, Crawford, and Labette Counties. This region is in general quite level and heavily cultivated. Extensive strip mining in this region in the early twentieth century greatly altered the landscape. Some of the reclaimed mined areas today offer interesting birding opportunities. The **Ozark Plateau** region covers tens of thousands of square miles in Arkansas, Missouri, and Oklahoma, but this region barely extends into Kansas and covers only fifty-five square miles in the southeastern corner of Cherokee County. Its thin soil is underlain by solid rock and is therefore mostly uncultivated with considerable oak-hickory woodland. Plants and animals that are rare or absent elsewhere in Kansas are found in this region. Extending northward from Oklahoma into the Osage Cuestas in a narrow strip, mostly in Chautauqua, Elk and Woodson Counties, is the scenic **Cross Timbers** region (also known as the Chautauqua Hills), which is typified by rocky, rugged slopes and unique stands of upland blackjack and post-oak woodland. West of the Glaciated Region and the Osage Cuestas is the **Flint Hills** region, which contains the largest remaining expanse of tallgrass prairie in the world. As in the Ozark Plateau, thin soils above a solid rock substrate make cultivation of the land difficult to impossible, which has allowed an extensive area of this grassland ecosystem to survive into the present day. Mature gallery forests are found along the rivers and streams in the Flint Hills. Continuing westward, the tallgrass prairie gives way to the mixed-grass prairies and visually appealing rock formations of the **Smoky Hills** region in the north-central part of the state, which extends from the Nebraska state line south to the Arkansas River. The **Arkansas River Lowlands** and the **Wellington-McPherson Lowlands** adjoin the entire course of the river between the Colorado and Oklahoma borders. Within these two lowland regions is a subregion of stabilized sand dunes known as the Sand Hills, which are both visually appealing and biologically diverse. The extensive and well-known wetlands of Cheyenne Bottoms and Quivira National Wildlife Refuge (NWR) lie within the Arkansas River Lowlands. In south-central Kansas south of the Arkansas River Lowlands is the **Red Hills** region, an area of ferrous soils, rugged rocky buttes, and deep wooded ravines. The **High Plains** region covers the western third of Kansas. This region is nearly entirely flat, semiarid, and heavily cultivated but has some outstanding oases of bird habitat. Within the High Plains region in southwest Kansas south of the Arkansas River Lowlands is a subregion known as Sandsage Prairie, a semiarid habitat with abundant sage, yucca, prickly-pear cactus, and other desert plant species.

The combined factors of elevation, precipitation, and physiography combine to create a subtle but steady gradient of change in Kansas bird populations from east to west. From the Missouri border west to the Flint Hills, all or nearly all bird species are typical of the eastern United States. As you travel farther westward in the state, more species of the western United States gradually begin to appear, becoming most prevalent near the Colorado border.

Throughout the state are numerous streams and river systems that generally flow from west to east. They provide corridors of woodland habitat for many species of birds. Beginning in the 1930s, and especially in the 1950s and 1960s, major dams were constructed on many of these rivers by the US Army Corps of Engineers and the Bureau of Reclamation. The large reservoirs that were created by these dams today attract many species of waterfowl, gulls, terns, and other aquatic species that were formerly scarce or absent in Kansas. Additionally, many of these reservoirs have extensive wetlands that provide important habitat for shorebirds, wading birds, and others.

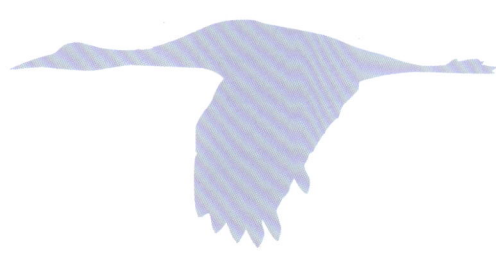

Calendar of Kansas Bird Activity

As the seasons advance throughout the year, Kansas bird populations are composed of an ever-shifting tapestry of species. This Kansas birding calendar provides a summary of the most significant trends and likely birds that can be expected during each month of the year. This is intended as a broad and general overview. Weather conditions and variations in region and habitat can greatly affect the abundance and distribution of individual species from year to year and place to place. For more detailed information, consult the individual species accounts.

January

As the year begins in Kansas, the birds of winter hold full sway. The large reservoirs that have not frozen over host large, massed flocks of Common Mergansers and Common Goldeneyes, with attendant flocks of hovering Herring and Ring-billed Gulls hoping to snatch fish from them. Large flocks of geese are present at some locations, especially in the south and east. Bald Eagles are at their greatest numbers, especially on major rivers and reservoirs. Red-tailed Hawks are abundant statewide, and by the end of the month, the males have begun courting females. Rough-legged Hawks and Northern Harriers patrol over open country, especially in grasslands and prairies. On the endless expanses of farmland in the west and central parts of the state, huge flocks of Lapland Longspurs and Horned Larks swirl restlessly over the wheat and milo fields, often pursued by Merlins or Prairie Falcons. This is the best time to look for Northern Shrikes in northern and western Kansas. In brushier habitats, Harris's and American Tree Sparrows join Dark-eyed Juncos in social foraging flocks. In the western third of Kansas, Harris's become scarce and are largely replaced in these mixed sparrow flocks by White-crowned Sparrows. Huge roosts of Red-winged Blackbirds assemble at wetlands in southern and central Kansas. Noisy flocks of Rusty Blackbirds roam along the wooded rivers of eastern Kansas, while flocks of Brewer's Blackbirds congregate at cattle-feeding areas statewide.

Several bird species of the northern boreal and Rocky Mountain forests invade Kansas in substantial numbers during some winters but are nearly absent in others. January is often the best month to look for these unpredictable wanderers, which include Red-breasted Nuthatches, Pine Siskins, Red Crossbills, and Purple Finches. Especially in the Red Hills but also elsewhere in the west, Mountain Bluebirds sometimes invade in large flocks to consume cedar berries, often joined by Cedar Waxwings and American Robins.

The days lengthen as the month draws to a close, and species such as Carolina Wrens and Northern Cardinals begin to sing more frequently in response. By the end of the month, Great Horned Owls are vocally engaged in courtship, and many are already incubating their eggs. Their telltale ear tufts can be seen protruding from large stick nests high in the bare trees.

February

The final month of winter is in many ways like January, but as the month progresses, the first hints of spring can be detected. Killdeer are among the earliest migrants to return north, arriving by the first week or two of the month. Their strident calls are a sure sign that winter's time is short. In central and western Kansas, large, vocal flocks of Sandhill Cranes begin moving north, staging in areas such as Quivira NWR and the playa areas of Meade County. Waterfowl populations are also on the move by the third week of the month. Large flocks of geese that have spent the winter in Kansas begin to migrate north, along with Common Mergansers and Common Goldeneyes that have spent the winter on the large reservoirs. As they depart, other waterfowl species begin to arrive. Northern Pintails are one of the first of the spring waterfowl to arrive, followed by American Wigeons, Ring-necked Ducks, Redheads, Canvasbacks, Lesser Scaups, and Green-winged Teal. Bald Eagles and other wintering raptors begin to thin out as they move northward to their breeding grounds. Other wintering species that begin to migrate north out of Kansas in late February are Herring and Ring-billed Gulls, American Tree Sparrows, and Lapland Longspurs.

Red-tailed Hawks are now nesting statewide. Most of the large stick nests in tall trees that are not occupied by Great Horned Owls now have Red-tails sitting on them, their heads peering over the rim. Along the woodland rivers of southeast Kansas, Red-shouldered Hawks begin to call noisily as they establish territories. Both Greater and Lesser Prairie-Chickens begin booming activity on their prairie leks during the latter half of the month. More songbirds begin to sing, especially Black-capped and Carolina Chickadees, American Robins, Eastern and Western Meadowlarks, Red-winged Blackbirds, Harris's Sparrows, and Dark-eyed Juncos. The first American Woodcocks arrive in eastern Kansas, and their "timberdoodle" displays can be observed at dusk late in the month. At the very end of February, the first Eastern Phoebes appear along the southern border.

March

Many of the migration trends that began in February accelerate during March. Wintering species continue to depart northward, and the hardier migrants begin to arrive from their southern wintering grounds.

The first small flocks of migrating American White Pelicans and Double-crested Cormorants appear at the large wetlands and reservoirs, along with Common Loons and most of the grebe species. Sandhill Crane numbers peak in mid-March, when they depart to join the huge flocks staging on the Platte River in Nebraska. The first Turkey Vultures reappear, lazily floating on the winds over the fields and prairies. On calm evenings, eastern Kansas birders visit locations such as Shawnee Mission Park and Linn County Park to view displaying American Woodcocks. The booming activity of prairie-chickens increases, and viewing their leks at dawn is a popular trip for birders, especially in the Flint Hills and Smoky Hills regions for Greater Prairie-Chickens and in the southwestern counties for Lesser Prairie-Chickens. In the Flint Hills and other areas where prairie burns are conducted, American Golden Plovers sometimes appear in large flocks to feed in the burned areas. At the large wetlands, shorebirds begin to arrive in a trickle early in the month. By the end of the month, thousands of individuals and numerous shorebird species will be represented. Mourning Doves return in large numbers, joining the small wintering population. Large numbers of migrating Northern Flickers are typically observed in March. By midmonth, Eastern Phoebes arrive and begin singing along streams, especially near bridges. Say's Phoebes also arrive, staking claims to their nesting territories in central and western Kansas. The first male Purple Martins arrive to nesting sites. The first Tree Swallows arrive at ponds and wetlands. The huge winter flocks of Horned Larks have departed, and the remaining resident individuals are paired up and beginning to nest in barren winter wheat fields and prairies. Northbound flocks of American Pipits appear at shorelines and in muddy farm fields, and Sprague's Pipits begin to arrive in prairie habitats. In the Flint Hills, Smith's Longspurs have shed their nondescript winter plumage for the more colorful summer plumage and will soon depart for their tundra nesting grounds. In the west, flocks of equally colorful Chestnut-collared and Thick-billed Longspurs are on the move, and this is the best time to look for them in shortgrass habitats. For several transient species of sparrows, March marks the peak of the spring migration. These include Savannah and Vesper Sparrows in open country and Fox Sparrows in brushy areas of eastern Kansas. In central and eastern Kansas, Field Sparrows return to their nesting grounds, and the pleasing song of territorial males is heard in a variety of brushy habitats. Huge flocks of Common Grackles and Red-winged Blackbirds arrive during March, often swarming bird feeders. Latest to arrive of the northbound waterfowl, Blue-winged Teal and Northern Shovelers are present in large numbers by the end of the month, just as most of the migrant goose and diving duck species are departing northward. In the final days of March, the first few egret and heron species return to the large rookeries in southern Kansas and immediately begin courtship and nesting activity. In eastern and southern Kansas, several migratory warmer-season woodland bird species also arrive late in the month. Some of these, such as Ruby-crowned Kinglets and Hermit Thrushes, are migrants on their way to more northerly breeding grounds; others will remain to nest, including Blue-gray Gnatcatchers, Black-and-white Warblers, and Louisiana Waterthrushes.

April

Nearly all the remaining wintering birds that did not depart in March depart northward in April. These include most wintering waterfowl, raptors, gulls, sparrows, and finches. The spring migration continues to increase in intensity, with new arrivals each week, especially after warm fronts have moved through the state from the south.

American White Pelicans and Double-crested Cormorants arrive in force, and flocks numbering in the thousands can be seen at the large reservoirs and wetlands. Several raptor species are also at their migration peak in April, including Swainson's Hawks, Broad-winged Hawks, and Ospreys. While the greatest diversity of shorebird species occurs in May, several of them are at their most numerous in April. Many birders make their first trip of the year to Quivira and Cheyenne Bottoms in mid- or late April to shake off the winter doldrums. This is also the month when many sparrow species migrate through Kansas. Chipping and Clay-colored Sparrows arrive in the middle of the month and are abundant and vocal in both rural and urban areas. Lincoln's and White-throated Sparrows are found in more wooded habitats. Song Sparrows begin to sing more frequently. Large flocks of Yellow-headed Blackbirds are frequently observed, especially around cattle herds and feedlots in central and western Kansas. In the last week of April, other migrants begin to arrive, including Least Flycatchers, Olive-sided Flycatchers, and Swainson's Thrushes. In eastern woodlands, several breeding warbler species arrive early in the month, and their songs can frequently be heard as they establish nesting territories. Among them are Black-and-white Warblers, Louisiana Waterthrushes, Northern Parulas, Yellow-throated Warblers, and Prothonotary Warblers. By midmonth, additional warbler species appear, particularly Yellow-rumped and Orange-crowned Warblers. In the closing days of the month, several others, such as Nashville, Tennessee, and Yellow Warblers, begin to arrive.

Most of the permanent resident bird species, including Red-shouldered Hawks, Ferruginous Hawks, Wild Turkeys, Bobwhites, Belted Kingfishers, woodpeckers, both kinds of chickadees, Tufted Titmouses, White-breasted Nuthatches, Carolina Wrens, Eastern Bluebirds, American Robins, Eastern Towhees, Northern Cardinals, and both kinds of meadowlarks begin nesting in April. The dawn chorus of singing robins, cardinals, and others increases in intensity as the month progresses. Both Greater and Lesser Prairie-Chickens continue to display on their leks throughout the month.

Many Kansas summer resident species arrive in April. Among them are Mississippi Kites; Chuck-will's-widows Whip-poor-wills; Great Crested Flycatchers; Eastern and Western Kingbirds; Scissor-tailed Flycatchers; Red-eyed, Bell's and Warbling Vireos; all the swallow species; Northern House Wrens; Gray Catbirds; Brown Thrashers; Lark and Grasshopper Sparrows; and Brown-headed Cowbirds. Herons and egrets continue to return to their huge nesting colonies in Wichita and elsewhere. In the wetlands of central Kansas, arriving summer residents include American and Least Bitterns, White-faced Ibises, King and Virginia Rails, Snowy Plovers, Black-necked Stilts, and Wilson's Phalaropes. Newly arrived Upland Sandpipers grace the Flint Hills and other grasslands with their ethereal "wolf-whistle" calls. Common Poorwills and Rock Wrens return to their specialized rocky slope habitats in the west. The beautiful display flights of Lark Buntings and Cassin's Sparrows brighten the arid western plains. In the last week of the month, a new set of summer songbirds arrives to establish nesting territories. These include Eastern Wood-Peewees; White-eyed and Yellow-throated Vireos; Wood Thrushes; Summer Tanagers; Indigo Buntings; and Orchard, Bullock's, and Baltimore Orioles.

May

For many Kansas birders, the first two weeks of May represent the most exciting and rewarding birding of the entire year. This is when the greatest variety and numbers of northbound migrants can be seen, especially among such popular bird families as plovers, sandpipers, vireos, warblers, grosbeaks, and buntings. The Kansas Ornithological Society holds its annual spring meeting in the first weekend of May in most years, usually in a location where interesting migrants are likely to be found. Many of the local Audubon chapters also schedule field trips to favored local birding hotspots in early May.

Cheyenne Bottoms, Quivira, and other wetlands across the state now teem with huge flocks of migrating shorebirds. Most of them are in full breeding plumage, including distinctively marked species such as Black-bellied Plovers, American Golden-Plovers, Whimbrels, Ruddy Turnstones, Dunlins, and Wilson's Phalaropes. Migrating Peregrine Falcons often lurk nearby, appearing seemingly out of nowhere to dive on the feeding shorebird flocks in dramatic plunges out of the sky. As the month passes, numbers of shorebirds diminish slowly and steadily. Last to arrive in numbers are the White-rumped Sandpipers, a sure sign that the shorebird flight is nearly over. Restless flocks of Franklin's Gulls and Black Terns roam over wetlands, lakes, and prairies statewide as they move north to their nesting grounds.

This is also the "warbler time," when birders comb the woodland hotspots of eastern Kansas, hoping for a large fallout of these feathered jewels of color. In one of the eastern hotspots such as Marais des Cygnes, Perry Reservoir, or Schermerhorn Park, twenty or more warbler species can be found on an exceptional day. The peak of the warbler migration usually falls between May 5 and May 15, although many are seen in the days before and after this period. Also at the peak of their migration are many species of flycatchers, vireos, thrushes, tanagers, grosbeaks, and buntings. By the third week of the month, the majority of these migrants have passed northward, but birders searching for stragglers are often rewarded with rare or unexpected species. In western Kansas locations such as the Cimarron National Grassland, Meade State Park, and Historic Lake Scott State Park, some of these same "eastern" species are encountered in May. They are joined by western migrants that are either rare or absent in central and eastern Kansas. These include species such as Dusky Flycatchers, Townsend's Warblers, Lazuli Buntings, Black-headed Grosbeaks, Western Tanagers, and Bullock's Orioles.

Arriving with these waves of northbound migrants in May are the last of the species that will remain to nest in Kansas in the coming months. These include wetland species such as Common Gallinules and Least Terns. The last of the breeding insectivorous species also arrive in May. These include Yellow-billed Cuckoos, Common Nighthawks, Wood Thrushes, American Redstarts, Yellow-breasted Chats, Scarlet Tanagers, Blue Grosbeaks, Indigo Buntings, Painted Buntings, Henslow's Sparrows and Dickcissels. By the final week of May, the spectacle of migration has spent itself, giving way to the vibrant energy of breeding activity.

June

After the frenetic activity of May, June is a comparatively quiet time of year for birding in Kansas. There is virtually no migration activity either northward or southward, and so the population of birds is composed entirely of nesting species. Early in the month, the dawn chorus of singing birds reaches its crescendo in forests, prairies, and wetlands across the state. Early-morning visits to local birding locations, regardless of location or habitat, reward birders with excellent views of singing territorial males. This is a good time to learn the songs and calls of the nesting species in your local area. Remember that most singing activity tapers off rather abruptly after about ten o'clock in the morning. During June, breeding-bird surveys are conducted statewide for the United States Geological Survey by volunteer birders. These surveys are conducted across the United States and Canada and are one of the most valuable databases of information on bird populations in North America.

At Quivira, Cheyenne Bottoms, and other wetlands, flocks of White-rumped Sandpipers linger into the first week of the month, representing the last echoes of the spring migration. A few nonbreeding shorebirds of other species are usually found during June. These are sometimes late migrants, but many are simply subadult individuals that will not continue north to the arctic nesting grounds. Similarly, many of the large reservoirs host a few nonbreeding Double-crested Cormorants, American White Pelicans, gulls, and terns.

July

High heat and humidity dominate the weather during July. Most breeding species have fledged their first brood. Some will begin their second or third broods during July, and these species continue to sing during the early-morning hours throughout the month. In agricultural areas, Dickcissels and meadowlarks continue to sing throughout the day. Yellow-billed Cuckoos, Great Crested Flycatchers, Rock Wrens, Bell's Vireos, Warbling Vireos, Red-eyed Vireos, Indigo Buntings, and Eastern Towhees are among the other species that continue to sing even in the hottest weather. Many other woodland and grassland species, even though they are still present, fall silent and therefore become much more difficult to find.

It is the middle of summer, but by mid-July, the first signs of the southbound "fall" migration are already apparent. At the large reservoirs, small flocks of postbreeding gulls and terns begin to appear. The first migrating Upland Sandpipers can be heard giving their choppy, chortling flight calls at night as they migrate overhead. Earliest of the southbound songbirds, a few Least Flycatchers also appear. Kansas hummingbird fans have learned to have their feeders full and ready by early July. Ruby-throated Hummingbirds that have nested to the north begin to arrive at nectar feeders throughout Kansas. Statewide, but especially in western Kansas towns such as Dodge City, Elkhart, Garden City, and Larned, they are often joined by rare western hummingbird species, especially Rufous Hummingbirds. Some locally nesting songbirds, especially those that arrived early in the spring, begin to migrate south in July. These include Yellow-throated and Kentucky Warblers, Louisiana Waterthrushes, and Orchard Orioles.

The most notable sign of the approaching fall season is the return of southbound shorebirds to the large central wetlands. Early in July, the first Lesser Yellowlegs, Long-billed Dowitchers, and small "peep" sandpiper species appear. By the end of the month, thousands of shorebirds can be found at Quivira, Cheyenne Bottoms, Neosho, and other wetlands. At this time of year, most of the adults are still in breeding plumage. During the fall migration, many juvenile shorebirds are observed, and this is a good time to learn their field marks, which often differ from those of the adults.

Trips to Quivira, Cheyenne Bottoms, and other Kansas wetlands during July are rewarding for other reasons. The young of most nesting marsh species have either hatched or fledged, and the adults are active as they engage in their parental duties. Look for Least Bitterns flying low over the cattails as they forage and carry food to the nest for their young. Adult King and Virginia Rails can sometimes be sighted leading their broods of fuzzy black chicks along the edge of the cattails. Common Gallinules are easier to find as well and are often sighted swimming near the edges of cattails. Mixed flocks of swallows are a familiar sight at the large wetlands in late July, with many immature birds among them. These are postbreeding individuals that are slowly migrating southward, taking advantage of the abundant supply of insects. These flocks gradually increase in size, sometimes numbering in the thousands. They remain well into August before moving on.

August

August marks the end of the breeding season, and the pace of migration continues to pick up as the month progresses. By August, most species have fledged their last broods. Many immature birds are seen, often in family groups with adults. American Goldfinches and Sedge Wrens are the only species still nesting to any significant degree in August. Once the young have fledged from the large heron rookeries in Wichita, Hutchinson, and elsewhere, the birds begin to disperse. Postbreeding herons and egrets wander widely throughout Kansas, sometimes well north of their breeding areas. They often congregate at large wetlands, reservoirs, and other locations with abundant food.

Migration continues to increase in intensity throughout August. Shorebird numbers reach their peak early in the month, and these birds remain numerous through the remainder of the month. This is the time to look for Buff-breasted Sandpipers at sod farms, as most are sighted during August. Cool fronts approaching from the north trigger large flights of Upland Sandpipers in the night skies. Flocks of Franklin's and Ring-billed Gulls and Black and Forster's Terns appear at wetlands and large reservoirs. This is also the peak of migration for hummingbirds, and hummer-feeder watchers are entertained throughout the month. Large roosts of southbound Purple Martins gather, often in urban areas. These flocks sometimes number in the tens of thousands. Late in the month, several other species begin to gather in conspicuous foraging flocks as they prepare to migrate southward. These include Mississippi

Kites, Eastern Kingbirds, Western Kingbirds, Dickcissels, Red-winged Blackbirds, and others.

Although comparatively few birders are in the field looking for them, flycatchers, vireos, warblers, orioles, and other migrants returning from their northern breeding grounds are moving through Kansas throughout August, especially during the last ten days of the month. Songbirds migrating in the fall do not sing and are generally more furtive and retiring than in the spring, but birders who spend time in the field in late August at songbird hotspots are usually rewarded with a good day of birding.

September

September is a month that many birders in Kansas look forward to with great anticipation. The variety of species that can be seen is greater than at any other time of year except May. Many of the birds of summer are still present early in the month. Additionally, a large percentage of the autumn bird migration from the northern states and Canada takes place during September. These factors combine to create exceptional birding for those who spend time in the field. While cool fronts from the north will sometimes trigger substantial flights of migrants, these are rarely as dramatic as the fallouts that are observed in late April and early May. On the other hand, fall migration is generally steady throughout the month, and birders who make a series of visits to local birding areas generally accrue a lengthy and varied list of birds.

September migration activity varies among the various bird families. Some waterfowl begin to return from the north, starting with Blue-winged Teal, Northern Shovelers, and Lesser Scaups early in the month, followed by Northern Pintails, American Wigeons, and Redheads in the last week of the month. Herons and egrets are numerous and widespread as the month begins, but by the end of the month, they have either become scarce or departed altogether. Many raptors are also on the move. Mississippi Kites gather in ever larger flocks prior to their abrupt departure in midmonth. Most fall observations of Broad-winged Hawks occur in September. Swainson's Hawks begin arriving from the north, sometimes in large, soaring flocks. The first few returning Northern Harriers and accipiters appear. Shorebirds and terns are still numerous as the month begins, but their numbers and species diversity both drop noticeably as the weeks pass. Franklin's and Ring-billed Gulls are increasingly numerous at the large wetlands and reservoirs. Hummingbirds are still common and widespread at the beginning of September but mostly have departed by month's end.

Songbird migration in September is quite diverse. Insectivorous species are abundant early in the month. The greatest variety of warblers and vireos is seen at this time. By the end of the month, most of these species will be absent. A few sparrow species such as Chipping, Clay-colored, and Vesper arrive early in the month, especially in the west. As October nears, several other sparrow species, especially Lincoln's and White-crowned, also begin to appear.

During September, birders often visit western Kansas hotspots such as Cimarron National Grassland, the Arkansas River Scenic Drive in Hamilton County, Historic Lake Scott State Park, and the St. Francis River Walk. This is the best time of year to observe western migrants such as Dusky Flycatchers, Western Tanagers, and Black-headed Grosbeaks. These species and other western migrants routinely drift eastward onto the plains during their southward migration. As a group, these western species tend to be more numerous in the fall than in the spring. Many birders visiting this part of the state in September are hoping for these sought-after western birds. Many of the rarest of these western species were omitted from this book because of their scarcity even in this part of Kansas, but each year, a few of them are found in the western counties. Of equal interest in western Kansas are migrant eastern species that also drift onto the High Plains during migration. These are often encountered in the small towns that provide islands of trees and water amid the vast semiarid "wheat desert." After driving through long miles of wheat fields and arid plains, a stop at parks and cemeteries in towns such as Colby, Garden City, Goodland, Oberlin, Sharon Springs, and Tribune can provide an excellent variety of migratory birds in early September.

October

October often seems like a quiet transitional period between the earlier warm-season migrants of September and the later cool-season migrants that will arrive in November. However, there is still substantial migratory activity.

Flocks of migrating Swainson's Hawks peak in numbers early in the month until significant cool fronts trigger their abrupt mass departure. Those lucky enough to be in the right place when this happens can observe hundreds or even thousands of these beautiful hawks as they slowly wheel southward in large "kettles." Noteworthy numbers of many other migratory hawk and falcon species are seen in October, sometimes migrating with the large flocks of Swainson's Hawks.

In the third week of the month, the first large flocks of diving ducks arrive somewhat abruptly at lakes and wetlands. They are often accompanied by the first loons and grebes of the fall season. Sandhill Cranes begin to arrive in central and western Kansas in the latter half of October. At the central wetlands, shorebird diversity continues to slowly diminish, but it is still possible to see fifteen or more species on a typical day early in the month. Terns also thin out rapidly, and by the end of the month, they have largely departed. Franklin's Gulls are present in huge numbers, especially at the large reservoirs, where hundreds of thousands are sometimes recorded in late October. These are accompanied by large numbers of Ring-billed Gulls and smaller numbers of Herring and Bonaparte's Gulls. A few hardy Eastern Phoebes and Scissortails represent the only flycatchers remaining in the state. Other bird families such as vireos, swallows, thrushes, and warblers have for the most part migrated southward, but each of these families has a few species that are still present as October begins. Nearly all of these lingerers have departed by the end of the month.

Sparrows are one of the most numerous bird families present in Kansas during October. Early- and late-migrating sparrows as well as lingering summer species are present. Adding to this mix of species, several winter resident sparrows such as Harris's Sparrows and Dark-eyed Juncos begin to arrive late in the month. As a result, birders visiting areas such as the Baker Wetlands, Kyle Marsh, Neosho Wildlife Area (WA), or Slate Creek Wetlands can potentially observe twenty sparrow species during a morning's walk in October. Most highly sought are several wetland sparrows, including LeConte's, Nelson's and Swamp Sparrows, which are easier to find and more numerous in the fall than in the spring, especially in the last half of October. While searching for these wetland sparrows, be alert for Sedge and Marsh Wrens, which often inhabit the same habitat at this time of the year.

November

Cool fronts give way to genuine cold fronts in November, and the last of the warm-weather migrants are now gone. But this is the best month of the year for viewing many of the late-migrating species. Many of these are more numerous and easier to see in late fall than during their flight north in early spring. The last of the winter resident species that have not already arrived will do so by the end of the month.

Quivira and Cheyenne Bottoms are popular destinations during November. The herons, egrets, rails, and terns are long gone, and only a handful of shorebirds remain, but early in the month, usually after a major cold front, massive flocks of geese, ducks, and Sandhill Cranes arrive. They remain throughout the month and into December in most years. Sandhill Cranes usually number between 50,000 and 100,000 individuals. Goose populations often exceed 500,000 birds, including all five species shown in this book. Bald Eagles sometimes gather to prey on them. These massive flocks of waterfowl and cranes disperse each day to forage in adjacent agricultural areas, returning near dusk to roost for the night. Birders often gather on the Wildlife Loop to watch these waves of birds descend into the Big Salt Marsh at sunset. If you are lucky, you might spot a Whooping Crane at Quivira or Cheyenne Bottoms in early November. Your odds are not favorable, but each November, several sightings of them are made as they pass through Kansas at these wetlands. Cheyenne Bottoms, Kirwin NWR, and Neosho WA can produce similar birding spectacles at this time of the year, but because of the location of the public hunting areas near the dike roads, they offer a less advantageous wildlife viewing venue during the waterfowl hunting season in November.

Large lakes and reservoirs are also popular birding destinations during November. Waterfowl migration is at its autumn peak, although the early-fall species such as Blue-winged Teal and other dabbling ducks are becoming scarce. Diving ducks such as Redheads and Lesser Scaups are abundant. Late in the month, flocks of Common Mergansers and Common Goldeneyes make their appearance. They are the last of the waterfowl to arrive. Common Loons can be seen in the dozens at the large reservoirs. Eared, Horned, and Pied-billed Grebes are also widespread and frequently seen on water impoundments throughout the state. Western Grebes are seen far less often but can appear on lakes anywhere in Kansas during November. Depending on weather fronts, huge flocks of Franklin's Gulls usually remain in Kansas for the first few days of November before abruptly departing for their wintering grounds in the Pacific Ocean off the coasts of Chile and Peru. As they depart, the numbers of Herring and Ring-billed Gulls are growing rapidly. By the end of the month, Ring-billed Gulls often number in the thousands and Herring Gulls in the hundreds of individuals at major reservoirs such as Cheney, John Redmond, Milford, Tuttle Creek, and Wilson.

Sparrows represent most of the songbirds that are still migrating in significant numbers in November. While some sparrow species are present in lower numbers than in October, others are at their peak, and a variety of sparrow species can still be found through midmonth. By the end of the month, most of those that remain are winter resident species, primarily American Tree, Song, Fox, Harris's, White-crowned, and White-throated Sparrows; Spotted Towhees; and Dark-eyed Juncos.

Several winter resident species that began to trickle into Kansas during October have all arrived by the end of the month. This is a diverse group of birds. Raptors include Bald Eagles, Northern Harriers, Rough-legged Hawks, and Merlins. Others are Long-eared and Short-eared Owls, Yellow-bellied Sapsuckers, Red-breasted Nuthatches, Brown Creepers, Winter Wrens, Golden-crowned Kinglets, Hermit Thrushes, and Rusty Blackbirds. Some of these late-migrating winter residents are restricted largely to the western half of Kansas, including Ferruginous Hawks, Prairie Falcons, Northern Shrikes, Townsend's Solitaires, and Mountain Bluebirds.

December

In general, the December bird population across Kansas is composed of the winter resident species described above for January, but if the weather has been mild, a few fall migrants are still on the move early in December. Many waterfowl species remain through the month, especially in the southern half of the state. Examples of other mostly migratory birds that sometimes linger into December are Common Loons, Horned Grebes, Double-crested Cormorants, Virginia Rails, American Coots, Wilson's Snipes, Eastern Phoebes, Marsh Wrens, Brown Thrashers, and Orange-crowned and Yellow-rumped Warblers. American Crows gather in immense roosts of thousands of birds in Wichita and elsewhere, taking advantage of the absence of predators and dispersing into surrounding agricultural areas to feed during the day. At wetland areas such as Quivira NWR and Slate Creek Wetlands, Red-winged Blackbirds and other blackbird species also gather in huge flocks, sometimes numbering in the hundreds of thousands. These flocks literally blacken the sky as they leave the roost in the morning and return at dusk.

The primary focus of many birders in Kansas during December is participation in one or more of the Christmas Bird Counts, which are held throughout the state from mid-December through early January. The Christmas counts are part of a continentwide effort administered by the National Audubon Society. Thousands of counts are held across the United States and Canada during a three-week period beginning on December 14 each year. Each count takes place within a circle fifteen miles in diameter. Birders spend an entire day counting all the birds they hear and see within the circle. Their findings are then entered into a national database. With over one hundred years of count data to draw on, this database is the most extensive source of information on bird populations in the world. Over fifty counts are conducted each year in Kansas, some of which have been conducted for seventy-five consecutive years.

Canada Goose

Branta canadensis

Field Identification: This common goose is recognized by nearly everyone. The head and neck are black, and there is a prominent white chin strap. The body is brown above and white below. Some individuals are very large; others can be almost as small as the closely related Cackling Goose. The Canada Goose always has a longer neck and a larger, more tapered head and bill than the Cackling Goose. Size: Length 36–45 inches; wingspan 53–60 inches.

Habitat and Distribution: Statewide on wetlands, marshes, lakes, ponds, city parks, golf courses, and other urban areas. Feeds in agricultural areas and on urban lawns. Canada Goose has been recorded in all 105 Kansas counties.

Seasonal Occurrence: Present year-round. Breeding has been confirmed in all 105 counties. Pairs of adults with family groups of downy yellow young are frequently observed in the summer throughout the state. It is an abundant migrant, especially in central and eastern Kansas. Huge winter flocks can be observed in both urban and rural areas.

Field Notes: Canada Geese are common in most of Kansas, but this was not always the case. In the 1980s, the Kansas Department of Wildlife and Parks introduced them in many parts of the state in order to establish a breeding population that was almost nonexistent at that time. They now are so numerous that they are considered a nuisance species at many urban parks and golf courses.

Branta hutchinsii

Cackling Goose

Field Identification: A miniature version of the familiar Canada Goose, the Cackling Goose differs by its much shorter neck, more rounded head, and small, stubby bill. Some Canada Geese can be as small, but they always have a more tapered head and proportionately longer bill. Size: Length 25 inches; wingspan 43 inches.

Habitat and Distribution: Statewide on wetlands and lakes, often in urban areas. Feeds in grainfields. Has been recorded in all but one Kansas county.

Seasonal Occurrence: Migrant and winter resident. Arrives in October and departs in March. Cackling Goose is not a breeding species in Kansas.

Field Notes: The Cackling Goose was formerly considered to be one of several small subspecies of the Canada Goose. It has been reclassified as a full species by the American Ornithological Society. Cackling Geese are frequently found with other geese but tend to gather in segregated flocks. They prefer shallower water than Canada Geese. Cackling Goose populations have increased in recent years. Flocks numbering in the thousands are observed regularly.

Greater White-Fronted Goose

Anser albifrons

Field Identification: This goose is smaller than most Canada Geese. The body is mostly brown, with prominent black speckling on the belly of adult birds. Note the bright orange legs and pinkish bill. It is named for the white plumage at the base of the bill. Size: Length 28 inches; wingspan 53 inches.

Habitat and Distribution: Found statewide on or near marshes, lakes, and farmlands, where they feed in grainfields. The goose is common to abundant in the central and eastern counties but less likely to be seen in the western third of the state. Has been recorded in 100 Kansas counties.

Seasonal Occurrence: Migrant and winter resident. Arrives in October and departs in March. Most numerous in migration, but many remain in winter if open water is present.

Field Notes: These geese are often referred to as "speckle-bellies" by hunters. They occur in large flocks at locations such as Neosho WA and Quivira NWR. Flocks can be distinguished by their calls, which are more rapid and high-pitched than those of other geese.

immature

Snow Goose

Anser caerulescens

white morph (Snow Goose)

Field Identification: There are two color morphs known as the Snow Goose and the Blue Goose. Some birds are intermediate in plumage between these two morphs. The bill and legs are pink. The bill mandibles have a black edge where they meet, forming a "grin patch," which helps to distinguish this species from the similar Ross's Goose. Size: Length 28–31 inches; wingspan 53–56 inches.

Habitat and Distribution: Found statewide on wetlands, lakes, and farmlands, where they feed in grainfields. Has been recorded in all 105 Kansas counties.

Seasonal Occurrence: Migrant and winter resident. Arrives in October and departs in March. Locally abundant in migration, especially in late fall and early spring. Large numbers remain in winter where open water is present.

Field Notes: Snow Geese have become abundant in North America in recent decades. Especially in late fall and early winter, concentrations in the hundreds of thousands can be seen at locations such as Quivira NWR. They roost in wetland areas at night and during the day disperse to surrounding agricultural areas to feed. Few sights are more inspiring than watching wave after wave of these geese returning at sunset on a calm autumn evening.

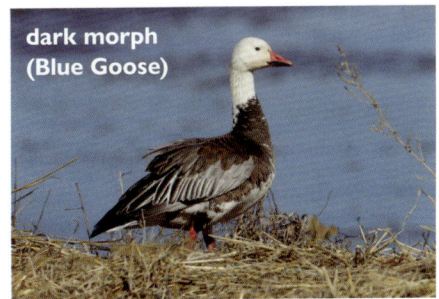

dark morph (Blue Goose)

Ross's Goose

Anser rossii

Field Identification: Looks like a small version of the Snow Goose. Like the Snow Goose, it has light and dark color morphs, although the dark morph is rare. It is distinguished from the Snow Goose by its smaller size, more rounded head, shorter neck, and small, stubby bill, which lacks the black "grin patch" of the Snow Goose. Size: Length 23 inches; wingspan 45 inches.

Habitat and Distribution: Statewide on lakes, ponds, and marshes. Often feeds in agricultural fields. Has been recorded in all but two Kansas counties.

Seasonal Occurrence: Migrant and winter resident. Arrives in late October and departs in March.

Field Notes: Ross's Geese were considered rare until the 1990s. As populations of the more abundant Snow Goose climbed in recent decades, so have those of the Ross's Goose. Flocks of Snow Geese often have some Ross's mixed with them. Single white geese seen with flocks of Canada Geese often prove to be Ross's.

Cygnus buccinator

Trumpeter Swan

adult with cygnets

Field Identification: Trumpeters are the largest waterfowl species in North America. Adults are all white, with a prominent all-black bill. Juveniles are dingy gray white, with some pink coloration on the bill. Typically, this species has a straight neck posture. Another swan species that is sometimes reported in Kansas is the introduced **Mute Swan**, which sometimes escapes from captivity. Mute Swans can be easily identified by their bright orange bill and curved neck posture. Size: Length 60 inches; wingspan 80 inches.

Habitat and Distribution: Wetlands, lakes, and farm ponds. Has been recorded in 81 Kansas counties, mostly in central and eastern Kansas.

Seasonal Occurrence: Migrant and winter resident. Seen from November through March.

Field Notes: Trumpeter Swans have been the beneficiaries of a vigorous reintroduction effort in several northern states, including Iowa, Minnesota, and Wisconsin. Most if not all of the birds seen in Kansas are from these reintroduced populations. Frequently they have neck collars to assist researchers in identifying individuals in the field.

Mute Swan

Tundra Swan

Cygnus columbianus

Field Identification: Tundra Swans are a bit smaller than Trumpeter Swans in both length and wingspan. The neck is slightly shorter and more slender than that of the Trumpeter Swan. The most reliable field mark is the shape of the head and the bill configuration. The Tundra Swan has a more rounded head and more curve on the top of the beak. The place where the bill meets the forehead is U-shaped, as opposed to the V-shape of the Trumpeter Swan. A yellow spot at the base of the bill in front of the eye is diagnostic but not always present. This can be difficult to discern at a distance. Immatures look a lot like immature Trumpeter Swans. Size: Length 49 inches; wingspan 75 inches.

Habitat and Distribution: Wetlands and lakes, usually in rural areas. Tundra Swans can be found at Quivira NWR in most years in late fall. Has been recorded in 64 Kansas counties, mostly in central and eastern Kansas.

Seasonal Occurrence: Migrant and winter resident. Arrives in November and departs in March. Wintering birds remain where open water is available.

Field Notes: Tundra Swans are more likely to be found in Kansas now than they were for most of the twentieth century. Frequently they are seen with Trumpeter Swans.

© Judd Patterson

Dendrocygna autumnalis

Black-bellied Whistling-Duck

Field Identification: This distinctively marked duck is unmistakable. It has a bright red bill contrasting with a pale gray head. The neck and upper breast are a burnt-orange color, and the belly is black. It has long bright pink legs and is frequently seen standing on shorelines. When in flight, the wings have a bold black-and-white pattern and the long legs extend beyond the tail. The sexes are identical. Juveniles have the same plumage characteristics but more muted colors. Size: Length 21 inches; wingspan 30 inches.

Habitat and Distribution: While it is to be looked for at the major wetlands, this species is often seen in city parks. It also seems to have an affinity for cattle feedlots. Its numbers are rapidly increasing in Kansas. Currently this species been recorded in 43 counties in eastern and central Kansas,

Seasonal Occurrence: Migrant and summer resident. Nesting has been confirmed in five south-central counties. Most records fall between April and September. In early spring, small flocks can show up just about anywhere.

Field Notes: This species has dramatically increased its range in Kansas and throughout the US over the past 20 years. Flocks exceeding 20 birds have been recorded in Pawnee County and elsewhere and have included multiple broods of fledglings. This species seems destined to become a more common Kansas bird in the future.

31

Wood Duck

Aix sponsa

Field Identification: The ornate colors of the male make it unmistakable. Females are soft gray overall, with white spotting on the flanks and a large white teardrop-shaped area around the eyes. Both sexes have a relatively long tail. Distinctive wheezy calls are given when individuals are alarmed or excited. Size: Length 18 inches; wingspan 30 inches.

Habitat and Distribution: Found on rivers, streams, and small ponds with trees along the banks. Has been reported in all 105 Kansas counties. It is most numerous in the eastern half of the state but is steadily becoming more numerous westward.

Seasonal Occurrence: Permanent resident. Most birds are present from March through November, with the greatest numbers seen in April and October as migrants move through. Widespread in summer. Breeding has been confirmed in 54 counties, with probable breeding in nearly 30 additional counties. Small numbers remain in winter where they can find open water.

Field Notes: This is one of the most colorful birds found in Kansas and one of the few waterfowl species that nests widely in the state. It readily adapts to urban areas. Wood Ducks nest in tree cavities as much as 50 feet above the ground. They were nearly exterminated by habitat loss in the early twentieth century but have since made a strong recovery throughout their range.

female

Mareca strepera

Gadwall

Field Identification: The male of this relatively nondescript duck has a pale head and neck, contrasting with a darker breast and gray flanks. Swimming males show a prominent black rump, which is the most useful field mark. Females are mostly light brown, with white fringes on the feathers and a pale throat. Flying birds of both sexes clearly show a neatly defined patch of white on the trailing edge of the wings. This is often visible on swimming birds as well. Size: Length 20 inches; wingspan 33 inches.

Habitat and Distribution: Found statewide on shallow ponds and marshes. Has been recorded in all 105 counties.

Seasonal Occurrence: Primarily a migrant and winter resident, but small numbers remain to nest in summer. Breeding has been confirmed in 13 counties, mostly in central Kansas. Migrants and wintering birds are seen from September through May, with the greatest numbers seen between mid-October and mid-April. Some remain throughout the winter where water remains open.

Field Notes: Ducks are often separated into "dabbling ducks," which feed in shallow water by tipping vertically to reach aquatic vegetation, and "diving ducks," which are usually found in deeper water, where they dive completely below the surface to feed. During migration, Gadwall are one of the most numerous dabbling duck species in Kansas. They tend to forage in somewhat deeper water than most of the other dabbling duck species.

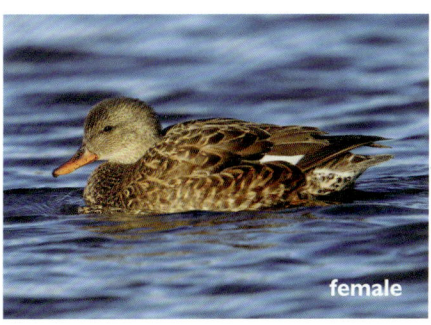

female

American Wigeon

Mareca americana

Field Identification: Males have a pinkish-brown breast and flanks, with gray cheeks, a broad green eye-stripe, and a prominent white crown. Females have more muted pinkish brown breast and flanks, with an entirely gray head speckled with black. On flying males, note the broad white wing panel. Size: Length 20 inches; wingspan 32 inches.

Habitat and Distribution: Statewide on shallow lakes, ponds, and wetlands. Has been recorded in all 105 counties.

Seasonal Occurrence: Migrant and winter resident. Some may remain at major wetlands into the summer months. There are confirmed breeding records from three counties. In fall, it arrives in September and departs in May. Common in migration. The greatest numbers are seen in March and October. Uncommon winter resident in areas where water remains open.

Field Notes: This is one of our more colorful ducks. They are often seen in small groups near flocks of other ducks. They sometimes graze on grass and winter wheat with flocks of geese.

female

Anas platyrhynchos

Mallard

Field Identification: This duck is recognized by nearly everyone. The male has an iridescent green head, bright yellow bill, chestnut breast, and light gray body. Females are mottled brown with grayish head, dark eye-line, and a yellow bill with a black saddle on the upper mandible. In flight, look for blue secondaries bordered with white on both males and females. In some city parks with domestic duck populations, Mallards will hybridize with them, producing a variety of intermediate-plumaged offspring. Size: Length 23 inches; wingspan 35 inches.

Habitat and Distribution: Found statewide on lakes, ponds, streams, and wetlands of all sizes. Mallards are not selective and readily adapt to almost any habitat with water, including flooded fields. Large flocks are seen at reservoirs and marshes in the nonbreeding season. They sometimes graze in corn and milo fields like geese. Has been recorded in all 105 counties.

Seasonal Occurrence: Permanent resident. Most numerous during migration and in winter, but many remain to nest, often in urban areas. There are confirmed breeding records from 77 counties.

Field Notes: This dabbling duck is the most abundant and widespread waterfowl species in Kansas. Nests are usually located near ponds or streams and sometimes several hundred yards from the nearest open water. Mallards are fully habituated to humans and can be found in densely populated urban areas. Females with family groups of up to 12 young are often seen in the summer months.

female

Northern Shoveler

Spatula clypeata

Field Identification: The most prominent field mark on both sexes is the unusually long and broad bill. Males have a solid green head, white breast, and chestnut flanks. Females are uniformly mottled brown. In flight, both sexes show a large blue wing-patch like that of the Blue-winged Teal. The patch is slightly paler on females. Size: Length 19 inches; wingspan 30 inches.

Habitat and Distribution: Statewide on wetlands and shallow bodies of water, including sewage lagoons. Has been recorded in all 105 counties.

Seasonal Occurrence: Migrant and summer resident. Spring migrants are seen from March through May, fall migrants from July through November. Occasionally nests in central and western Kansas. Confirmed breeding has been reported in 10 counties, mostly in central and western Kansas. Some remain through the winter in southern Kansas where water remains open.

Field Notes: This is an abundant duck in Kansas, especially during migration in central and western areas of the state. The large bill is used to strain out small aquatic life. Sometimes these ducks form large, rotating groups that feed together in a big swirling pinwheel, with heads almost submerged in the water. They often associate with Blue-winged Teal. Formerly rare in winter, in recent years flocks of 100 or more birds have wintered in southern Kansas.

female

Anas crecca

Green-winged Teal

Field Identification: This is a small, fast-flying dabbling duck. The male has a chestnut head with a broad green eye-stripe, a prominent vertical white slash mark on the flanks, and a buffy-yellow patch at the base of the tail. Females are dark brown and lack the white around the eye and the long bill of other female teal. In flight, both sexes show a green trailing edge on the wing. Size: Length 14 inches; wingspan 23 inches.

Habitat and Distribution: Statewide on wetlands, rivers, and ponds. Has been recorded in all 105 counties.

Seasonal Occurrence: Common migrant and uncommon winter resident. Fall arrival is in September, spring departure in May. Spring migrants are most abundant in March and early April. Fall migrants are most abundant from late October through late November. A few remain in the summer months. Confirmed breeding has been reported from five counties scattered across the state. Probable breeding has been reported from five additional counties.

Field Notes: These ducks are sometimes seen in flocks of several hundred birds. They fly in compact groups that execute deft maneuvers. When excited or alarmed, the flocks make distinctive high-pitched peeping calls.

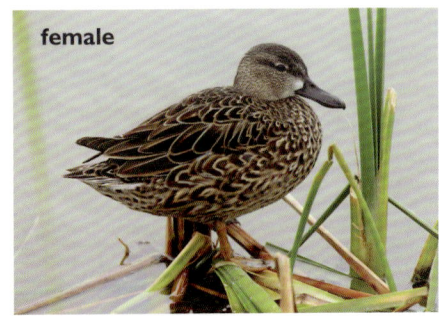

female

Blue-winged Teal

Spatula discors

Field Identification: The Blue-winged Teal is a small dabbling duck. In spring, the male has a dark blue-gray head with a prominent white crescent mark at the base of the bill. Females have brown feathers with pale edges, a thin dark line through the eye, and a thin white eye-ring. Males in late summer and early fall look like females. At all seasons, both sexes show a large blue wing-patch in flight. The three teal species can be distinguished from most other waterfowl species at some distance by their noticeably smaller size and fast flight. Size: Length 15 inches; wingspan 23 inches.

Habitat and Distribution: Found statewide on wetlands, shallow lakes, and ponds. There are records from all 105 counties.

Seasonal Occurrence: Migrant and summer resident. Arrives in April and departs in October. During migration,

they can be found on even the smallest bodies of water. Some remain to nest in summer. Confirmed breeding has been reported from 35 counties, probable breeding from an additional 18 counties. This is one of the waterfowl species more likely to be observed in the summer months in Kansas.

Field Notes: This is one of the most abundant and widespread waterfowl species in Kansas. In the spring, they tend to arrive later than other ducks, and in the fall, they are among the first to depart for the south.

female
© Judd Patterson

Spatula cyanoptera

Cinnamon Teal

Field Identification: Males are a rich rusty-red color for most of the year. The bill is large and spatulate but not as much as that of the closely related Northern Shoveler. Females and early-fall males resemble female Blue-winged Teal except for the size and shape of the bill. The bright red eye is a useful way of distinguishing them from Blue-winged Teal in this fall plumage. In flight, they show blue patches on the wings like those of Blue-winged Teal. Size: Length 16 inches; wingspan 22 inches.

Habitat and Distribution: Wetlands and shallow ponds, especially in the western half of the state. They are seen annually at Cheyenne Bottoms and Quivira NWR. Most likely to be seen in western Kansas, but spring migrants are reported annually in the eastern counties. There are records from 86 counties.

Seasonal Occurrence: Migrant and rare summer resident. Observed most often

in April and May, but there are records from the fall months. In some years, a few pairs remain to nest at Cheyenne Bottoms and in southwestern Kansas. Confirmed breeding has been reported from only two counties, probable breeding from three other counties.

Field Notes: It is always a treat to find one of these teal in Kansas. Check large flocks of Blue-winged Teal, and you might be lucky enough to pick one of these birds out of the flock. This close relative of the Blue-winged Teal some-times hybridizes with it, producing birds with intermediate plumage characteristics.

female

Northern Pintail

Anas acuta

Field Identification: Both sexes of this large, slender dabbling duck have dark gray bills, long necks, and long, pointed tails, longest in the male. Males have a brown head, gray back and flanks, white breast, and long white stripe on the sides of the neck. Females have a plain, evenly brown head and neck and a brown body with strong V-shaped markings on the feathers. Flying birds can be identified by their long wings and tails. Size: Length 21 inches; wingspan 34 inches.

Habitat and Distribution: Statewide on wetlands, lakes, and ponds. There are records from all 105 counties.

Seasonal Occurrence: Primarily a migrant. A few remain in Kansas in both summer and winter. Spring migrants are seen from February through May, fall migrants from September through November. Occasionally remains to nest, most often at Cheyenne Bottoms. Confirmed breeding has been recorded in 13 counties, mostly in the western half of Kansas. A few remain through the winter where water remains open.

Field Notes: One of the first signs of spring in Kansas is the arrival of large Northern Pintail flocks during the last half of February. Some of these flocks number in the thousands. Many regard this as the most elegant and graceful of the waterfowl family.

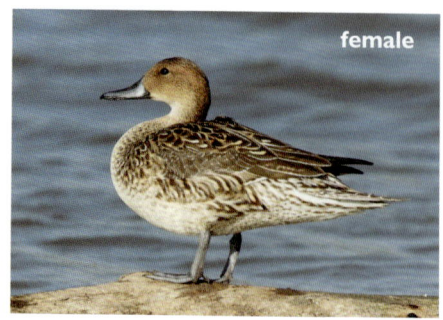

female

Aythya valisineria

Canvasback

Field Identification: Both sexes have a distinctively sloped head and a large, curved black bill. Males have a white body, black breast and tail, and red head. Females have a grayish body and brown head. The underwings of both sexes are white. Canvasbacks can be identified at some distance by the bright white plumage of the males. Size: Length 21 inches; wingspan 29 inches.

Habitat and Distribution: Statewide on marshes, ponds, and lakes. There are records from all 105 counties.

Seasonal Occurrence: An uncommon migrant. Present in winter where water remains open. Arrives in fall in October, departs in April. Occasionally nests at Cheyenne Bottoms and Quivira NWR. There are no other nesting records from elsewhere in Kansas.

Field Notes: This duck feeds by diving below the surface of the water to forage. Most diving ducks feed on shellfish and other aquatic animals, but the diet of Canvasbacks consists mostly of aquatic vegetation. Their population has declined because of the loss of nesting habitat in the prairie pothole region of the northern Great Plains. Flocks of several hundred birds can be seen in March and October.

female

Redhead

Aythya americana

Field Identification: Males superficially resemble Canvasbacks, but their body is gray instead of white, the head is a brighter shade of red, and the bill is pale blue with a dark tip. Females are a uniform soft brown, paler on the face. Redheads have a more rounded head and less prominent bill than Canvasbacks. Size: Length 19 inches; wingspan 29 inches.

Habitat and Distribution: Statewide on lakes and marshes. Has been recorded in all 105 counties.

Seasonal Occurrence: Common migrant, uncommon winter resident. Fall migrants arrive in October and remain numerous through November. Small numbers remain through winter. Spring migrants arrive in late February and depart in April. Spring migrants are most abundant in March. Each year a few pairs remain to nest at Cheyenne Bottoms and Quivira NWR, occasionally elsewhere. Breeding has been confirmed in five south-central counties.

Field Notes: This colorful duck is often seen in large flocks on deepwater lakes and reservoirs in spring and fall. It is the only diving duck likely to be seen during the summer in Kansas, especially along Quivira NWR's Wildlife Drive, where adults with downy young can often be seen during July.

female

Aythya collaris

Ring-necked Duck

Field Identification: Males have a dark head, breast, and back with pale gray flanks. A prominent field mark on the male is the white vertical slash mark on the side of the breast. Females are mostly dark brown with lighter gray cheeks and a distinct white eye-ring. On both sexes, a distinct white ring separates the black tip from the dark gray base of the bill. Size: Length 17 inches; wingspan 25 inches.

Habitat and Distribution: Statewide on wetlands, lakes, and ponds. Has been recorded in all 105 counties.

Seasonal Occurrence: Common migrant, uncommon in winter. Arrives in October and departs in April. Not present in summer.

Field Notes: These ducks are usually seen in smaller flocks than the other diving ducks. They are also more likely to be observed on smaller ponds. The species is named for the chestnut collar on the male that is difficult to see in the field, even at close range.

female

Lesser Scaup

Aythya affinis

Field Identification: Males have a dark iridescent purple head and breast, gray back with whitish flanks, and black tail. They are "black on the ends and light in the middle." Males can be easily separated even from a distance from Ring-necked Ducks, which have black backs. Females are a uniform dark brown except for a bright white area surrounding the base of the bill. Distinguishing Lesser Scaup from Greater Scaup is discussed in the next account. Size: Length 17 inches; wingspan 25 inches.

Habitat and Distribution: Statewide on large reservoirs, lakes, ponds, and marshes. Has been recorded in all 105 counties.

Seasonal Occurrence: Common migrant and uncommon winter resident. Rare summer resident. Fall migrants arrive in October and are abundant in November. Some remain through the winter. Spring migrants are seen from late February through early May and are most abundant in March. The only nesting records are from Cheyenne Bottoms, Quivira NWR, and Slate Creek WA, and only one of these has been fully confirmed.

Field Notes: Large rafts of these diving ducks can be seen in migration, often with other diving duck species, especially Redheads.

female

Aythya marila

Greater Scaup

Field Identification: Greater Scaup and Lesser Scaup are closely related and similar in appearance. Male Greater Scaup have a greenish sheen on the head. Male Lesser Scaup have a purple sheen on the head. Male Greater Scaup have a smoothly rounded head, as opposed to the more angular-looking and slightly peaked head of Lesser Scaup. The head of Greater Scaup appears larger in proportion to the body than does the head of Lesser Scaup. Greater Scaup have a broader bill that expands near the tip. The bill of Lesser Scaup is a bit narrower and does not widen at the tip. In general, male Greaters appear whiter on the flanks than do Lessers, but this is subjective when looking at a single bird. Female Greaters are larger overall, with a larger and more rounded head. The white area at the base of the bill is typically larger on Greater Scaup. Size: Length 18 inches; wingspan 28 inches.

Habitat and Distribution: Found across most of the state on large reservoirs, lakes, ponds, and marshes. Has been recorded in 91 counties. Most records are from central and eastern Kansas. Much less common in the west.

Seasonal Occurrence: Migrant and local winter resident. Fall migrants arrive in early November. Some remain through the winter at favored locations. Spring migrants depart northward during April.

Field Notes: Once considered uncommon, this species is increasing in numbers. Winter flocks sometimes exceed 100 birds in the Wichita area and at some eastern Kansas reservoirs.

female

Common Goldeneye

Bucephala clangula

Field Identification: Males are mostly white. The have a puffy black head with a prominent white oval at the base of the bill. Females have a soft gray body with a warm brown head. Both sexes show large white wing-patches in flight. The wings make a metallic whistling sound when these birds are in flight. Size: Length 19 inches; wingspan 26 inches.

Habitat and Distribution: Statewide on lakes and reservoirs with deep water. Has been recorded in 101 counties.

Seasonal Occurrence: Common migrant and winter resident. Fall arrival is in mid-November; spring departure is in mid-March. In spring, a few stragglers remain through late April. This is the last duck species to arrive in the fall and the earliest to depart north again in the spring.

Field Notes: Common Goldeneyes thrive in cold weather. When harsh winter weather drives other waterfowl south, Common Goldeneyes remain present on rivers and large reservoirs until no patch of open water remains. In February, the males begin to court females with a dramatic display. They throw their heads back and give a loud nasal vocalization, then snap the head forward with great emphasis.

female

Bucephala albeola

Bufflehead

Field Identification: This is the smallest duck found in Kansas. Males have a white body with a black back. When seen in good light, the black portion of the head reflects a rainbow of iridescent colors. Females are mostly dark gray but have a large round white spot on the head. Both sexes have large white wing-patches that are visible in flight. Size: Length 13 inches; wingspan 21 inches.

Habitat and Distribution: Statewide on wetlands, lakes, ponds, and reservoirs. Has been recorded in all 105 counties.

Seasonal Occurrence: Common migrant. Uncommon but regular in winter where open water remains. Arrives in fall in late October, departs in spring in April. Fall migrants are most abundant in November, spring migrants in March.

Field Notes: The name is derived from the disproportionately large head, which is vaguely suggestive of a buffalo. Buffleheads typically linger later in both spring and fall migration than most other diving duck species. They do not gather in large flocks but occur in smaller groups, which forage on the edges of large rafts of other diving duck species.

female

Hooded Merganser

Lophodytes cucullatus

Field Identification: This is the smallest of the mergansers, almost as small as a teal. Their flight is fast and agile. The male has a black back, reddish-brown flanks with two black bars near the breast, and a black head with a large white patch that can be raised into a puffy crest. The female is dark gray, with a smaller frosty brown crest. Immature males are similar in appearance to females. Size: Length 18 inches; wingspan 24 inches.

Habitat and Distribution: Statewide at wetlands and lakes as well as streams in or near wooded areas. Has been recorded in all 105 counties.

Seasonal Occurrence: Migrant and winter resident. Rare in summer. Most are observed between late October and mid-April. There are confirmed breeding records from eight counties, mostly at eastern wetlands such as Benedictine Bottoms, Marais des Cygnes, and Neosho WA. A few immature birds are found each year in the summer months at scattered locations across the state.

Field Notes: This is one of the most strikingly attractive duck species in Kansas. Males displaying for females raise and lower their prominent crests with dramatic flourishes. Mergansers are diving ducks that feed primarily on fish. They have narrow, pointed beaks with serrated edges that allow them to catch and hold their prey. This is one of the waterfowl species that remains common through the winter. Flocks at favored wintering areas can exceed 100 individuals.

female

Mergus merganser

Common Merganser

Field Identification: These mergansers appear long and slender when swimming and in flight. The male is mostly white, with dark back and tail, green head, and orange bill. Females are gray with a brown head and short crest, clean-cut white throat, and orange bill. Size: Length 25 inches; wingspan 34 inches.

Habitat and Distribution: Large reservoirs and deepwater lakes, mostly in the eastern two-thirds of the state. Has been recorded in 96 counties.

Seasonal Occurrence: Migrant and winter resident. Fall arrival is in November, spring departure in March. Like the Common Goldeneye, it is among the last of the duck species to arrive in the fall and among the earliest to depart in the spring.

Field Notes: These are familiar diving ducks during winter months, when they form large flocks on large reservoirs. These flocks may number in the tens of thousands of birds. If the lakes freeze, they move south into Oklahoma and Texas but return immediately when open water is again available. They feed on fish, primarily gizzard shad. They are social and usually swim and fly in tightly grouped flocks.

female

Red-breasted Merganser

Mergus serrator

Field Identification: This merganser also has a long and slender appearance, both on the water and in flight. The male has a green head with a shaggy crest, white collar, dark reddish-brown breast, gray flanks, and black back. The female resembles the female Common Merganser except for the evenly brownish-gray throat and more prominent crest. Both sexes have longer, thinner bills than Common Mergansers. Size: Length 23 inches; wingspan 30 inches.

Habitat and Distribution: Statewide on wetlands, reservoirs, and deep lakes. Has been recorded in 88 counties. Possible anywhere but most likely to be seen in central and eastern Kansas.

Seasonal Occurrence: Migrant. Spring migrants are seen in March and April, fall migrants mostly in November. Occasionally seen in winter.

Field Notes: This is the least observed merganser in Kansas, and finding one is always noteworthy. It is never found in the huge flocks that Common Mergansers form. Most sightings are of just a few individuals, although flocks of up to 30 birds can sometimes be seen in November at locations such as Cheney Reservoir, John Redmond Reservoir, and Wilson Reservoir.

female

© Judd Patterson

Oxyura jamaicensis

Ruddy Duck

breeding male

Field Identification: Males have black caps and white cheeks and are colorful in full breeding plumage, when their bills are bright powder blue and their bodies brick red. In winter plumage, males retain the same head pattern but have a dark gray back, white underparts, and dark bills. Females at all seasons have a dark line crossing the white cheek below the eye. Ruddy Ducks often swim with their tails held stiff and upright. Size: Length 15 inches; wingspan 18 inches.

Habitat and Distribution: Statewide on marshes, reservoirs, and lakes. Has been recorded in all 105 counties.

Seasonal Occurrence: Common migrant and uncommon winter resident. Winters locally on lakes with open water, mostly in the southeastern third of the state. Fall arrival is in October, spring departure in May. Fall migrants are most abundant in late October and November. Spring migrants are most abundant in late March and early April. In some years, a few pairs remain to nest. There are confirmed breeding records from seven counties, including several southwestern counties. Probable breeding has been reported in five additional counties. In summer, most likely to be seen at Cheyenne Bottoms and Quivira NWR.

Field Notes: In winter, the dark plumage blends well with dark water, and they can be difficult to see until the white cheeks are visible. Their flight is direct and rapid.

female

nonbreeding male

51

Long-tailed Duck

Clangula hyemalis

nonbreeding male

Field Identification: Long-tailed Duck identification can be challenging because they have three distinctive seasonal plumages. Almost all that are seen in Kansas are winter adults or first-winter birds. The summer plumage is seldom seen in Kansas. The adult winter male is mostly white with black breast, a large black patch on the side of the head, and a long streaming tail. Adult winter females and first-year males are largely white with a brownish cap, wings, and back and brownish patch on the side of the head. Size: Length 16 inches; wingspan 28 inches.

Habitat and Distribution: Reservoirs and lakes. Has been recorded in 53 counties, largely in central and eastern Kansas but also in eight southwestern counties.

Seasonal Occurrence: Migrant and winter resident. Observed mostly between November and March.

Field Notes: This species and the three scoter species are "sea ducks." While these species are usually found on ocean coasts, inland sightings have grown in frequency over the past 25 years. Sea duck species are most often seen at larger reservoirs, especially along the faces of dams, but there are numerous records from smaller impoundments. Sometimes wintering flocks of 5–10 individuals can be seen at Clinton State Park and other large reservoirs.

nonbreeding female

Melanitta perspicillata

Surf Scoter

Field Identification: Adult males are mostly all black with distinctive white patches on the back of the neck and forecrown and a huge yellow-and-white bill. Females are mostly dark brown, with a vertical white patch behind the bill and a round white spot behind the eye. First-winter immature birds resemble females but are lighter brown with whitish flanks and belly. Size: Length: 20 inches; wingspan 30 inches.

Habitat and Distribution: Reservoirs and lakes. There are records from 50 counties, largely in central and eastern Kansas but also in eight southwestern counties.

Seasonal Occurrence: Migrant and rare winter resident. Most individuals seen in Kansas are females or immatures. Arrives in November and departs in March. Most likely to be seen in November and December.

Field Notes: It's always a good birding day when you find a scoter in Kansas. Most sightings of these sea ducks are of one or two individuals at the large reservoirs, but they have also been observed on much smaller bodies of water.

female

White-winged Scoter

Melanitta deglandi

© David Seibel

Field Identification: This is the largest scoter. Adult males are entirely black with a teardrop of white surrounding the eye and a sharply defined white patch on the secondaries that can usually be seen on swimming birds. Females are dark brown with white patches at the base of the bill and on the sides of the head. First-year males resemble females except that the flanks and belly are whitish. The white wing-patch is present in all plumages but is not always visible. On female and immature birds, the white patch at the base of the bill is important for identification. On Surf Scoters, this is a well-defined vertical oval. On White-winged, it is rounder, smaller and fainter. Size: Length: 21 inches; wingspan 36 inches.

Habitat and Distribution: Reservoirs and lakes. Has been recorded in 41 counties, mostly in central and eastern Kansas but also in six southwestern counties.

Seasonal Occurrence: Migrant and winter resident. Most are observed between mid-November and March.

Field Notes: As with the Surf Scoter, most Kansas sightings are of females and first-winter males.

© David Seibel

Melanitta americana

Black Scoter

Field Identification: Adult males are entirely black except for the bright orange knob at the base of the bill. Adult females are mostly a solid dark brown except for the white facial plumage below the eye and the neck. First-winter males resemble adult females but are not as dark overall, and the white plumage is not as cleanly defined. The bill structure of Black Scoters is not as large and prominent as that of the other two scoter species. Size: Length 19 inches; wingspan 36 inches.

Habitat and Distribution: Reservoirs and lakes. Has been recorded in 37 counties in central and eastern Kansas.

Seasonal Occurrence: Migrant and winter resident. Most observations fall between November and March.

Field Notes: This was formerly considered to be the rarest of the scoter species in Kansas, but it is now reported annually. All scoters have increased in numbers in Kansas in the twenty-first century, but none more than Black Scoters.

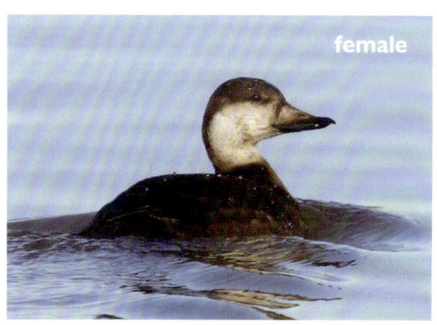

female

Common Loon

Gavia immer

Field Identification: This large swimming bird is seldom seen in flight. It dives and remains submerged for long periods. It rides low in the water and has a large dagger-shaped bill. In spring breeding plumage, look for the black head, striped white collar, checkered black-and-white back, and white breast and belly. In winter, the head and back are an even grayish brown, and the throat and chin are white. Size: Length 32 inches; wingspan 46 inches.

Habitat and Distribution: Found in large reservoirs and deep lakes. Has been recorded in 82 counties. Most records come from the eastern two-thirds of the state, but there are also records from several western counties.

Seasonal Occurrence: Spring and fall migrant. Spring migrant from mid-March through April; fall migrant from late October through early December. It is most numerous in early November.

During winter, a few occasionally remain at major reservoirs with open water.

Field Notes: Three other loon species have been reported in Kansas, all of which were considered too rare to be included in this book. Most Common Loon sightings are of single birds, but at reservoirs such as Wilson, Perry, or Cheney, it is possible to see up to 30 in a single day. If you are lucky, you may hear their wild and haunting primordial calls.

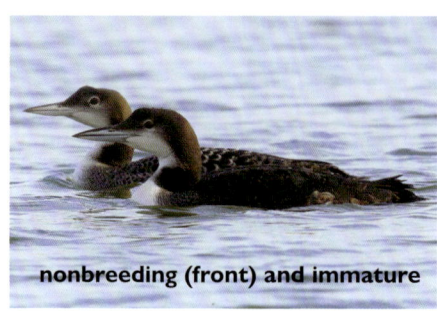

nonbreeding (front) and immature

Podilymbus podiceps

Pied-billed Grebe

Field Identification: Pied-billed Grebes are often mistaken for a small duck. They are gray and brown overall, with a black chin. In spring and summer, the pale, chicken-like bill has a black ring near the tip. In winter, the black chin and ring on the bill are lost. This species is seldom seen in flight and prefers to dive when disturbed. Size: Length 13 inches; wingspan 16 inches.

Habitat and Distribution: Reservoirs, lakes, ponds, and wetlands. Has been recorded in all 105 counties.

Seasonal Occurrence: Seen year-round, but populations fluctuate significantly with the season. Most numerous during spring and fall migration. Spring migration occurs from early March through early May, and fall migrants are observed from early September through early December. In summer, some remain to nest in wetlands throughout the state. Confirmed breeding has been reported from 26 counties. A few remain through the winter, especially in southern Kansas.

Field Notes: This is the most common grebe in Kansas. Almost any body of water can attract this species, even in urban areas. When submerging, they frequently sink out of sight like a submarine, and they sometimes swim with just their head remaining above water. At Cheyenne Bottoms and Quivira NWR, family groups can be seen during the summer. The young have striped faces. Their loud, echoing song is one of the familiar summer sounds of the marshes.

nonbreeding

immature

Horned Grebe

Podiceps auritus

Field Identification: In size and shape, this grebe resembles a small duck. Adults in breeding plumage have a bright rufous neck and flanks, a black head with flared yellow "horns," and red eyes. The bill is short and pointed. In winter, they are all black and white, with a broad white cheek patch and clean-cut black cap. In general, this grebe rides lower in the water and is more flat-headed in appearance than the similar Eared Grebe. Size: Length 14 inches; wingspan 18 inches.

Habitat and Distribution: Reservoirs, lakes, and wetland areas with large areas of open water. Has been recorded in 94 counties. Most common in central and eastern Kansas.

Seasonal Occurrence: Spring migrant from March through May; fall migrant in October and November. A few linger into December in mild years. Not present in summer.

Field Notes: Observing these grebes in their brightly colored plumage in spring is always a treat. They are most numerous in late fall. These grebes sometimes join large flocks of diving ducks during migration.

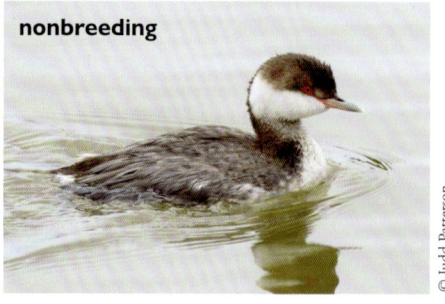

nonbreeding

© Judd Patterson

Podiceps nigricollis

Eared Grebe

Field Identification: In spring and summer, look for the black head, neck, and back; large feathered yellow "ears" on the sides of the head; and bright red eyes. In winter plumage, they can be confused with the Horned Grebe. On winter Eared Grebes, look for the dark plumage of the crown extending to the side of the face, often giving them a smudgy look, as opposed to the cleanly defined black cap and white cheek of the Horned Grebe. They often show a peaked crown, which is unlike the flatter head of the Horned Grebe. The Eared Grebe generally rides higher on the water, especially on the rear of the body. Look for the "big butt." Size: Length 13 inches; wingspan 16 inches.

Habitat and Distribution: Statewide, but most likely to be found in central and western Kansas. Has been recorded in 103 counties. Prefers shallow ponds and marshes but also seen regularly on large reservoirs.

Seasonal Occurrence: Spring and fall migrant. Some remain to nest in central and western Kansas, where confirmed breeding has been reported from four counties. They often nest in large, loose colonies. Northbound spring migrants can be seen from March into early June with a peak in late April. Fall migrants begin to arrive in September and remain common through November. A few remain into December in some years.

Field Notes: This grebe prefers shallow marshy habitats more than the closely related Horned Grebe.

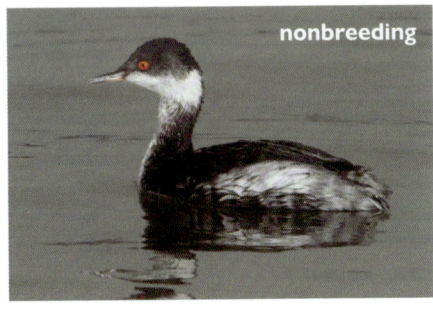

nonbreeding

Western Grebe

Aechmophorus occidentalis

Field Identification: A large grebe with a long, slender neck and a long, thin greenish-yellow bill. The crown, hindneck, and upperparts are all dark gray or black. The white throat and breast are cleanly separated from the darker plumage. In breeding plumage, the black plumage on the crown extends down to surround the eyes. Size: Length 25 inches; wingspan 24 inches.

Habitat and Distribution: Statewide at large reservoirs and wetlands. Has been recorded in 83 counties throughout the state.

Seasonal Occurrence: Spring migrant in April and May; fall migrant in October and November. Occasionally seen during the winter. Confirmed breeding has been recorded in two counties, with most breeding records from Cheyenne Bottoms.

Field Notes: These handsome birds are often seen in small flocks of 3–10 individuals in late fall on large reservoirs. In some years, a few remain to nest at Cheyenne Bottoms when water conditions are favorable. If you are lucky, you may see their dramatic courtship ritual, which involves both adults dancing across the water in unison with curved necks outstretched.

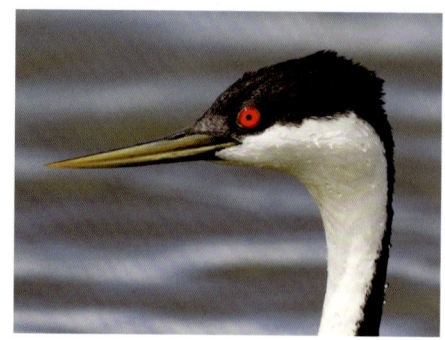

Aechmophorus clarkii

Clark's Grebe

Field Identification: This species closely resembles the Western Grebe. The bill is a brighter yellow color than the greenish-yellow bill of the Western Grebe. In breeding plumage, the eye is completely surrounded by white. In winter plumages, the head patterns of the two species can appear nearly identical. Clark's usually has brighter white flanks than Western. The bill color is the best field mark on winter-plumaged birds. Size: Length 25 inches; wingspan 24 inches.

Habitat and Distribution: Large lakes and reservoirs. Reported much less often than Western Grebe. Has been recorded in 33 counties, mostly in central and western Kansas.

Seasonal Occurrence: Spring migrant in April and May; fall migrant in October and November. There is one confirmed breeding record from Cheyenne Bottoms in Barton County and probable breeding from two other counties in western Kansas.

Field Notes: At Cheyenne Bottoms, courtship between Clark's and Western Grebes has been observed in some years, and this could have produced hybrid offspring.

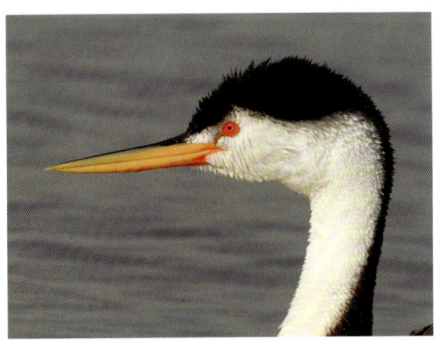

Common Gallinule

Gallinula galeata

Field Identification: This secretive bird of dense marshes is closely related to rails. It swims like a duck but walks like a shorebird. Breeding adults are uniformly slate gray with a scarlet bill. The bill is grayer on juveniles and winter adults. In all plumages, look for the long white horizontal mark on the sides, a field mark not shared by the similar and much more abundant American Coot. Size: Length 14 inches; wingspan 21 inches.

Habitat and Distribution: Eastern Kansas wetlands. Most recent sightings come from Baker Wetlands, Cheyenne Bottoms, and Quivira NWR, although it has been recorded in 29 central and eastern counties.

Seasonal Occurrence: Summer resident. Present from early April through October. There are confirmed breeding records from eight counties.

Field Notes: A walk along the levees at Baker Wetlands or a drive along the dikes at Cheyenne Bottoms and Quivira in summer may yield a glimpse of one of these enigmatic birds lurking at the edge of cattails. Most sightings are brief, as they swiftly swim or walk into dense cattails.

Fulica americana

American Coot

Field Identification: Coots are relatives of the rails, but in size, shape, and habits, they are more like waterfowl. The plumage is uniformly slate gray, darkest on the head. On adults, the bill is bright white and chicken-like. Juveniles are lighter gray with a grayish bill. This is a vocal species with a variety of resonant calls. Coots typically bob their heads while swimming, creating a comical impression. Size: Length 16 inches; wingspan 24 inches.

Habitat and Distribution: Statewide at marshes, lakes, ponds, and streams. Has been recorded in all 105 counties.

Seasonal Occurrence: Migrant and summer resident. Rare winter resident. Abundant during migration. Spring migration is from early March through early May, fall migration from mid-September through late November. Fairly common in summer at large wetlands. Confirmed breeding has been recorded in 25 counties scattered throughout the state. Least numerous in winter, but a few will remain at favored locations where water remains unfrozen.

Field Notes: This is an abundant aquatic bird in Kansas. Flocks numbering in the thousands are seen during migration. Coots are not shy and frequently occur in urban areas. To become airborne, coots run across the water to gain momentum.

American White Pelican

Pelecanus erythrorhynchos

Field Identification: This huge white waterbird has a long and broad orange bill that separates it from swans, the only species with which confusion is even remotely possible. The trailing half of the wings is black, visible only on birds in flight. Size: Length 62 inches; wingspan 108 inches.

Habitat and Distribution: Marshes, lakes, and reservoirs, mostly in the eastern two-thirds of Kansas. Has been recorded in all 105 counties.

Seasonal Occurrence: Common and conspicuous migrant in spring and fall, especially in April and October. During the summer and winter months, a few linger at the central wetlands and major reservoirs. There are no breeding records from Kansas.

Field Notes: This is the largest bird found in Kansas. The large, pouched bill is used to scoop up fish in shallow water. Sometimes a group of these birds will feed communally, forming a long line of birds that swim toward the shore, driving and scooping fish in front of them. White Pelicans migrate in large flocks, soaring on air thermals in surprisingly graceful formations. These migrating flocks are sometimes seen far from water.

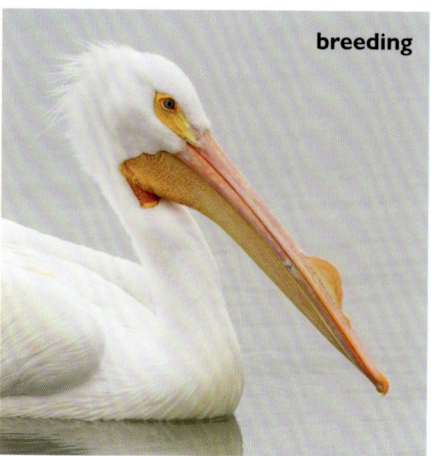

breeding

Nannopterum auritum

Double-crested Cormorant

© Judd Patterson

Field Identification: This is a large, dark waterbird, similar in habits and appearance to waterfowl. Adults are mostly black, with orange throat and bill. Birds less than two years old are brownish and paler on the throat and breast. It usually swims with the head angled upward, imparting a somewhat comical appearance. Size: Length 33 inches; wingspan 52 inches.

Habitat and Distribution: Wetlands, reservoirs, and lakes statewide. Has been recorded in all 105 counties.

Seasonal Occurrence: Can be seen year-round in Kansas. Abundant migrant in spring and fall. Smaller numbers are seen in summer and winter. Spring migration is from early March through late May with a peak in early April. Fall migration is from August through November with a peak in late October. There are confirmed breeding records from nine counties, most often at nesting colonies in north-central Kansas locations such as Cheyenne Bottoms, Glen Elder Reservoir, and Kirwin NWR. There are always a few nonbreeding individuals in summer. In many years, winter flocks numbering in the hundreds occur at lakes in Wichita and elsewhere in southern Kansas.

Field Notes: These birds are a familiar sight at wetlands and lakes of all sizes during migration. They fly in V-formations like geese and are fond of roosting on islands or sandbars surrounded by water. They are disliked by some fishermen, who are concerned that they are depleting game fish populations, but studies have shown that most of their diet consists of food fish species, such as gizzard shad.

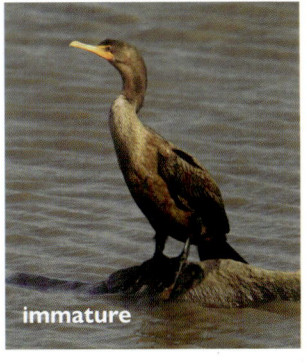

immature

65

Neotropic Cormorant

Nannopterum brasilianum

Field Identification:
Neotropic Cormorants are similar in appearance to Double-crested Cormorants but are smaller and slimmer, with proportionately longer tails. The most prominent mark for identification is the area of bare yellow skin at the base of the bill. On Neotropic adults, this comes to a sharp point behind the eye and is bordered with white feathers. On Double-crested Cormorants, this yellow area at the base of the bill does not come to a sharp point and is not bordered with white. Immatures can be difficult to separate from Double-crested but always have longer tails, slimmer bodies, and overall darker appearance. Size: Length 25 inches; wingspan 40 inches.

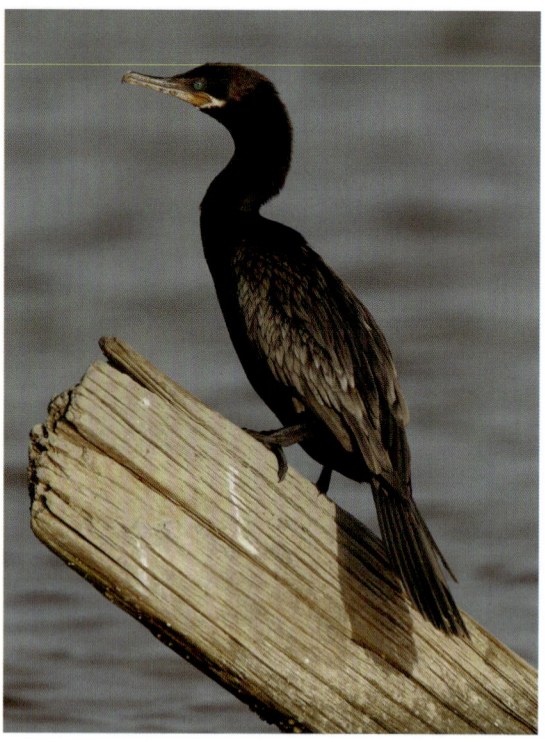

Habitat and Distribution: Marshes, reservoirs, and lakes. Has been recorded in 37 counties, all in central and eastern Kansas except for one record from Morton County.

Seasonal Occurrence: Migrant and summer resident. Most records fall between mid-March and early September, but there are records from all months of the year. Confirmed breeding has been recorded at Cheyenne Bottoms, and breeding has been suspected but not confirmed at several other locations. A few are sometimes present during the winter at John Redmond Dam, Arkansas City, and Wichita.

Field Notes: Singles or small groups are often found roosting amid larger flocks of Double-crested Cormorants. In 2011, there were records from 14 counties. By 2022, this had increased to 37 counties. Sightings continue to increase each year, likely due to our warming climate.

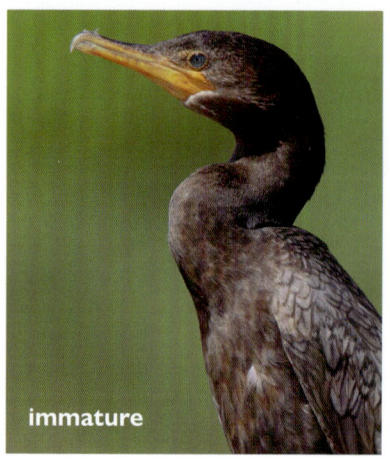

immature

Grus americana

Whooping Crane

Field Identification:
This crane is all white, like an egret, but much larger and more heavy-bodied, with a red crown. In flight, look for the black outer wing feathers, lacking on any heron or egret species. Immatures often show brownish feathers on the body and neck. Size: Length 52 inches; wingspan 87 inches.

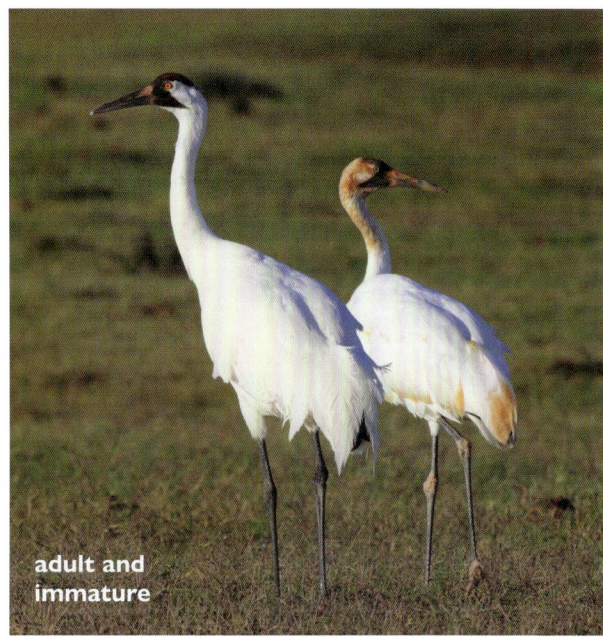

adult and immature

Habitat and Distribution:
Wetlands and grainfields adjacent to wetlands. There are records from 61 counties. Their typical migration flyway is a narrow band crossing the state from north to south. Most Kansas records are from Kirwin NWR, Quivira NWR, and Cheyenne Bottoms, all of which lie within this flyway. Migrating birds regularly stray east and west of this corridor.

Seasonal Occurrence:
Migrant in spring and fall. Most spring records are from April; fall records are from late October and early November. Occasionally single individuals linger at Quivira into December.

Field Notes:
This well-known endangered species has a wild population of several hundred individuals, most of which pass through central Kansas each year. If weather conditions are favorable for migration, their stay is brief, and they fly over the state without stopping. Occasionally a cold-weather front will incite them to remain for a few days as they wait for the winds to shift. Small numbers are seen each year at Cheyenne Bottoms and Quivira. They are generally more numerous and linger longer in the fall than in the spring.

Sandhill Crane

Antigone canadensis

Field Identification:
This large heron-like species is almost always seen in flocks. Sandhill Cranes are mostly gray except for the red crown. Immature birds lack the red crown and show more brownish color on the back. In flight, the neck is held straight out from the body, unlike herons, which always curl the neck back while flying. Their loud, chortling calls can be heard up to a mile away. Size: Length 46 inches; wingspan 77 inches.

Habitat and Distribution: Found at wetlands and in grainfields. Conspicuous migrating flocks can be seen anywhere. Has been recorded in 99 counties. Most often observed in central and western Kansas.

Seasonal Occurrence: Spring migrant in March and early April; fall migrant in October and November. In some years, wintering flocks numbering in the thousands are observed in Barber County and Meade County and at Quivira NWR.

Field Notes: Concentrations of Sandhill Cranes sometimes number in the tens of thousands during migration, especially in late fall but also in early spring. When wind direction and weather conditions are favorable, multiple migrating flocks of 50–100 cranes can be seen during a day trip across central and western Kansas. They usually roost together at night at favored staging areas and fan out to forage on waste grain during the day, returning near dusk. For a true wildlife spectacle, visit Quivira's Big Salt Marsh near sundown on an early-November evening. Watch as wave after wave of these huge birds return to the roost along with thousands of Snow and White-fronted Geese.

Ardea herodias

Great Blue Heron

Field Identification: This is the largest heron species in Kansas. Most of the plumage is blue-gray. The head is mostly white with black plume feathers. The bill is spike-shaped, long, and thick. In flight, the neck is folded back in an S-shape, unlike the straight neck of cranes. Size: Length 46 inches; wingspan 72 inches.

Habitat and Distribution: Found throughout the state in or near water at almost any stream or body of water. Has been recorded in all 105 counties.

Seasonal Occurrence: Permanent resident. Most move south during the coldest winter months, but a few remain through the winter. Found statewide in summer. Breeding has been confirmed in 90 counties.

Field Notes: This is the most widespread and conspicuous heron in Kansas. It is frequently seen in urban areas. Great Blue Herons nest communally in colonies, which are used year after year. Their nests are large stick platforms placed high in mature cottonwood and sycamore trees.

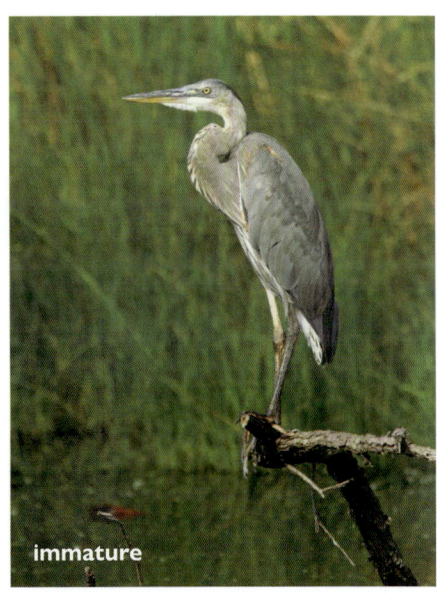

immature

Little Blue Heron

Egretta caerulea

Field Identification: Adults are similar in size and shape to Snowy Egrets but are a slate-blue color on most of the body, with a dark reddish-purple head and neck. Adults have a blue bill with a black tip. Juveniles are completely white and are most easily distinguished from the other white herons by their pale pinkish bill with a black tip. At about one year of age, the white plumage starts giving way to the dark adult plumage. This transitional plumage is sometimes referred to as "calico." Size: Length 24 inches; wingspan 40 inches.

Habitat and Distribution: Wetlands, streams, and lakeshores, mostly in the eastern two-thirds of Kansas. Has been recorded in 84 counties.

Seasonal Occurrence: Summer resident. Present from April through October. Confirmed breeding has been documented in 10 south-central and eastern counties.

Field Notes: This heron nests in large colonies that often include other heron species, including Great Egrets, Snowy Egrets, Western Cattle Egrets, and Black-crowned Night Herons. In late summer, postbreeding birds wander widely and can be found across the state. It sometimes feeds along wooded streams and shores, where other egrets are not typically found.

immature

© Judd Patterson

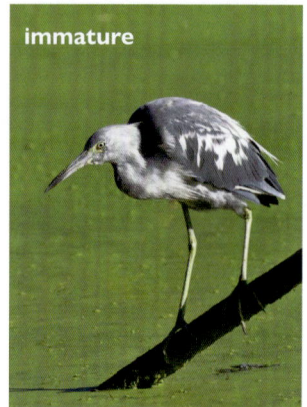

immature

© David Butel

Ardea alba

Great Egret

Field Identification: This is the largest of the white herons. The colors of the bill and legs are the most helpful field marks to easily differentiate the white heron species. Great Egrets have a large yellow bill, and the long legs are completely black. In spring and early summer, they have showy plume feathers, called aigrettes, on the breast and wings. Size: Length 39 inches; wingspan 51 inches.

Habitat and Distribution: Seen at lakes, ponds, streams, flooded fields, and wetlands. Mostly found in central and eastern Kansas but has been seen west to the Colorado border. Has been recorded in 100 Kansas counties.

Seasonal Occurrence: Summer resident. Arrives in late March, and most individuals depart in October. A few linger into November and early December. In late summer, postbreeding wanderers can turn up anywhere in the state. There are confirmed breeding records from seven south-central and eastern counties.

Field Notes: Egrets are herons that are white. Great Egrets are more numerous and widespread in Kansas today than they were 40 years ago. They typically nest in large mixed-species heron rookeries. Postbreeding wanderers sometimes gather in large numbers when abundant prey is present, especially at Cheyenne Bottoms, Glen Elder WA, and Quivira NWR. Like the Great Blue Heron, they feed on a variety of aquatic prey, such as fish and frogs, and readily hunt at almost any location with water.

courtship display

71

Snowy Egret

Egretta thula

Field Identification: This all-white heron is about two-thirds the size of the Great Egret. Showy head plumes appear in spring and early summer. The bill is all black and slenderer than that of the Great Egret. The legs of the adult are black with bright yellow feet. Immatures also have all-black bills, and their legs are a mixture of black and yellow. When excited, the bare skin on the face can turn from bright yellow to bright pink. Size: Length 24 inches; wingspan 41 inches.

Habitat and Distribution: Shallow wetlands, shorelines, and streams. Found most often in the eastern half of the state but can turn up anywhere in Kansas. Has been recorded in 92 counties. Least likely to be seen in the northwestern corner of the state.

Seasonal Occurrence: Summer resident. Arrives in early April and departs in late September. Breeding has been confirmed in six counties in south-central and southwestern Kansas. In late summer, postbreeding wanderers are seen statewide.

Field Notes: This egret usually favors shallower water than the Great Egret. Sometimes feeding birds raise their wings over their head and run erratically through the water as they attempt to capture fish. In late summer, hundreds of Snowy Egrets often gather at Quivira NWR, where they offer excellent views from the Wildlife Drive.

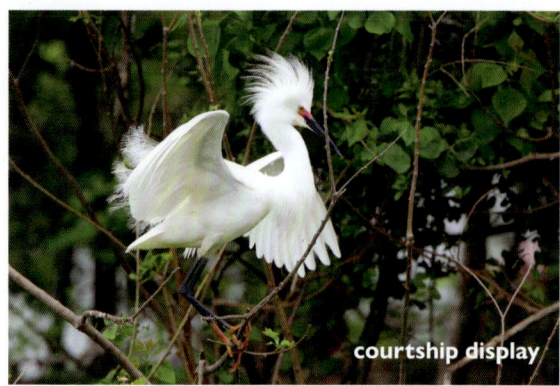

courtship display

Ardea ibis

Western Cattle Egret

Field Identification: This is the smallest of the white herons and the most likely to be seen away from water. In spring and early summer, adults have patches of light orange on the head, back, and breast. Compared with the other white herons, the Western Cattle Egret has a shorter neck, legs, and bill. The bill is all yellow. The legs are orange-yellow in the breeding season, darker the rest of the year. Size: Length 20 inches; wingspan 36 inches.

Habitat and Distribution: Found mostly in the eastern two-thirds of Kansas, but there are records from all but one Kansas county. Often seen in pastures and other grassy areas following livestock or mowers. Also common at marshes and other aquatic areas.

Seasonal Occurrence: Summer resident. Present from April through October. There are confirmed breeding records from seven counties scattered across the state.

Field Notes: This old-world species arrived in Brazil from Africa in the nineteenth century. From there, it expanded its range to include much of North and South America. It first arrived in Kansas in the 1960s and now is common and widespread in the warm seasons. In late summer, small flocks roam widely across Kansas. They feed on insects stirred up by the movement of cattle and often descend on recently mowed fields to feed on insects, earthworms, snakes, and small mammals.

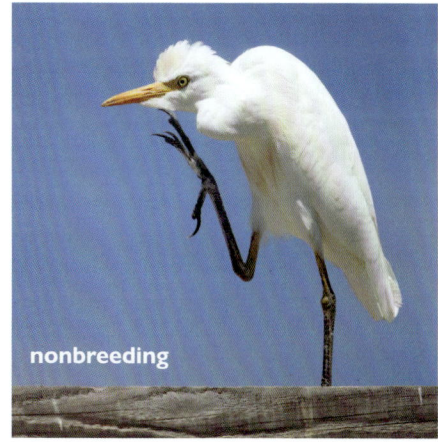

nonbreeding

Green Heron

Butorides virescens

Field Identification: This small, chunky heron has a greenish-blue back and crown, dark chestnut-brown neck and face, and orange legs. Juvenile birds have broad white and brown streaking on the breast. Flying birds can be mistaken for a crow at first glance but can be identified by their long bill and trailing legs. Size: Length 28 inches; wingspan 26 inches.

Habitat and Distribution: Found throughout Kansas. Has been recorded in 103 counties. Favors small, wooded streams but also frequents marshes and lakeshores.

Seasonal Occurrence: Summer resident. There are confirmed breeding records from 36 counties and probable breeding records from an additional 20 counties, mostly in central and eastern Kansas. Present from April through early October.

Field Notes: Green Herons are an inconspicuous and usually solitary species. They crouch motionless for long periods on low-hanging branches near the bank and wait for an opportunity to catch minnows or other prey. They utter a squawk and fly when startled. At urban parks and fishing lakes, they have become accustomed to humans and can be easily observed at close range.

immature

Botaurus exilis

Least Bittern

Field Identification: This is the smallest heron in Kansas, and the most secretive. Look for the black back and crown and the white breast with broad streaks. The rest of the plumage is mostly orangish brown or buff. In flight, large buff patches on the upper wings are apparent. It moves through dense wetland vegetation by grasping plants with its feet. Like the American Bittern, it will freeze with bill held upright to avoid detection. Size: Length 13 inches; wingspan 17 inches.

Habitat and Distribution: Dense cattail marshes of central and eastern Kansas. Stands of cattails near small impoundments are often sufficient to attract this species. Has been reported from 55 counties.

Seasonal Occurrence: Summer resident. Present from May through October. Nesting has been confirmed in 10 counties located in south-central and northeastern Kansas, and it likely nests in many others. Rare or absent in the west.

Field Notes: Although relatively common at marshes such as Baker Wetlands, Cheyenne Bottoms, and Quivira NWR, the Least Bittern is difficult to observe. The best opportunity to see one is during late-summer months, when adults are actively feeding nestlings. At these times, adults are seen flying low over the cattails as they come and go from the nest. Carefully check vegetation just above the water surface, where they perch motionless waiting for prey. Their song is a clue to their presence. It is a rhythmic, descending series of monotone cooing notes.

immature

American Bittern

Botaurus lentiginosus

Field Identification: This is an elusive brownish heron of dense cattail marshes. Large and chunky, it is mostly a subdued brown color, with broad vertical stripes on the neck and breast, a long greenish-yellow bill, and bright yellow eyes. Size: Length 28 inches; wingspan 42 inches.

Habitat and Distribution: Marshes with extensive stands of cattail or other dense wetland vegetation. Occurs statewide during migration. Has been reported in 88 counties. Much less common in western Kansas than in the rest of the state.

Seasonal Occurrence: Migrant and summer resident. Present from early May through mid-October. Breeding has been confirmed in five counties, mostly in the eastern half of Kansas. Nests annually at Cheyenne Bottoms and Quivira NWR, less frequently at smaller wetlands in eastern and central Kansas.

Field Notes: American Bitterns conceal themselves in tall cattails and wetland grasses by stretching vertically with the bill pointed skyward and freezing in position, rendering themselves virtually invisible within the dense vegetation. Their "song" is a deep guttural pumping sound that is often the best clue to their presence. It is heard most often at night or near dawn and dusk. Like many wetland birds, it has suffered from the loss of wetland habitat. Carefully watch the cattails and dense sedges while driving the roads at Cheyenne Bottoms and Quivira. You might be rewarded with a good look of one of these unique birds.

Nycticorax nycticorax

Black-crowned Night Heron

Field Identification:
This short, stocky, large-headed heron has a black crown and back. The rest of its body is mostly white and gray. Its bill is thicker than that of most heron species. Juveniles are frequently seen away from adults. They are mostly brown, heavily streaked below and spotted with white above. Size: Length 25 inches; wingspan 44 inches.

Habitat and Distribution: Statewide at marshes and other aquatic habitats. Has been reported in 101 Kansas counties.

Seasonal Occurrence: Migrant and summer resident. Present from April through October. There are confirmed breeding records from 17 counties in south-central and southwestern Kansas. This heron is unique in that it has nested in several southwestern counties where other heron species are scarce or absent. In some years, a few remain for the winter in southern Kansas.

Field Notes: This distinctive heron forages mostly at night, although it is frequently seen during the day. Groups of up to 20 birds will gather at productive feeding areas. It is most easily seen at Cheyenne Bottoms and Quivira NWR, but it is a widespread species in Kansas. This is one of the herons most likely to be found in southwestern Kansas, where it nests in shelterbelts adjacent to foraging areas. At the Elkhart wastewater lagoons in semiarid southwestern Kansas, flocks of up to 50 birds are seen each year, especially in September.

immature

77

Yellow-crowned Night Heron

Nyctanassa violacea

Field Identification: This species is similar in proportion to the Black-crowned Night Heron but is slimmer and longer-necked, with a bill that is thicker and shorter. Adults are mostly blue-gray, with a striking black-and-white facial pattern. Juveniles resemble juvenile Black-crowns but have much smaller white spots on the back that give them a speckled appearance. The body shape and bill structure of juveniles are the same as those of adults. Size: Length 24 inches; wingspan 42 inches.

Habitat and Distribution: Mostly in the southern half of the state, at marshes and along shorelines. Has been reported from 75 counties. Rare or absent from the northwest.

Seasonal Occurrence: Summer resident. Confirmed breeding has been reported in 17 counties scattered across the state. Present from April through October.

Field Notes: Unlike the more social Black-crowned Night Heron, this species is generally seen feeding alone. Its favorite food is crayfish, and it usually hunts for them in shallow pools and streams. These herons are not shy around humans, and on many occasions, they have nested in mature trees located in suburban yards, sometimes adjacent to major highways. They readily feed in ditches along busy streets and roads if hunting is productive.

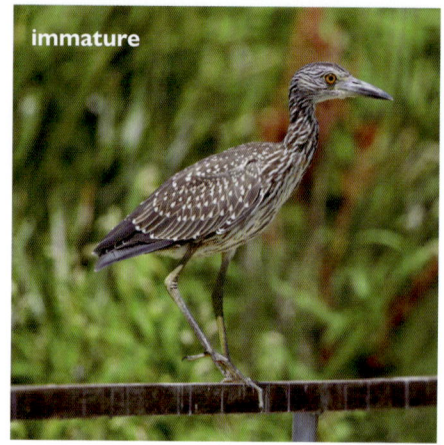

immature

Plegadis chihi

White-faced Ibis

Field Identification: This is a dark heron-like bird with a long, curved bill. Adults in breeding season are chestnut brown on the head, neck, and breast, with iridescent green wings and back. The facial skin of breeding adults is bright red and is bordered with bright white plumage. Juveniles and winter adults are much less colorful, but if seen closely, the facial skin is still pinkish and the eyes are always red. It is easily identified in flight by the dark color, trailing legs, and prominently curved bill. Size: Length 23 inches; wingspan 36 inches.

Habitat and Distribution: Found statewide at wetlands, lakeshores, and flooded fields. Has been recorded in 102 Kansas counties.

Seasonal Occurrence: Migrant and summer resident. Present from April through October. Breeding has been confirmed at Cheyenne Bottoms and Quivira NWR as well as playa wetlands located in Meade and Finney Counties.

Field Notes: Nesting in Kansas is restricted mostly to Cheyenne Bottoms and Quivira. Small flocks appear in many areas statewide during migration. These birds often feed in tall wetland vegetation, and despite their relatively large size, they often go undetected until they take flight. When disturbed, they take flight with unique nasal grunting calls.

nonbreeding

Glossy Ibis

Plegadis falcinellus

© Judd Patterson

Field Identification: Breeding adults differ from White-faced in having gray facial skin with a narrow and bright powder-blue border that is "pinched" behind the eye. The eyes are always brown. Nonbreeding birds are difficult to separate from White-faced, but with a good view, you can see the gray facial skin with the pale blue border still visible and the brown eyes. Size: Length 23 inches; wingspan 36 inches.

Habitat and Distribution: Wetlands and flooded fields. Much less likely to be seen than the White-faced Ibis. There are records from 18 Kansas counties in central and eastern Kansas.

Seasonal Occurrence: Migrant and summer wanderer. Present from April through September.

Field Notes: Both Glossy and White-faced Ibis have increased their range in the past 30 years. Glossy Ibis first appeared in Kansas in 1992, and sightings have steadily increased since then. Most are sighted at Cheyenne Bottoms and Quivira NWR and are typically single birds seen with flocks of White-faced. In addition to being similar in appearance, the two species hybridize. Consequently, some dark ibis, especially in nonbreeding plumage, cannot be reliably identified in terms of species. When in doubt, it is probably a White-faced Ibis.

nonbreeding

Rallus elegans

King Rail

Field Identification:
Rails are chicken-like birds of the marshes. King Rails are the largest rail species in the US and are twice the size of any other rail in Kansas. Adults are a deep rufous-orange on the throat and breast, with contrasting black and white bars on the flanks and belly, and a long spike-shaped bill. The loud grunting call is often the first clue to its presence. Size: Length 15 inches; wingspan 20 inches.

Habitat and Distribution: A resident of densely vegetated marshes. Has been reported from 45 counties, mostly in central and eastern Kansas.

Seasonal Occurrence: Summer resident, present from March through October. There are confirmed breeding records from 15 counties. Most contemporary nesting records are from the large central wetlands, but this species will also nest at small wetlands that have the preferred habitat conditions. In 2020 a pair raised a brood of young at a small marshy pond amid tallgrass prairie in Cowley County.

Field Notes: King Rails are in decline across much of the interior US because of wetland destruction. The population at Quivira NWR probably numbers less than 50 pairs, which makes it one of the most robust populations remaining anywhere west of the Mississippi River. A drive through the Big Salt Marsh at Quivira may provide the sharp-eyed observer with a glimpse of one. Carefully watch the edges of the cattails for these well-camouflaged birds.

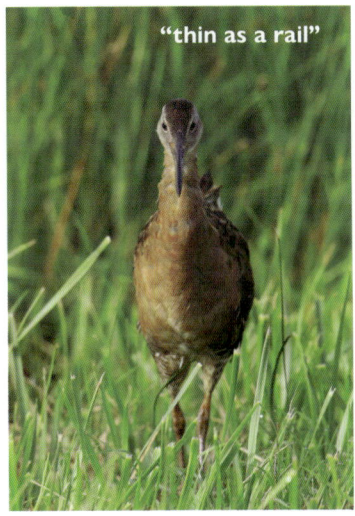

"thin as a rail"

Virginia Rail

Rallus limicola

Field Identification: The Virginia Rail resembles the King Rail in proportions and plumage but is about half the size. The face is gray; the bill and legs are orangish red. The most common call is a peculiar series of oinking sounds. Size: Length 10 inches; wingspan 13 inches.

Habitat and Distribution: Statewide in wetlands, especially dense cattail marshes. There are records from 80 counties.

Seasonal Occurrence: Found year-round, but populations fluctuate greatly. Uncommon migrant throughout, especially in May and September. Confirmed breeding has been documented in nine counties scattered across the state. During some winters, a few remain in areas with open water and thick stands of cattails.

Field Notes: This rail is more numerous and widespread than the King Rail. Like all rails, it is difficult to observe. It is common in summer at Cheyenne Bottoms, Quivira NWR, and other wetlands throughout the state. As with King Rail, patiently watching the edges of cattail stands is the best strategy for seeing one. During migration, these rails can turn up in unexpected locations such as suburban lawns and farm fields.

Porzana carolina

Sora

Field Identification: This rail is similar in size to the Virginia Rail but with a shorter and thicker bill, gray plumage on the throat and breast, and an area of black at the base of the bill. The barring on the flanks is less distinct than that on other rails. When Soras are flushed into flight, the long legs dangle below the body. Size: Length 9 inches; wingspan 14 inches.

Habitat and Distribution: Typically seen in marshes, but during migration, it is sometimes found in upland thickets far from water. Has been observed in 90 counties. Least common in the northwest.

Seasonal Occurrence: Migrant and local summer resident. Spring migrants are seen in April and May, fall migrants during September and October. A few remain to nest. There are confirmed breeding records from seven counties, mostly at the major wetland complexes.

Field Notes: Soras often make their presence known by their vocalizations. The call is a high-pitched descending whinny or a one- or two-note squeaking call. Soras are more apt than other rails to feed in the open, although they are still secretive birds. Walking along the drier edge of marshy areas will sometimes flush these birds into brief flight.

Black Rail

Laterallus jamaicensis

Field Identification: This tiny rail is heard far more often than it is seen. It is almost entirely charcoal gray speckled with white and has a chestnut nape. The distinctive *kik-ee-doo* call can be heard late at night from May through July. Size: Length 6 inches; wingspan 9 inches.

Habitat and Distribution: Shallow wetlands with stable water levels that are dominated by sedges. Has been recorded in 19 counties. Most likely to be found at Quivira NWR but also found in much smaller wetlands, generally in south-central Kansas. Sometimes turns up in unexpected locations during fall migration.

Seasonal Occurrence: Migrant and summer resident. Arrives in late April and departs in October. There are confirmed breeding records from seven counties.

Field Notes: This secretive and little-known species is the most rarely seen bird listed in this book. It has been included largely because of its exceptional mystique. Many birders in Kansas have never seen or heard one. Even at the national level, the inland populations of Black Rail are poorly understood. There is another population of these rails in southeastern Colorado in wetlands associated with the Arkansas River. These isolated Colorado and Kansas populations represent most of the known inland range of this species in the US.

Pluvialis squatarola

Black-bellied Plover

Field Identification: This is the largest plover in Kansas. Plovers have proportionately shorter and thicker bills than most sandpipers. In spring, the Black-bellied is checkered white and black on the back, with jet-black underparts bordered by snowy white. Winter birds are mostly mottled gray above and white below. On flying birds in all plumages, look for black axillaries, or "wing-pits," and a prominent white wing-stripe. These field marks are absent on the similar American Golden-Plover. Size: Length 12 inches; wingspan 29 inches.

Habitat and Distribution: Statewide in wetland areas. There are records from 86 counties. Usually seen at mudflats, playas, plowed fields, and open shorelines. The largest numbers are seen at Cheyenne Bottoms and Quivira NWR.

Seasonal Occurrence: Spring and fall migrant. Most spring migrants are seen in April and May and most fall birds

during September and October. A few linger into November.

Field Notes: When in full breeding plumage, this is one of the most striking shorebird species in Kansas. Sometimes sizable flocks can be seen at the larger wetlands, but most sightings are of small groups or single birds. Their plaintive flight call is a memorable sound of the marshes.

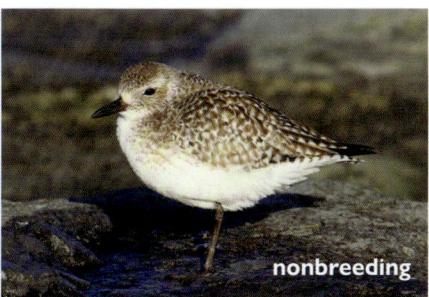

nonbreeding

American Golden-Plover

Pluvialis dominica

Field Identification: This species is similar in appearance to the Black-bellied Plover. Breeding-plumaged birds seen in spring have rich gold-and-black checkering on the back. The undertail of breeding-plumaged birds is completely black, unlike the Black-bellied, which has a white undertail. Winter birds are like Black-bellied but have a more noticeable white stripe above the eye and lack the black "wing-pits." In both spring and fall, they often show a transition between breeding and winter plumage, and flocks often contain individuals with a variety of plumages. Size: Length 10 inches; wingspan 26 inches.

Habitat and Distribution: Statewide at wet mudflats, wetlands, sod farms, and burned prairies. Has been recorded in 86 counties. Rarest in western Kansas.

Seasonal Occurrence: Migrant in spring and fall and most numerous during late March and early April. Seen less often in the fall, when many migrate south along the Atlantic seaboard instead of through the interior of the continent.

Field Notes: Flocks of up to several hundred birds can be seen on freshly burned prairies in the Flint Hills during the early spring. Watch for their pale bodies contrasting with the blackened prairies.

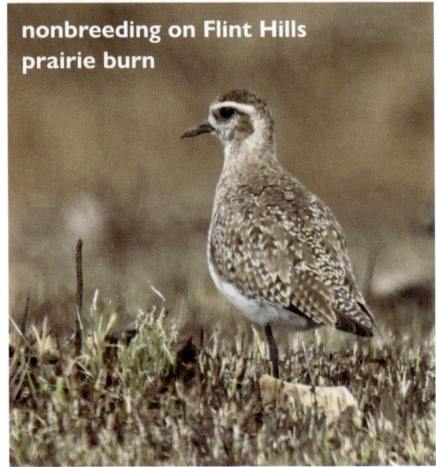

nonbreeding on Flint Hills prairie burn

Anarhynchus nivosus

Snowy Plover

Field Identification: A small, pale plover, the Snowy Plover is all white below, with sandy upperparts and a black bill. The male in breeding plumage has black patches on its head and shoulders. These patches are brown on the females and on the winter males. Size: Length 6 inches; wingspan 17 inches.

Habitat and Distribution: In Kansas, most are found on the white alkali flats of Quivira NWR and some at Cheyenne Bottoms. In early spring, migrants regularly turn up throughout the state, including in many eastern counties. These sightings are believed to be birds that have migrated beyond their target destinations. Has been recorded in 52 counties.

Seasonal Occurrence: Summer resident. Most arrive in early April and depart by early September. Nests annually at Cheyenne Bottoms and Quivira, irregularly elsewhere. There are confirmed breeding records from nine counties in central and western Kansas.

Field Notes: Central Kansas represents the northeastern edge of the breeding range for this small plover in North America. Their pale colors are an adaptation to the alkali flats environment in which they live. They are often difficult to see when they are motionless. The nest is a simple, shallow scrape in dry flats. Be careful when walking the roadsides at Quivira, because the eggs are even better camouflaged than the adults!

Piping Plover

Charadrius melodus

Field Identification: Sandy brown like the Snowy Plover, the Piping Plover is differentiated by its narrow black breastband, orange legs, and orange bill with black tip. In flight, the upper tail is all white at the base with a black tip, while on the Snowy Plover, the upper tail is sandy brown with white edges. Size: Length 7 inches; wingspan 19 inches.

Habitat and Distribution: Mudflats, river sandbars, and sandy shorelines statewide, but rare in the west. Most often seen at the large wetlands. Has been recorded in 54 counties, mostly in central and eastern Kansas.

Seasonal Occurrence: Migrant and local summer resident. Spring migrants are seen between mid-April and mid-May. Fall migrants are seen between late July and late August. In some years, a few pairs have nested on the Kansas River between Manhattan and Lawrence. In drought years, it has also nested on exposed flats at Cedar Bluff and Webster Reservoirs. Confirmed breeding at least once has been documented in five counties.

Field Notes: This species is considered threatened or endangered throughout its range. Kansas is located on the flyway for the inland population, which nests mostly in the northern Great Plains. This is one of the more sought-after shorebird species for avid birders in Kansas.

Charadrius semipalmatus

Semipalmated Plover

Field Identification: This small plover is dark brown above and white below. It has a single black band on the breast and a black mask. Killdeer are similar in appearance but are more than twice as big and have two breastbands. Piping Plover is also similar but has much paler upperparts. The legs are orange, and the bill is orange with a black tip. Size: Length 7 inches; wingspan 19 inches.

Habitat and Distribution: Statewide at wetlands and marshes, almost always on open mudflats. Has been recorded in 98 counties.

Seasonal Occurrence: Spring and fall migrant. Spring migrants are seen in April and May, fall migrants from late July through late September.

Field Notes: Look for this species on the wettest mudflats. At large wetlands, the darker brown Semipalmated Plover is found on the equally dark wet mud nearest to the open water, while the pale-colored Piping and Snowy Plovers prefer the drier, lighter-colored mud located farther from the water's edge.

immature

Killdeer

Charadrius vociferous

Field Identification: This is an abundant and familiar large plover of rural and urban areas. It is similar in color and pattern to the Semipalmated Plover but is larger and longer-legged, with two black breastbands. In flight, look for the orange-and-black tail and broad white wing-stripe. It is vocal at all seasons, and its two-note *killdeer* call is a familiar sound of farm country. Size: Length 11 inches; wingspan 24 inches.

Habitat and Distribution: Common statewide in unvegetated areas at urban parks, plowed fields, wet areas, gravel roads, and parking lots. Has been recorded in all 105 counties.

Seasonal Occurrence: Found year-round, but seasonal abundance varies significantly. Migrant and summer resident. Most numerous in migration. Spring migration takes place from late February through late April. Fall migration peaks in late July and early August but continues through October.

Common in summer throughout the state. Breeding has been confirmed in 100 counties statewide. A few hardy individuals remain through all but the harshest winters.

Field Notes: The arrival of the first northbound migrant Killdeer in mid-February is one of the earliest signs of spring in Kansas. Killdeer frequently nest on gravel parking lots and roadsides. They often feign injury with a broken-wing display to distract humans and other predators from the chicks or nest site. In the summer, most are seen singly or in family groups, but during migration, they often occur in flocks.

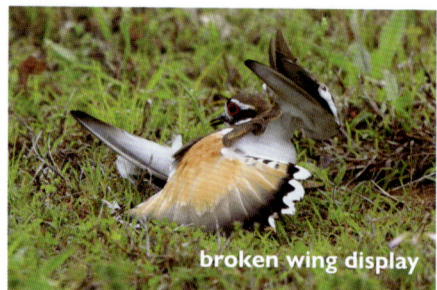

broken wing display

Himantopus mexicanus

Black-necked Stilt

Field Identification: This large, slender, strongly patterned black-and-white shorebird has extremely long, bright pinkish-red legs. It is impossible to confuse it with any other bird species. Size: Length 14 inches; wingspan 29 inches.

Habitat and Distribution: Wetlands of central and western Kansas. Often observed on playa wetlands in southwestern Kansas. Wandering birds appear throughout the state, especially during spring migration. Has been recorded in 73 counties.

Seasonal Occurrence: Migrant and summer resident. Arrives in early April and departs in late September. There are confirmed breeding records from nine counties in central and southwestern Kansas, including playa lakes in Finney, Ford, Gray, Meade, and Seward Counties.

Field Notes: In Kansas, this conspicuous bird is easily found at Cheyenne Bottoms and Quivira NWR. A summer cruise along the Wildlife Drive at Quivira should provide numerous looks at these distinctive birds and their young. When perceived threats approach the nest, they become agitated and fly about, giving a variety of alarm sounds. In recent decades, they have become more widespread and numerous during migration, and sightings throughout the state have increased.

American Avocet

Recurvirostra americana

Field Identification: Avocets are an easily identified shorebird with exceptionally long blue-gray legs and a thin, upcurved bill. They have a bold black-and-white pattern on the body. The head and neck are cinnamon brown in summer, pale gray in winter. Size: Length 18 inches; wingspan 31 inches.

Habitat and Distribution: Statewide at lakes, playa wetlands, and marshes. Has been recorded in all 105 counties.

Seasonal Occurrence: Migrant and summer resident. Migrants arrive in early April and depart in late October. Most abundant during migration. Many remain to nest. There are confirmed nesting records from 16 counties in central and western Kansas. Most numerous as a nesting species at Cheyenne Bottoms and Quivira NWR. Like the Black-necked Stilt, it has nested at playa wetlands in western Kansas, especially in the southwest.

Field Notes: This species is related to the Black-necked Stilt. Avocets typically feed by sweeping the bill back and forth across the surface of the water. Sometimes an entire flock will be found feeding together in this manner. During migration, it is not uncommon to see them swimming in lakes and ponds in deep water. Avocets show agitation and loudly vocalize when their nests or young are approached.

nonbreeding

Tringa melanoleuca

Greater Yellowlegs

Field Identification:
This is one of the large sandpipers. Look for the long legs and the slender overall build. The best field mark is the bright yellow leg color. In breeding plumage, it has vertical black bars on the flanks that are not present in the fall. It is distinguished from the closely related Lesser Yellowlegs by its larger size as well as the longer, slightly upcurved bill. The two- or three-note call is a familiar sound of the marshes. Size: Length 14 inches; wingspan 28 inches.

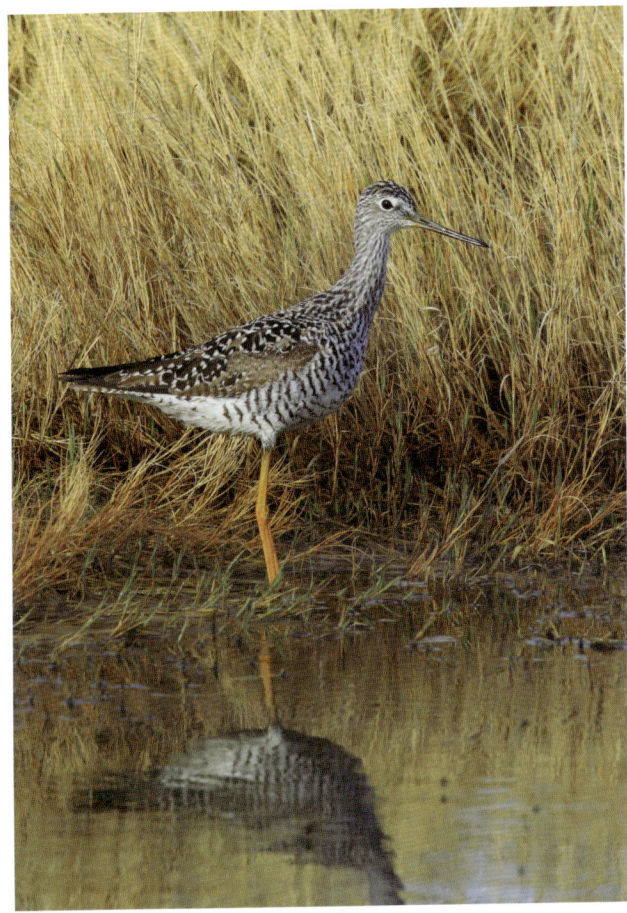

Habitat and Distribution:
Statewide at wetlands and shorelines. This is one of the most common migratory shorebirds in Kansas and has been observed in all 105 counties.

Seasonal Occurrence: Spring and fall migrant. This is one of the earliest shorebirds to arrive in spring and remains later into the fall than most other shorebirds. Spring migrants are seen from early March through early June. Fall migrants start returning in July, and a few are seen well into November. Some are seen throughout the summer at major wetlands. During mild winters, a few linger at wetlands if open water is available.

Field Notes: Greater Yellowlegs are one of the more common shorebird species in Kansas. They tend to arrive earlier in the spring and linger longer in the fall than the closely related Lesser Yellowlegs.

nonbreeding

93

Lesser Yellowlegs

Tringa flavipes

Field Identification: Lesser Yellowlegs are similar in appearance to Greater Yellowlegs but are smaller. They also have a shorter, straighter bill. In breeding plumage, the vertical markings on the flanks are not as long or heavy as those on Greater Yellowlegs. The call notes are less strident and are usually given in one or two notes, in contrast to the two- or three-note call of the Greater Yellowlegs. On both species, look for the all-white tail and rump in flight. This is helpful in distinguishing them from most other flying shorebirds. Size: Length 11 inches; wingspan 24 inches.

Habitat and Distribution: Found statewide at wetlands, flooded fields, and a variety of shoreline habitats. Has been recorded in all 105 counties.

Seasonal Occurrence: Spring migrant from March through June; fall migrant from July through early November.

Field Notes: This is one of the most common migratory sandpipers in Kansas. At the peak of migration, hundreds can be seen at large wetlands. It is more of a generalist than most other shorebird species. It readily adapts to marginal habitats if any shallow water is present.

nonbreeding

© Judd Patterson

Tringa solitaria

Solitary Sandpiper

Field Identification: The Solitary Sandpiper is closely related to the yellowlegs, which it resembles in many ways. The legs are shorter and more greenish. Look for the fine white speckling on the back and a prominent eye-ring. In flight, the vertical center of the upper tail is black with prominent black horizontal bars. When flushed, it gives a loud, clear two-note call, accented on the second note. Sometimes it bobs its head while walking. Size: Length 8 inches; wingspan 22 inches.

Habitat and Distribution: Statewide. Favors small temporary pools but is often seen at large wetland areas and other shoreline habitats. Has been recorded in all 105 counties.

Seasonal Occurrence: Migrant. Seen in spring from mid-April through May and in fall from July through October.

Field Notes: True to its name, this species is rarely seen in groups of more than two or three birds. It is often inconspicuous until flushed, flying away while giving its clear, two-note call. This is among the earliest species to return south in late summer.

Willet

Tringa semipalmata

Field Identification: This large shorebird is heavy-bodied with a long, stout bill and gray legs. It is mostly grayish, with a scalloped or patterned look in breeding season that gives way to a plainer appearance in winter plumage. When it takes flight, strongly patterned black-and-white wings are seen. Size: Length 15 inches; wingspan 26 inches.

Habitat and Distribution: Statewide at large wetlands, small prairie pools, lakeshores, and riverbanks. Has been recorded in 103 Kansas counties.

Seasonal Occurrence: Spring migrant from April through June and fall migrant from July through October.

Field Notes: This large sandpiper is one of several shorebirds that show a strong preference for temporary rain pools. Despite the subtle plumage colors, it is conspicuous due to its large size and striking wing pattern in flight.

nonbreeding

Actitis macularius

Spotted Sandpiper

Field Identification: This is a small shorebird of streams and ponds. It has a short neck and an elongated, tapered body. It walks with an odd bobbing gait. When flushed, it flies low across the water with stiff wingbeats, usually uttering a two-note alarm call. Summer birds have heavy black spotting below, bright orange legs, and a thin white eye-line. Winter birds lose the spots and have a dull brown wash on the breast. Size: Length 7 inches; wingspan 15 inches.

Habitat and Distribution: Statewide. Favors rivers and streams more than any other Kansas shorebird, although it is also found at ponds and lakes. Has been recorded in all 105 counties.

Seasonal Occurrence: Migrant and summer resident. Most numerous during migration. Spring migrants are seen from mid-April through May.

Fall migrants are seen from July through late September. Some remain to nest where appropriate habitat exists, and confirmed nesting has been documented in 30 counties.

Field Notes: This unique sandpiper is almost always found alone or with only a few others. Small flocks may form during migration. It is widespread throughout Kansas.

nonbreeding

Upland Sandpiper

Bartramia longicauda

Field Identification: A large shorebird found on the open prairies, the Upland Sandpiper appears round-headed, with large, dark eyes on a pale face. It has a potbellied look, long yellowish legs, and a relatively short bill. Size: Length 12 inches; wingspan 26 inches.

Habitat and Distribution: Native prairies and hayfields throughout the state. Has been recorded in all 105 counties.

Seasonal Occurrence: Migrant and summer resident. Arrives in April and departs in early September. Fairly common during fall migration in late July and August. Nests mostly in the eastern two-thirds of Kansas. Conspicuous in the Flint Hills in summer. There are confirmed nesting records from 49 counties and probable nesting records from an additional 20 counties.

Field Notes: This is truly an upland bird, seldom seen near water. A summer drive through the Flint Hills and other prairie regions usually produces good looks at this species. They are fond of perching on top of wooden fence posts. The unique song is often referred to as the "wolf whistle" and is one of the signature sounds of the Kansas prairies. Their nesting range extends north to Alaska. Birds from these northern populations are already moving south by mid-July. On a quiet night in late summer, you can hear their distinctive four-note flight calls as they pass overhead. In July and August, large numbers sometimes gather in freshly mowed hayfields to feed on insects.

Numenius americanus

Long-billed Curlew

Field Identification: This is the largest shorebird in Kansas, with an absurdly long and deeply curved bill. It is mostly mottled brown above, unmarked and paler on the belly. In flight, the bright cinnamon underwings are prominent. Size: Length 23 inches; wingspan 35 inches.

Habitat and Distribution: Migrant and local summer resident. Found in shortgrass prairies and irrigated crop circles. Occasionally reported at Cheyenne Bottoms, Quivira NWR, and other wetlands. Has been recorded in 65 counties, mostly in the western two-thirds of Kansas.

Seasonal Occurrence: Migrant and summer resident. Spring migrant from March through May; fall migrant from August through September. A few pairs nest in shortgrass prairies of far western Kansas, mostly in Morton and Stanton Counties. Confirmed breeding has been recorded in five counties. The small breeding population is present throughout the summer months.

Field Notes: This is one of the most distinctive shorebirds in the world. Overhunting and habitat alteration during European settlement significantly reduced the breeding range and population in Kansas. Northbound migrants can turn up anywhere in the western counties during March and April. In early spring, large flocks are sometimes found in irrigated crop circles west of Garden City in Finney and Kearney Counties. If you are in the right place at the right time, you can see these impressive flocks, which in many years represent a measurable percentage of the entire world population.

Whimbrel

Numenius phaeopus

Field Identification: This large, brownish-gray shorebird with a long, deeply curved bill is smaller than the Long-billed Curlew, and the bill is not as long. The distinct black and white stripes on its head separate it from the plainly marked Long-billed Curlew. It also lacks the cinnamon wing linings of the Long-billed Curlew. Size: Length 17 inches; wingspan 32 inches.

Habitat and Distribution: Found at wetlands and on shorelines throughout the state. Has been recorded in 49 counties.

Seasonal Occurrence: Migrant in spring and fall. Most spring sightings are in April and May. Fall records are fewer but have occurred between July and October.

Field Notes: This species is one of the harder-to-find migratory shorebirds in Kansas. Most sightings have come from Cheyenne Bottoms and Quivira NWR, but flocks of up to several dozen have been reported elsewhere in the state. Look for them in flooded grass, mudflats, and recently burned grasslands as well as at wastewater treatment facilities in western Kansas.

Limosa haemastica

Hudsonian Godwit

nonbreeding

Field Identification: A large sandpiper with a long, two-toned bill, the Hudsonian Godwit in breeding plumage is deep rusty red on the breast and belly and darker above, with a pale head. Some seen in early spring are still in winter plumage, which is mostly a uniform light gray. In flight, the tail is black with a clean-cut white rump, and a narrow white wing-stripe is visible. Size: Length 15 inches; wingspan 29 inches.

Habitat and Distribution: Prairie marshes, playa wetlands, and burned prairies. Most are seen in the central and eastern parts of the state, far fewer in the western counties. Has been recorded in 82 counties.

Seasonal Occurrence: Spring migrant in April and May. In fall, most migrate south along the Atlantic coast, and only a few stragglers return south across the center of the continent.

Field Notes: Most of the entire world population of this species passes through Kansas in spring. Flocks numbering in the hundreds can sometimes be seen at Cheyenne Bottoms and Quivira NWR when habitat and feeding conditions are favorable.

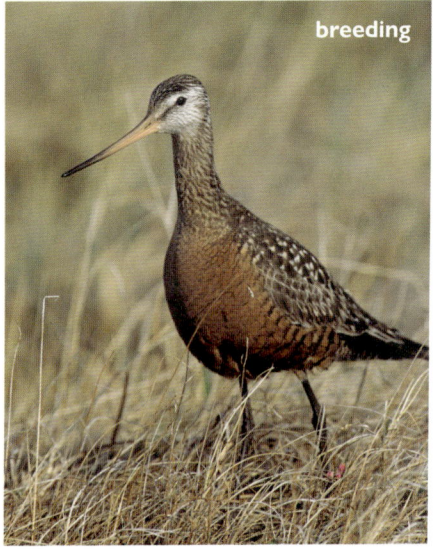

breeding

Marbled Godwit

Limosa fedoa

Field Identification: In comparison with the Hudsonian, the Marbled Godwit is larger and heavier-bodied and has a longer bill. Adults in breeding season are barred brown and black, becoming a plainer brown by late summer. In flight, the bright cinnamon underwings are visible. Size: Length 18 inches; wingspan 30 inches.

Habitat and Distribution: Found at marshes and other shoreline habitats. Has been recorded in 87 counties statewide. More numerous in the western two-thirds of Kansas.

Seasonal Occurrence: Migrant. Spring migration is in April and May and fall migration from July through early October.

Field Notes: This large shorebird is more likely to be seen on lakeshores than the Hudsonian Godwit. Any godwit seen in late summer and fall is probably this species, since nearly all Hudsonian Godwits migrate south along the Atlantic coast in fall. The two godwit species are often seen together during the spring migration, as they typically forage in water of similar depth.

Arenaria interpres

Ruddy Turnstone

Field Identification: Adults in spring are patterned with bright chestnut and black on the back and wings and have an intricate black-and-white facial pattern and orange legs. Fall birds have the same basic pattern, but it is less colorful and poorly defined. Size: Length 9.5 inches; wingspan 21 inches.

Habitat and Distribution: Found in wetlands and along shorelines, often those with rocks, gravel, or woody debris. Has been reported from 50 counties, mostly in central and eastern Kansas. There are records from a few counties in the western quarter of the state.

Seasonal Occurrence: Migrant in both spring and fall. Arrives later in spring than most other shorebird species. Most spring migrants are seen in May and fall migrants from August through October.

Field Notes: Ruddy Turnstones have slightly upturned bills which they use to flip over rocks and other shoreline debris in search of food. Turnstones are most often observed at Cheyenne Bottoms and Quivira NWR. When water and shoreline conditions are favorable at Cheyenne Bottoms, several can be seen in a single day along the rocky riprap adjacent to the dike roads.

nonbreeding

Sanderling

Calidris alba

Field Identification: Birds in winter plumage are pale gray above, sometimes with black checkering on the back and an obvious black shoulder mark. Summer birds have a rich rusty head, breast, and upperparts. The bill is proportionately longer and stouter than on other small sandpipers. In flight, look for the broad, prominent white wing-stripe.

As with many other shorebird species, individuals in winter and summer plumage can often be seen side by side during migration. Size: Length 8 inches; wingspan 17 inches.

Habitat and Distribution: Prefer shallow mudflats and sandy shorelines. Often seen at wastewater treatment impoundments in western Kansas. Has been recorded in 83 counties.

Seasonal Occurrence: Migrant. Spring migrants are seen in April and May. This is one of the shorebirds that tends to arrive later in the spring than other species, with a noticeable peak in the last half of May. Fall migrants are seen from July through late October.

Field Notes: This is one of the most common shorebirds along the ocean coasts but is one of the scarcer migratory shorebird species in Kansas and other interior states. Several of the small shorebird species in the genus *Calidris* are often collectively referred to as "peeps" by birders. They share many field marks and can be difficult for the beginning birder to tell apart, especially since they are often seen at a distance, are very active when feeding, and occur in mixed flocks. To identify Sanderlings, scan large feeding flocks of small sandpipers for birds that are perceptibly larger than the other species of "peeps" and either noticeably pale white or, alternatively, bright rusty red in color.

nonbreeding

Calidris pusilla

Semipalmated Sandpiper

Field Identification: This is one of the small "peep" sandpiper species. The dark legs separate it from the Least Sandpiper. The bill is shorter and straighter than that of the similar Western Sandpiper. When standing, the wingtips do not extend beyond the tail feathers, separating it from Baird's. In spring, it is mottled brownish above and streaked on the breast. Winter birds are gray above with a grayish breast. Size: Length 6 inches; wingspan 14 inches.

Habitat and Distribution: Mudflats, flooded fields, and shorelines statewide. Has been recorded in all 105 counties.

Seasonal Occurrence: Spring migrant from early April through early June; fall migrant from early July through October.

Field Notes: Semipalmated is probably the most abundant peep in Kansas, and hundreds or even thousands can be seen on mudflats at Cheyenne Bottoms and Quivira NWR. They are commonly observed at small wetlands across the state.

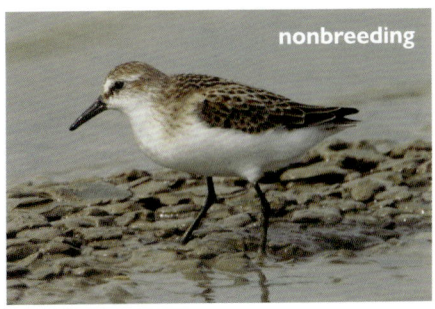

nonbreeding

Western Sandpiper

Calidris mauri

Field Identification: This small "peep" is often difficult to differentiate from the similar Semipalmated Sandpiper. Adults have a longer bill than Semipalmated, with a slight droop at the tip. In spring breeding plumage, it is brightly rufous on the shoulders, cheeks, and crown. Spring adults have streaked flanks, a field mark that Semipalmated lacks. In winter plumage, the most useful field mark is the longer bill with a drooped tip. Juveniles seen during fall migration look a lot like juvenile Semipalmated Sandpipers but always have the longer bill with a drooped tip. Size: Length 6.5 inches; wingspan 13 inches.

Habitat and Distribution: Statewide, but most numerous in the western half of Kansas. Has been recorded in 88 counties.

Seasonal Occurrence: Migrant. Seen in spring from April through early June and in fall migration from late July through October. This species is more abundant in fall than in spring migration.

Field Notes: This species is most easily found at the large central wetlands.

nonbreeding

Calidris minutilla

Least Sandpiper

Field Identification: This is the smallest of the "peep" sandpiper species and is the smallest sandpiper species in the world. The Least Sandpiper is smaller and darker above than other peeps and is the only peep with yellow legs. Spring birds have rufous markings on the upperparts; winter birds are a dull grayish brown. Winter individuals have a dark breastband of streaks that Semipalmated and Western do not have. Juveniles in fall are brighter reddish brown but always have a breastband and yellow legs. When standing, the wingtips do not extend beyond the tail feathers. Size: Length 6 inches; wingspan 13 inches.

Habitat and Distribution: Statewide at wetlands, shorelines, and river sandbars. Has been recorded in all 105 counties.

Seasonal Occurrence: Spring migrant from March through late May; fall migrant from July through November. Some remain into early winter in southern Kansas at Cheney Reservoir, along the Arkansas River, and occasionally elsewhere.

Field Notes: Along with the Semipalmated, this is one of the most abundant shorebirds in Kansas. Large numbers can be seen during migration, often in association with other species of peeps.

Baird's Sandpiper

Calidris bairdii

Field Identification: One of the largest "peep" sandpiper species, Baird's is a bit larger than the preceding three species, with a longer and more tapered body shape and a longer bill. When standing, the wingtips extend beyond the tail feathers. In spring, the back feathers have a scaly appearance, giving way to a dull grayish brown in winter. Size: Length 7.5 inches; wingspan 17 inches.

Habitat and Distribution: Statewide at wetlands and other shoreline habitats; found on pastures and sod farms more often than other peeps. Has been recorded from all 105 counties.

Seasonal Occurrence: Spring migrant from March through June, in fall from July through October. A few linger into November in most years.

Field Notes: This is one of the earliest shorebirds to arrive in Kansas on the journey north in spring, and in fall, they tend to linger longer than most other shorebirds. With time, it becomes easier to recognize their longer, tapered shape, especially when they are seen with other small sandpipers for comparison.

Calidris fuscicollis

White-rumped Sandpiper

Field Identification: One of the larger "peep" sandpiper species, the White-rumped is similar in size to Baird's but appears heavier-bodied. When standing, the wingtips extend beyond the tail feathers. Spring-plumaged birds have gray-and-rufous-patterned plumage on the back, with streaks on the breast and flanks. Baird's lacks these streaked flanks. In flight, it has an all-white rump. All other peep sandpipers in Kansas have white rumps divided by a black bar. Size: Length 7.5 inches; wingspan 17 inches.

Habitat and Distribution: Statewide at wetlands, prairie playas, and shorelines. Has been recorded in 101 counties.

Seasonal Occurrence: Spring migrant. Migrants are seen from late April through mid-June. Most are seen in late May and early June. Seen only rarely in fall.

Field Notes: In spring, most of the population of this species migrates north through the central plains to its arctic nesting areas, but it returns south along the Atlantic seaboard in fall. A large percentage of the world population passes through Kansas each spring. When you start seeing large numbers of White-rumps, you know the spring shorebird migration is nearly over, as they are among the last of the shorebird species to pass through Kansas on the long flight to the arctic tundra nesting grounds.

Pectoral Sandpiper

Calidris melanotos

Field Identification: A plump, barrel-chested, and medium-sized sandpiper, the Pectoral has heavy streaking on the breast with a clearly defined bottom edge and yellow legs. Juveniles in the fall are similar to adults but have brighter rufous tones on the back. Size: Length 9 inches; wingspan 18 inches.

Habitat and Distribution: Statewide at wetlands and shorelines. Like the Baird's Sandpiper, it is often found at sod farms, pastures, and other grassy habitats. Has been reported from all 105 counties.

Seasonal Occurrence: Spring migrant from March through May; fall migrant from July through October. A few linger into November in some years.

Field Notes: This sandpiper is usually less numerous than some of the other shorebirds but is seen on most trips to the major wetland areas during spring and fall migration.

Calidris alpina

Dunlin

© Judd Patterson

Field Identification: This stocky, medium-sized shorebird has a long and drooped bill. Adults in spring have a bright rufous back and a large black belly patch on the white underparts. In fall, the plumage is uniformly brownish gray on the head, upperparts, and breast, and the belly is white. In flight, it shows an extensive white wing-stripe, although the stripe is not as prominent as that of the Sanderling. Size: Length 8.5 inches; wingspan 17 inches.

Habitat and Distribution: Found on wetland mudflats, on sandy or muddy shorelines, and occasionally in flooded fields. Has been recorded in 64 counties, most of which are in central and eastern Kansas. Has been reported in only a few of the western counties.

Seasonal Occurrence: Spring migrant in April and May; fall migrant from July through late November. Most spring records are from May. Fall birds are most often seen in late October and early November.

Field Notes: Careful scanning of large flocks of shorebirds feeding on mudflats will often turn up a few Dunlins. They are hardy and often arrive later and linger later into the fall than most other shorebirds.

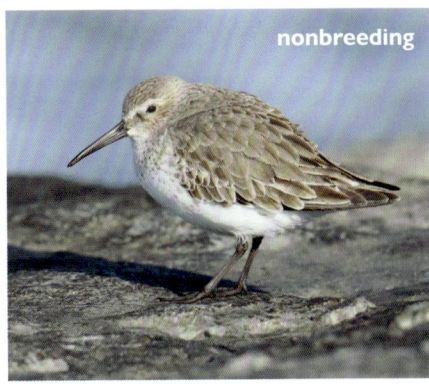

nonbreeding

Stilt Sandpiper

Calidris himantopus

Field Identification: The Stilt Sandpiper is a plump, medium-sized sandpiper with long legs and a heavy, drooped bill. Spring birds have dense bars on the breast and belly and a bright chestnut ear-patch. In fall, they are a uniform pale gray above with a prominent white line above the eye. Size: Length 9 inches; wingspan 18 inches.

Habitat and Distribution: Statewide at wetlands and shallow ponds. Has been recorded in 101 counties.

Seasonal Occurrence: Spring migrant from April through early June; fall migrant from July through October.

Field Notes: Stilt Sandpipers often associate with dowitchers, usually foraging in slightly shallower water. They feed in a distinctive "sewing machine" style, much as dowitchers do. The long legs allow them to feed in deeper water than many other sandpipers. They are one of the late-arriving species in spring and can often be seen in large numbers in late May and early June.

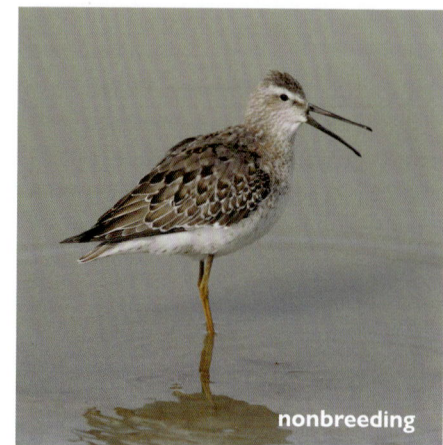

nonbreeding

Calidris subruficollis

Buff-breasted Sandpiper

Field Identification: This is a distinctive sandpiper with a pigeon-headed look. It has a scaly appearance on the back and is evenly buffy brown below, with a short plover-like bill and yellow legs. Snowy-white wing linings are prominent on flying birds. Size: Length 8 inches; wingspan 18 inches.

Habitat and Distribution: Found at prairie burns, sod farms, airports, short prairies, and plowed fields. Has been recorded in 73 counties, mostly in central and eastern Kansas. Rare in the west.

Seasonal Occurrence: Spring migrant in April and May; fall migrant from late July through early September.

Field Notes: This is one of the rarest shorebird species in North America. Their migration path passes directly through the Great Plains in both spring and fall. In spring, flocks numbering in the hundreds can be seen on recently burned prairies in the Flint Hills. Most fall sightings come during the month of August from sod farms, where they are fattening up for the long flight to the Argentine pampas. Sod farms in Douglas, Johnson, and Sedgwick Counties produce sightings of these birds each year. Spring migrants seldom linger and occasionally perform an impressive courtship display for lucky observers.

courtship display

© James W. Arterburn

113

Long-billed Dowitcher

Limnodromus scolopaceus

Field Identification: Dowitchers are large sandpipers with a chunky appearance and a prominent long, straight bill. In breeding plumage, they have a scalloped look on the back and reddish underparts. Winter birds are a nondescript gray but are easily identified by shape and bill. Juveniles have a faint reddish tinge on the breast. The narrow white wedge on the back is an excellent field mark on flying birds of both dowitcher species. Size: Length 11 inches; wingspan 19 inches.

Habitat and Distribution: Statewide at wetland areas. Has been recorded in 104 counties.

Seasonal Occurrence: Spring migrant from early March through late May; fall migrant from early July through early October.

Field Notes: This sandpiper can be found in large flocks during migration at wetlands with appropriate water conditions. It is often seen feeding in association with other sandpipers, especially Stilt Sandpipers. Dowitchers probe the mud in a methodical up-and-down motion that is often compared to the motion of a sewing machine. They typically feed in water that reaches their bellies.

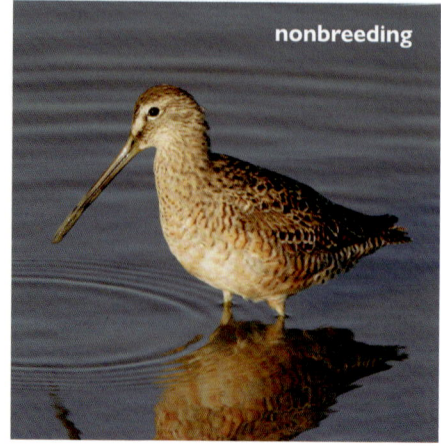

nonbreeding

Limnodromus griseus

Short-billed Dowitcher

Field Identification: This species strongly resembles the Long-billed Dowitcher. On breeding adults of the Great Plains subspecies, the black markings on the neck are all small spots. In comparison, Long-billed in this plumage has a combination of spots and bars on the neck. Juvenile Short-billed seen in early fall have bright golden markings on the back and wings, and identification of these is straightforward. In comparison, juvenile Long-billed have darker and less colorful markings on the back and wings. Winter-plumaged birds of the two species are similar. Birds in various stages of molt are often seen together, causing additional identification challenges, especially during the fall migration. Size: Length 11 inches; wingspan 19 inches.

Habitat and Distribution: Statewide at wetland areas. Has been reported from 59 counties, mostly in central and eastern Kansas.

Seasonal Occurrence: Spring migrant from late April through late May; fall migrant from mid-July through early September. Short-billed tends to arrive later in spring and earlier in the fall than Long-billed.

Field Notes: The two dowitchers are nearly identical, and not all can be confidently identified in terms of species, especially in late fall, when they are in winter plumage. Becoming familiar with their calls is useful, as these are distinct from each other. Most major field guide apps have recordings that will allow you to compare the vocalizations in the field.

nonbreeding

Wilson's Snipe

Gallinago delicata

Field Identification: This is a medium-sized sandpiper that resembles dowitchers. It has a long, straight bill; short legs; barred flanks; and bold striping on the head and back. When flushed, it gives a harsh two-note call while flying away swiftly and erratically on pointed wings. Size: Length 10 inches; wingspan 18 inches.

Habitat and Distribution: Statewide in densely vegetated, shallow wetlands and flooded grassy ditches. Sometimes seen on open mudflats and shorelines. Has been recorded in 104 counties.

Seasonal Occurrence: Spring migrant from late February through early May; fall migrant from August through November. Most numerous in fall in late October. Small numbers remain for the winter in areas with appropriate habitat and open water.

Field Notes: Although it is sometimes seen on open mudflats, this sandpiper typically goes unnoticed until it is startled into flight from its grassy wetland haunts. Its plumage provides excellent camouflage. This species and the American Woodcock are the only shorebirds hunted as game birds in Kansas.

Scolopax minor

American Woodcock

Field Identification: The American Woodcock is a medium-sized, plump shorebird with odd proportions. The legs are extremely short for a shorebird. It is pale buffy orange below, patterned in gray and black above, with a long, straight bill. It has black bars on its crown. Size: Length 11 inches; wingspan 18 inches.

Habitat and Distribution: Wet, brushy meadows in or near woodland areas. Has been recorded in 69 counties. Most records are from central and eastern Kansas, but stray migrants have been seen as far west as the Colorado state line.

Seasonal Occurrence: Migrant and summer resident. Most sightings are from late February through early April, when conspicuous courtship is taking place. Most continue north, but some remain to nest. There are confirmed nesting records from 18 counties in the eastern half of Kansas and west to Reno and McPherson Counties. Fall migrants are inconspicuous and are seen far less often, mostly in October and November.

Field Notes: This species is mostly nocturnal and as a result often goes undetected. It typically nests in the same areas from year to year, and birders enjoy visiting these areas on calm evenings in early March to observe their courtship flights. Migrating birds sometimes display in areas where they will not remain to nest. With twittering wings, the male flies up into the sky and performs a complex display flight while giving a loud nasal *peent* call every few minutes. Places to observe these displays include Shawnee Mission Park in Overland Park, Nemaha WA, and the nature trail at Harvey County West Park. During migration, Woodcocks often turn up in a variety of odd locations, including suburban yards.

Wilson's Phalarope

Phalaropus tricolor

Field Identification: This is a small shorebird with long legs and a thin, pointed bill. In spring, the female has colorful black and reddish markings on the head and neck. The male has similar patterns, but the colors are more subdued. Nonbreeding birds are pale gray above and white below. In all plumages, the broad white rump on flying birds is a good field mark. Size: Length 9 inches; wingspan 17 inches.

breeding female

Habitat and Distribution: Statewide at marshes, playas, and wastewater treatment ponds. Has been recorded in all 105 counties.

Seasonal Occurrence: Spring migrant from April through June; fall migrant from July through September. A few remain to nest at Cheyenne Bottoms and Quivira NWR. There are confirmed breeding records from three counties. Sometimes seen in breeding season on playa lakes in the southwestern counties.

Field Notes: Phalaropes are unique. Unlike most bird families, all female phalaropes are more colorful than their male counterparts. Most phalaropes migrate and winter on the open oceans, but Wilson's migrate through the interior of the continent and are a familiar sight on the prairie marshes in spring and fall. Wilson's Phalaropes are often seen swimming in large flocks where they spin around in circles to stir up aquatic insects and their larvae, which they pick from the surface of the water. Watching hundreds of these birds spinning on a prairie pool is mesmerizing.

breeding male

nonbreeding

Phalaropus lobatus

Red-necked Phalarope

Field Identification: Red-necked Phalarope is slightly smaller than Wilson's Phalarope. In all plumages, it has bold wing-stripes visible in flight and a streaked back, unlike the plain, unstreaked back of Wilson's. Females in breeding plumage have a distinctive chestnut-red color on the sides and front of the neck and a dark gray back with buffy stripes on the flanks. Breeding males have a similar but more muted color pattern. Nonbreeding adults are gray above and white below. The most prominent mark on nonbreeding birds is the dark cap and the bold black patch around the eye. Size: Length 7 inches; wingspan 15 inches.

breeding female

Habitat and Distribution: Wetlands, playa lakes, ponds, and water treatment lagoons statewide. Has been recorded in 61 counties. Most likely to be found in the central and western counties, but there are records from multiple eastern counties.

Seasonal Occurrence: Spring migrant mainly in mid- to late May. Fall migrant from mid-August through early October.

Field Notes: This shorebird is rare but regular in Kansas. Most spring sightings occur during the last two weeks of May. Most likely to be found in spring at Cheyenne Bottoms and Quivira NWR. In fall, the migration period is more prolonged and is spread more broadly across the state.

breeding male

nonbreeding

© David Seibel

119

Franklin's Gull

Leucophaeus pipixcan

Field Identification: This small gull of the prairies is darker gray on the back than the other common gulls in Kansas. The head is all black in spring. By autumn, much of the black is lost, leaving a smudgy area on the nape and around the eye. The wings have white trailing edges and black primary tips with white spots. One-year-old Franklin's are lighter above, with all-black primaries, and their tails have a sharply defined black terminal band. Gulls go through a complex series of plumages as they mature and have different plumages in summer and winter. To find Franklin's in large mixed-species flocks of gulls, look for smaller gulls darker than the rest. Size: Length 14 inches; wingspan 36 inches.

Habitat and Distribution: Lakes, marshes, and plowed fields throughout Kansas. Loose flocks of migrating birds fly low and are frequently observed. Has been recorded in all 105 counties but is far more common in central and eastern Kansas. It is scarce in western Kansas.

Seasonal Occurrence: Spring migrant in April and May; fall migrant from September through November. A few linger in summer at reservoirs and wetlands. A small nesting colony succeeded in fledging some young at Cheyenne Bottoms in 1989 for the only confirmed breeding record of this species in Kansas.

Field Notes: This species nests on prairie wetlands in the northern US and Canada and winters along the coasts of Chile and Peru. In spring, hundreds are sometimes seen foraging behind farm equipment working the soil. Concentrations of up to several hundred thousand migrants gather at many of the large reservoirs in October and November. These huge gatherings represent most of the entire world population and are one of the greatest wildlife spectacles in Kansas.

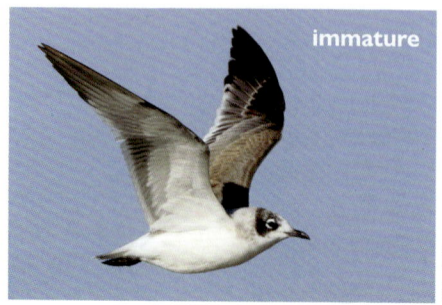

immature

© Judd Patterson

Chroicocephalus philadelphia

Bonaparte's Gull

Field Identification: Bonaparte's is slightly smaller and slimmer than Franklin's Gull and much paler gray on the back. The head is black in spring and white with a black ear spot in winter. On flying birds, the wings have prominent white outer wing feathers with black tips that separate them from all other Kansas gulls. Their flight style is lighter and more graceful than that of Franklin's Gull. Size: Length 13 inches; wingspan 33 inches.

© James W. Arterburn

Habitat and Distribution: Lakes, rivers, and wetlands. Has been reported in 91 counties, mostly in the eastern two-thirds of Kansas. Rare in western counties.

Seasonal Occurrence: Spring migrant from March through May; fall migrant from October through November. In winter, some remain at large reservoirs in eastern Kansas through December and January but are mostly absent by February.

Field Notes: Bonaparte's, Franklin's, Herring, and Ring-billed Gulls represent most of the gulls seen in Kansas. Bonaparte's is the least numerous of these four and is typically seen in small flocks at lakes and reservoirs. In late fall and early winter, flocks numbering into the hundreds sometimes occur at reservoirs in eastern Kansas.

nonbreeding

© David Seibel

immature

121

Sabine's Gull

Xema sabini

Field Identification: This small gull is the same size as Bonaparte's. The distinctive and graceful flight is more like that of a tern than a gull. Almost all seen in Kansas are juveniles during fall migration. Juveniles are brown on the crown, hindneck, and back. The feathers on the back have white edges, creating a scaly look. On flying birds, the upper side of the wing shows a striking pattern of black, brown, and white triangles. This is the only gull with a forked tail. On fall juveniles, the tail has a black band at the tip. Adults are almost never seen inland but have a black head and retain the unique wing pattern. Size: Length 13 inches; wingspan 35 inches.

immature

© David Seibel

Habitat and Distribution: Large reservoirs, wetlands, and wastewater treatment impoundments, especially in the west. There are records from 34 counties.

Seasonal Occurrence: Fall migrant. Can be seen from August through November. Most often seen from mid-September through mid-October.

Field Notes: This is the most oceanic of all gull species. It nests in the high arctic, migrates almost entirely off both the Atlantic and Pacific coasts, and winters at sea in southern oceans. Each fall, a small number of juvenile Sabine's migrate south across inland North America instead of by sea. A few of these inland migrants are seen each year in Kansas. In some years, they are reported from multiple locations in Kansas, occasionally in groups of several birds. Seeing this enigmatic and beautiful gull is a treat.

adult

© Judd Patterson

immature

© David Seibel

Larus delawarensis

Ring-billed Gull

Field Identification: This is the most abundant winter gull in Kansas. This medium-sized gull takes three years to reach full adult plumage. The white head of the adult is streaked brownish in winter. Adults are pale gray on the back and wing. Wing tips are black with white spots. The bill is yellow with a black ring near the tip, and the legs are also yellow. Consistent marks on one- and two-year-old individuals are the flesh-pink legs and a dark band at the end of the tail. Size: Length 18 inches; wingspan 48 inches.

Habitat and Distribution: Statewide at lakes, ponds, marshes, and rivers. There are records from all 105 counties. Least common in the west.

Seasonal Occurrence: Migrant and winter resident. Fall migrants begin to arrive as early as July, becoming numerous in late fall. Many winter to the south of Kansas, but large numbers remain throughout the winter at reservoirs and urban areas with open water. In spring, most have departed by May.

Field Notes: Other than the strictly migratory Franklin's Gull, this is the most often observed gull in Kansas for most of the year, especially in the winter months. From late October through March, Ring-bills are abundant at large reservoirs and landfills. Like Franklin's Gulls, they follow plows in the spring, foraging for insects and other prey. When major water releases are made at reservoirs, large numbers of Ring-bills gather to feed on stunned fish in the spillways.

1st year

2nd year

Herring Gull

Larus argentatus

Field Identification: This large, white-headed gull is similar to the Ring-billed Gull but is larger and more barrel-chested. The yellow bill of the adult is proportionately larger, and its legs are pink. The Herring Gull has a succession of plumages as it matures and does not attain adult plumage until it is five years old. The large size and pink legs are constant throughout these changes. Many immature birds are seen in Kansas. Most are one-year-old birds that are uniformly mottled dark brown on the body, with an entirely black tail. When they reach two years of age, the gray mantle color begins to appear and the black on the tail begins to recede. Gull identification is complex and in Kansas is epitomized by this species. Size: Length 25 inches; wingspan 58 inches.

Habitat and Distribution: Lakes, rivers, and wetlands. Has been recorded in 84 counties, mostly in central and eastern Kansas, occasionally west to the Colorado border.

Seasonal Occurrence: Migrant and winter resident. Most arrive in fall in October, and spring birds depart by April. Straggling individuals are occasionally observed in summer. There are records from all months of the year.

Field Notes: During winter, Herring Gulls may occur in flocks numbering in the hundreds at large reservoirs and near landfills, and a few are usually found within large flocks of Ring-billed Gulls.

1st year

2nd year

Larus hyperboreus

Glaucous Gull

Field Identification: This stocky, barrel-chested bird is one of our largest gulls and one of the "white-winged" gulls that have no black on the wings or tail in any plumage. Most seen in Kansas are one- or two-year-old birds that can appear to be all white, but one-year-old birds show some light gray or brownish mottling on the back and wings. On younger birds, which are typically seen in Kansas, the large bill is pink with a black tip, and the legs are also pink. Adults are seen less often in Kansas. They have a pale gray back, pale gray wings without black tips, and all-yellow bills. Size: Length, 26 inches; wingspan 60 inches.

2nd year

Habitat and Distribution: Large reservoirs, wetlands, and landfills. There are records from 32 counties in central and eastern Kansas.

Seasonal Occurrence: Winter visitor from December through March. Most likely to be seen in January and February.

Field Notes: These "winter wanderers" are found only in the winter months. In Kansas, they are attracted to large landfills, where they forage for food scraps and roost at nearby lakes and rivers. Look for Glaucous Gulls amid the large flocks of gulls that gather at winter locations such as Nelson Island in Overland Park and reservoirs such as Cheney, Clinton, Glen Elder, Perry, and Wilson. In these large flocks, Glaucous Gulls are usually easy to pick out because of their large size and all-white plumage.

adult

Iceland Gull

Larus glaucoides

Field Identification:
Think of the Iceland Gull as a smaller version of the Glaucous Gull. If you see a "white-winged" gull that is obviously smaller than a Glaucous Gull, it is likely this species. The overall body shape is proportionately more slender than that of the Glaucous Gull. Additionally, the bill structure is smaller than that of the Glaucous, and the folded wings extend farther beyond the tail. There are darker first- and second-year Iceland Gulls with moderately dark wingtips and tails. Some adults have limited black in the outermost primaries. These plumage characteristics can make separation of darker Iceland Gulls from Herring Gulls challenging. To be safe with your identification, stick to those Iceland Gulls that are pale in color. Size: Length 22 inches; wingspan 54 inches.

Field Notes: This is another of the "winter wanderer" gull species. Like the Glaucous Gull, it is most likely found amid large flocks of wintering gulls at landfills and large reservoirs. It occurs much less frequently than the Glaucous Gull.

Habitat and Distribution:
Large reservoirs, wetlands, and landfills. Frequently found at Cheyenne Bottoms. There are records from 30 counties in central and eastern Kansas.

Seasonal Occurrence:
Winter visitor. Seen from late November through February. Most likely to be seen in January and February.

adult

© David Seibel

Larus fuscus

Lesser Black-backed Gull

Field Identification: This large gull resembles the Herring Gull in structure, although it is slightly smaller and slimmer. At a distance, adults are dark charcoal gray on the back and wings. Adults have yellow legs, unlike the pink legs of the Herring Gull, and heavier streaking on the head, especially around the eye. The dark smudge around the eye is a consistent field mark on all ages of this species. Immature birds have a four-year succession of plumages. Separate one-year-old Lesser Black-backed from Herring by its dark eye smudge, more patterned back, and narrower black tailband compared with the all-dark tail of the one-year-old Herring Gull. Two-year-old birds have a heavily streaked whitish head with a dark smudge around the eye, and the dark gray feathers on the back begin to emerge. Size: Length 23 inches; wingspan 55 inches.

© Joseph Miller

Habitat and Distribution: Large reservoirs, wetlands, and landfills. There are records from 33 counties in central and eastern Kansas and a few records from farther west.

Seasonal Occurrence: Most likely to be seen between December and March.

Field Notes: This gull is another "winter wanderer" but is seen over a much greater length of the year than the others. There are records from all months except June in Kansas. Of the "winter wanderer" gulls, this is the most widespread and likely to be seen. Each year, it seems to become more common.

© James W. Arterburn
1st year

© Alvan Buckley
2nd year

Great Black-backed Gull

Larus marinus

Field Identification: This is the largest gull in the world. Look for the exceptional size, big head, huge bill, and standing posture taller than that of any other gull. Adults have a coal-black back and wings, an all-yellow bill, and pink legs. One-year-old birds have a distinct checkered pattern on the back and wings, a white head, and an all-black bill. Size: Length 30 inches; wingspan 65 inches.

Habitat and Distribution: Large reservoirs, wetlands, and open-air landfills. There are records from 16 counties in central and eastern Kansas.

Seasonal Occurrence: Most records are from December through March.

Field Notes: This is another of the "winter wanderers" and is the rarest Kansas gull included in this book. It is expected to become more frequently seen within winter gull flocks at the big lakes. Its huge size makes it conspicuous.

1st year

128

Hydroprogne caspia

Caspian Tern

Field Identification: This large tern is nearly the size of a Ring-billed Gull but is slimmer, with more angular, pointed wings. It has a pale gray back and upper wings. The undersides of the wingtips are black and visible in flight. Look for the swept-back black crown and the blood-red bill, which is much thicker than the bill of other terns. Size: Length 21 inches; wingspan 50 inches.

Habitat and Distribution: Lakes, reservoirs, and wetlands in eastern and central Kansas. Common east of the Flint Hills, scarce in the central counties, and rare in the west. There are records from 56 counties.

Seasonal Occurrence: Spring migrant from April through June; fall migrant from August through October.

Field Notes: Terns are closely related to gulls but are more graceful in both appearance and behavior. This is the largest tern found in Kansas. Its size alone separates it from other tern species. It is most likely to be found at large reservoirs in eastern Kansas, where it patrols above the surface of the water in search of fish. These terns sometimes roost on sandbars with flocks of gulls and can be overlooked.

129

Forster's Tern

Sterna forsteri

Field Identification: This is a medium-sized tern with frosty-white primary feathers and pale gray secondary feathers and back. It has long outer tail feathers. It has a black cap in spring and a well-defined black eye-patch in fall and winter. The bill is orange with a dark tip in spring, black in fall. Size: Length 13 inches; wingspan 31 inches.

Habitat and Distribution: Statewide at reservoirs, lakes, wetlands, and wastewater treatment facilities. Has been recorded in 103 counties.

Seasonal Occurrence: Spring migrant in April and May; fall migrant from July through November. Fall migration peaks in late September and early October. In years when habitat conditions are favorable, some will remain to nest at Cheyenne Bottoms and Quivira NWR. These are no other Kansas nesting locations.

Field Notes: This tern is a familiar sight throughout Kansas during migration. Like other terns, it forages by flying over water in a graceful fashion and suddenly diving into the water when it spots a fish or minnow. In contrast to the more eastern distribution of Caspian Tern, Forster's is common at the larger, shallow wetlands of central Kansas. Sewage lagoons in western Kansas, in the absence of other bodies of water, also attract them.

nonbreeding

Sterna hirundo

Common Tern

Field Identification:
The Common Tern is like Forster's in appearance, but despite the name, it is much less likely to be seen in Kansas. In spring breeding plumage, check the wings for an obvious dark wedge on the otherwise gray primary feathers. There is a light gray wash on the underparts and thin dark edges on the outer tail feathers. In comparison, Forster's has frosty white primaries, white underparts, and white outer tail feathers. In fall, both species lose most of the solid black cap. On winter-plumaged Common Terns, the retained black on the head wraps around the back of the head, while Forster's retains only a clearly defined black eye-patch. The bill of the Common Tern is deeper red than the orange bill of Forster's, and the black tip is smaller. Immatures are often seen in fall and are easily identified by a dark patagial bar on the wing that can be seen both in flight and on perched individuals. Size: Length 12 inches; wingspan 31 inches.

Habitat and Distribution: Lakes, reservoirs, and wetlands in central and eastern Kansas. Rare farther west. Has been recorded in 58 counties.

Seasonal Occurrence: Spring migrants occur in April and May with a peak in late May, fall migrants in September and October.

Field Notes: Common Terns are less numerous in Kansas than Forster's and are easy to overlook because they closely resemble them. Carefully study flocks of terns, and you might pick one of these out of the crowd. Fall immatures with the dark patagial bar are the easiest plumage to identify.

nonbreeding

Least Tern

Sternula antillarum

Field Identification: This is the smallest tern in Kansas and can be identified by the black outer primaries on the wings, gray back, and bright yellow bill. The white forehead contrasts with the black cap and eye-line. Sharp alarm notes are often heard as it flies overhead. Size: Length 9 inches; wingspan 20 inches.

Habitat and Distribution: Prefers sandbars, salt flats, and other unvegetated areas near water. It is seen in many parts of the state during migration and has been observed in 67 counties.

Seasonal Occurrence: Migrant and summer resident. Spring arrival is in April, and fall migrants depart in September. It nests in colonies at a few locations on or near the Arkansas and Kansas Rivers. There are confirmed breeding records from 16 counties, always on alkali flats or sandbars. It currently nests on the Big Salt Marsh at Quivira NWR, at Jeffrey Energy Center near Topeka, and at several locations in Wichita on or near the Arkansas River. Kansas nesting birds usually depart by August.

Field Notes: This tern is small but plucky. It was formerly on the endangered species list but was removed in 2021. It formerly nested on river sandbars throughout Kansas, but reduced flow and dam construction have led to habitat changes that have made many of its former nesting sites unsuitable. Because its nests are on sandbars, they are vulnerable to rising water, a variety of human-caused disturbances, and predation by coyotes and other mammals and birds.

© Judd Patterson

Chlidonias niger

Black Tern

Field Identification:
Black Terns are markedly different from all other tern species in Kansas. In breeding plumage, the head and body underparts are uniformly black. In flight, the bottom of the wings is contrastingly white. Winter-plumaged birds are white below and on most of the head, with a blackish nape and ear-patch. Size: Length 10 inches; wingspan 24 inches.

Habitat and Distribution:
Migrant statewide at marshes, lakes, and wastewater treatment impoundments and over prairies. Has been recorded in all 105 counties.

Seasonal Occurrence:
Spring migrant in April and May; fall migrant from July through October. Small colonies nest at Cheyenne Bottoms and Quivira NWR in years when habitat conditions are favorable.

Field Notes:
Large flocks of Black Terns are often observed migrating low over the Kansas prairies during the spring, acrobatically foraging as they journey north. Unlike other terns, they rarely dive for fish but instead feed almost entirely on insects captured on the wing. Their primary breeding grounds are in the northern US and Canada.

nonbreeding

133

Ring-necked Pheasant

Phasianus colchicus

Field Identification: Both sexes are easily identified by their large size and long, pointed tails. The loud, squawking call of the male is a familiar sound in rural areas. Size: Length 21 inches; wingspan 31 inches.

Habitat and Distribution: Found in or near grainfields as well as cattail marshes and brushy areas. Found over most of the state but is absent or rare in the southeastern counties and along most of the Missouri border. Has been reported in all 105 counties but is most abundant in the western two-thirds of Kansas.

Seasonal Occurrence: Permanent resident. Confirmed breeding has been reported from 77 counties. Does not nest in the southeastern or easternmost counties along the Missouri border.

Field Notes: Ring-necked Pheasants are a species native to China. They were introduced as game birds in the early 1900s and have thrived in Kansas. They are often seen along roadsides in farm country. They remain motionless when hiding, then burst into flight with whirring wings and a loud squawk. This is the most popular game bird in Kansas, and hundreds of thousands of them are taken each year by hunters.

female

Tympanuchus cupido

Greater Prairie-Chicken

Field Identification:
Plump and chicken-like in appearance. Most of the body is patterned with brown-and-white barring. On flying birds, the tail is short, rounded, and dark in color, unlike the longer, pointed tail of pheasants. Size: Length 17 inches; wingspan 28 inches.

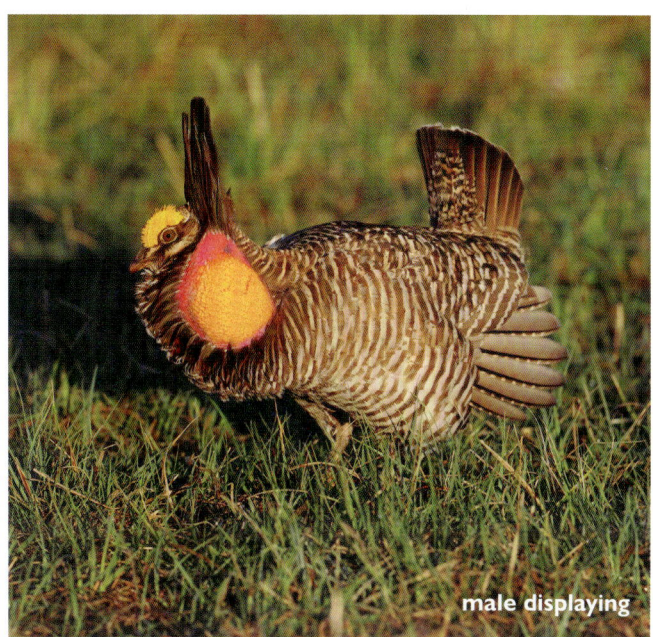
male displaying

Habitat and Distribution:
Found in tallgrass prairies in the Flint Hills and mixed prairies in north-central and northwestern Kansas, generally in or north of the Smoky Hill Valley. There are records from 81 counties, but it is no longer found in more than 20 of the eastern counties where it was once a resident.

Seasonal Occurrence: Permanent resident. There are confirmed nesting records from 46 counties, but many of these are old records from eastern counties where this species no longer occurs.

Field Notes: In early spring, males gather to compete for the attention of females at specific areas on the prairie called "booming grounds," or leks. The orange sacs on the sides of the neck are used to create a unique "booming" sound, which can be heard at a distance of a mile or more. Males face off in a series of ritual confrontations involving raising the long neck feathers, making vigorous clucking sounds, and jumping into the air. The female observes the activity and, if sufficiently impressed, will allow the dominant male to mate with her. In the Flint Hills, Greater Prairie-Chickens are declining in numbers due to habitat fragmentation and annual burning that leaves little grass cover for concealment and successful nesting. Populations in north-central Kansas are more stable.

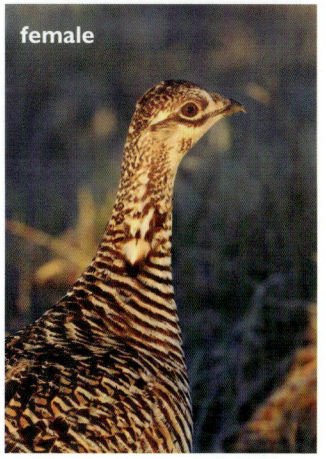
female

Lesser Prairie-Chicken

Tympanuchus pallidicinctus

Field Identification: The Lesser Prairie-Chicken is similar to the Greater Prairie-Chicken but is paler, with less contrasting brown-and-white barring. The neck sacs of the male are plum-colored instead of orange, and the "booming" is a rapid bubbly sound, quite unlike the slow, deep tones of the Greater Prairie-Chicken. Size: Length 16 inches; wingspan 25 inches.

male displaying

Habitat and Distribution: Sand sage and other semiarid prairies in the southwestern quadrant of the state, north to Trego County, and east to Barber County. Has been recorded in 36 counties.

Seasonal Occurrence: Permanent resident. There are confirmed nesting records from 28 counties.

Field Notes: This is a close relative of the Greater Prairie-Chicken. It favors sandy, semiarid prairies of southwestern Kansas, northwestern New Mexico, southeastern Colorado, and the Panhandle region of Oklahoma and Texas. From about one million birds at the time of settlement, the population has declined by over 97 percent, and their habitat has been reduced by over 90 percent. It is now listed as "endangered" in New Mexico and Texas and "threatened" in Colorado, Oklahoma, and Kansas. Because of its limited range and rarity, it attracts birders from across the US and around the world each year.

female

Meleagris gallopavo

Wild Turkey

Field Identification: This large game bird is unlikely to be mistaken for any other species. Males are larger than females and have a "beard" that grows from the middle of their chest. Size: Length 46 inches; wingspan 64 inches.

Habitat and Distribution: Statewide where woodlands and brushy areas are present, especially those near rivers and streams. Has been recorded in all 105 counties.

Seasonal Occurrence: Permanent resident. There are confirmed breeding records from 70 counties.

Field Notes: Early in the twentieth century, this formerly abundant species was almost eliminated from Kansas by overhunting. Populations have recovered in a major way since the Kansas Department of Wildlife and Parks began reintroducing the birds decades ago. It is a memorable sight to see males displaying for females in the spring. The tail is raised and fanned, the wings are held downward, and the body feathers are raised, giving the bird a unique appearance as it struts around. After the nesting season, large flocks, some numbering in the hundreds, gather and forage for waste grain and acorns. The large footprints of these birds in wooded areas are obvious clues to their presence.

Northern Bobwhite

Colinus virginianus

male and female

Field Identification: This is a small, plump game bird of brushy areas. Males are patterned with white and chestnut brown, with a white stripe above the eye and a white throat. Females' head markings are buff instead of white. Size: Length 10 inches; wingspan 13 inches.

Habitat and Distribution: Statewide in brushy and woodland edges adjacent to grasslands and grain crops. Has been recorded in all 105 counties. Easy to find in much of the state, but populations in the western counties are smaller and more localized.

Seasonal Occurrence: Permanent resident. There are nesting records from 75 counties. Least common as a nesting species in the northwestern corner of the state.

Field Notes: The clear, two-note call of the male is a familiar sound across Kansas in the spring and summer months. In the fall and winter, they gather into small flocks called coveys, which flush in unison when alarmed. Populations of Northern Bobwhites fluctuate significantly from year to year. Data from breeding bird surveys and other studies show that in recent years, populations of this species have declined significantly. This is thought to be mostly due to habitat loss.

Callipepla squamata

Scaled Quail

Field Identification: The most notable field mark is the "cottontop" crest on the head. Feathers on the underparts and hindneck have clean black edges, creating a unique scaled appearance. Size: Length 10 inches; wingspan 14 inches.

Habitat and Distribution: Sand sage prairies of southwestern Kansas, mostly south of the Arkansas River and west of Clark County. There are records from 13 counties in the southwestern corner of the state.

Seasonal Occurrence: Permanent resident. There are confirmed nesting records from seven counties.

Field Notes: Scaled Quail are sometimes called by their nicknames, "cottontop" and "blue quail." These are birds of southwestern deserts, with southwestern Kansas at the northwestern edge of their range. They are found around farms, abandoned farm equipment, auto salvage yards, and old junk piles in sage country. They frequent these places because of the shelter they offer from the elements and predators. Look for them on the Cimarron National Grassland in Morton County and along the River Road in Hamilton and Kearney Counties. Their loud, piercing calls are unlike those of the Bobwhite, and many people hearing them for the first time are surprised to learn that they are hearing a species of quail. They prefer to run rather than fly.

Turkey Vulture

Cathartes aura

Field Identification: This is a large, soaring bird of the summer skies. The two-tone gray-and-black wings are always tilted upward. It flaps its wings only reluctantly, preferring to soar on warm thermal air currents. Perched adults appear all black with a bare red head. Immatures have black heads. Size: Length 26 inches; wingspan 67 inches.

Habitat and Distribution: Found statewide. Can be seen soaring overhead anywhere. Most often observed along roads and highways, in river valleys, or near dams of large lakes and reservoirs. There are records from all 105 counties.

Seasonal Occurrence: Migrant and summer resident. Present from late March until late October. There are confirmed nesting records from 58 counties, mostly in central and eastern Kansas.

Field Notes: Turkey Vultures roam widely in search of carrion and are frequently seen attempting to feed on road-killed animals along highways. They often gather in large communal roosts at night, especially during migration. As air currents warm in the morning, the entire roost takes flight and begins foraging. While vultures look like hawks and eagles in appearance, DNA studies have shown that they are most closely related to the stork family.

immature

© David Seibel

140

Coragyps atratus

Black Vulture

Field Identification: The Black Vulture is a bit smaller than the Turkey Vulture. The bare head is black, but immature Turkey Vultures also have black heads. The tail is shorter and broader than that of the Turkey Vulture. Flying birds are easiest to identify. The wings are all black except for contrasting white primaries that form a white patch at the ends of the all-black wings. Size: Length 25 inches; wingspan 57 inches.

Habitat and Distribution: Wooded streams in southeastern Kansas. Its range is expanding. In 2011, there were records from 11 counties. By 2022, there were records from 32 counties. It is still most likely to be seen in southeastern Kansas, but it has been seen north to Jefferson County and as far west as Finney County. Future range expansion appears likely.

expanded in Kansas and will likely continue to do so. Look for them soaring with Turkey Vultures.

Seasonal Occurrence: There are records from all months of the year, but most likely to be found in spring and summer. There has been an increase in winter reports. Some winter roosts near Sedan in Chautauqua County have had up to 30 individuals. The only confirmed nesting record comes from Labette County.

Field Notes: Black Vultures were present in Kansas in the nineteenth century, but there were no twentieth-century records until 1998, when a few were found in Cherokee County in the southeastern corner. Since then, their range has

© Judd Patterson

Osprey

Pandion haliaetus

Field Identification: This large, unique bird of prey is always seen near water. It is distinguished by its black back and crown, white underparts, and white face with a prominent black eye-stripe. Flying birds have a noticeable crook in the long wings and show a black-and-white pattern below. Size: Length 23 inches; wingspan 63 inches.

Habitat and Distribution: Statewide along rivers, lakes, and wetlands. Less common in the western part of the state due to the scarcity of open water. Has been recorded in 100 counties.

Seasonal Occurrence: Migrant. Most are seen in spring during April and May and in fall during September and October.

Field Notes: Ospreys feed exclusively on fish. They soar over lakes and rivers in search of prey. When a fish is spotted near the surface, they plunge dramatically into the water, sometimes completely submerging, while capturing prey in their talons. They fly with the fish in their grasp, oriented with the head of the fish pointing in the direction of their flight. They perch on utility poles or trees to consume their catch. Ospreys are not shy and are frequently seen in busy urban areas.

Ictinia mississippiensis

Mississippi Kite

Field Identification:

This slender, graceful bird of prey is evenly gray, lighter on the head, and has a black tail. White secondary wing feathers are prominent on flying birds. Juveniles have broad streaks below and a banded tail and can be mistaken for accipiters or falcons. Size: Length 14 inches; wingspan 31 inches.

Habitat and Distribution:

Most common in south-central Kansas. Has been recorded in 101 counties. Originally restricted to south-central Kansas, it is now found statewide. Its historical habitat was open prairies with stands of trees. This species has adapted well to nesting in cities and towns.

Seasonal Occurrence: Summer resident. Arrives in late April or early May and departs in mid-September. In the fall, a few linger into early October. There are confirmed nesting records from 61 counties, mostly west of the Flint Hills but also including Douglas and Johnson Counties.

Field Notes: These raptors can be seen "kiting" by skillfully manipulating their flight feathers to navigate wind currents. They feed mostly on cicadas and other large insects but will readily take small vertebrate prey. In many south-central Kansas towns, several can be seen at once as they soar overhead. When the young are near fledging, the adults sometimes become quite aggressive and will dive on people who come too close to the nest. In September, these kites gather into flocks of up to 100 individuals as they prepare to migrate south for the winter.

immature

Bald Eagle

Haliaeetus leucocephalus

Field Identification: Eagles are the largest birds of prey in Kansas. Adult Bald Eagles are unmistakable. They have a dark body contrasting with a white head and tail. Younger birds are uniformly dark in their first year, gradually gaining more white plumage until they reach adulthood at four years of age. Size: Length 31 inches; wingspan 80 inches.

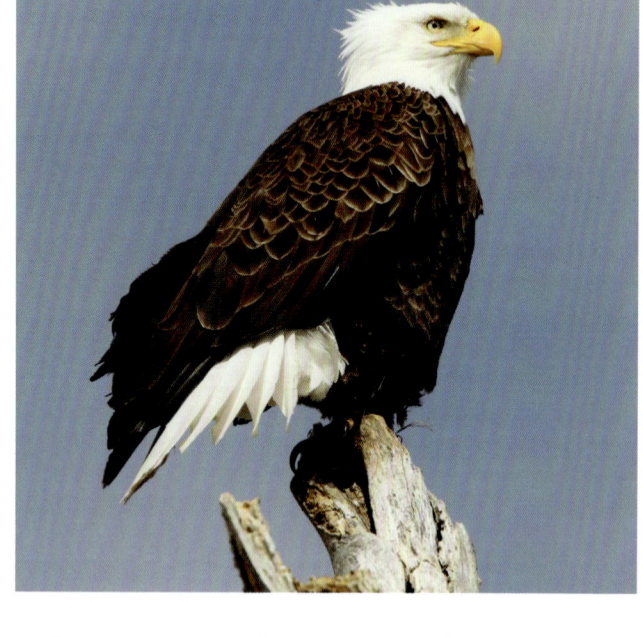

Habitat and Distribution: Statewide near lakes, rivers, reservoirs, and wetlands. It can also be seen in semiarid upland areas of western Kansas, especially during migration. Bald Eagles have been reported from 103 counties in Kansas. They are less common in the west.

Seasonal Occurrence: Migrant and winter resident, rare and local in summer. The greatest numbers are present from November through March, typically around large reservoirs and along rivers. Confirmed breeding has been recorded in 55 counties. As of 2023, there were about 200 active nests in Kansas, mostly near reservoirs and rivers in the eastern two-thirds of the state.

Field Notes: During the winter, several thousand birds move into Kansas when open water in northern states has frozen over. They may be found at large reservoirs and along major rivers. Look for large, dark birds perched in trees near the shore and standing on the ice at the edge of open water. They feed on both live and dead fish, waterfowl, and carrion.

1st year

© Roni and LaVern Allen

2nd year

Aquila chrysaetos

Golden Eagle

Field Identification: This is a large, dark-colored raptor of the western plains. Its size alone separates it from most other raptor species. Both adults and immatures have golden feathers on the head. Adults have a banded tail with a dark terminal band. Immatures have a white tail with broad black terminal band and oblong light patches in the wings. Size: Length 30 inches; wingspan 79 inches.

Habitat and Distribution: Seen most regularly in western Kansas. It prefers rugged open country but wanders widely and has been recorded in 99 counties. During migration and winter, a few are seen each year in the eastern counties.

Seasonal Occurrence: Mostly a migrant and winter resident. Rare and local nesting species. Most are seen from September through March. There are nesting records from nine counties in the Red Hills and in western Kansas.

Field Notes: Golden Eagles are majestic birds of the arid plains. They primarily eat mammals and are frequently seen near prairie dog towns, which unfortunately are disappearing from the Kansas landscape. Golden Eagles are sometimes confused with juvenile Bald Eagles, which are also all dark. Reports of Golden Eagles seen with Bald Eagles at large reservoirs in the winter almost always prove to be young Bald Eagles. Juvenile Bald Eagles always have some white near the body on the underwings. Golden Eagles have more restricted white patches on the underwings.

© Judd Patterson

145

Northern Harrier

Circus hudsonius

Field Identification: This is a low-flying hawk of open country. Flying birds are easily identified by the prominent white rump and long tail. Males are smaller and pale gray above, white below, with black wingtips. Females are larger and brown above, mottled with paler brown below. Harriers are usually seen cruising low over grasslands, marshes, and farmlands with wings tilted upward, rocking back and forth as they fly. Size: Length 18 inches; wingspan 43 inches.

Habitat and Distribution: Found statewide. Prefers marshes, prairies, and Conservation Reserve Program (CRP) grasslands. Also seen over grainfields. Has been recorded in all 105 counties.

Seasonal Occurrence: Found in Kansas year-round, but their abundance varies significantly by season. Widespread and common in winter. Most numerous from October through April. Some remain to nest in locations where appropriate habitat is present and prey is abundant. Confirmed nesting has been reported from 36 counties throughout the state.

Field Notes: The presence of this hawk is usually an indicator of nearby grasslands or marshes. The low flight allows it to surprise prey that is captured with a sudden plunge. The preferred prey is small rodents, but it also takes snakes and birds. This species often shares hunting grounds with Short-eared Owls, although they hunt at different times of the day.

female

© David Seibel

Accipiter striatus

Sharp-shinned Hawk

Field Identification: Separating Sharp-shinned Hawks from Cooper's Hawks is challenging. Both have long tails and short wings and fly with a few quick wingbeats, followed by a glide. Sharp-shinned is generally jay-sized, whereas Cooper's is crow-sized. In both species, females are larger than males. The female Sharp-shinned nearly overlaps the male Cooper's in size. Sharp-shinned appears to have a square-tipped tail, whereas Cooper's is round-tipped. In flight, Cooper's head projects beyond the wings, giving it a "flying cross" appearance. The head of a Sharp-shinned does not project much, creating a "flying mallet" appearance. The crown and nape of adult Sharp-shinned is a uniform dark gray, creating a "hooded" look, while Cooper's has a dark cap with a whitish nape, creating a "capped" appearance. The head of Sharp-shinned usually appears rounded, whereas Cooper's is typically flat-headed. Immature Sharp-shinned Hawks have heavy vertical streaks on the chest, while the streaks of immature Cooper's are thinner and sparser. Size: Length 13 inches; wingspan 23 inches.

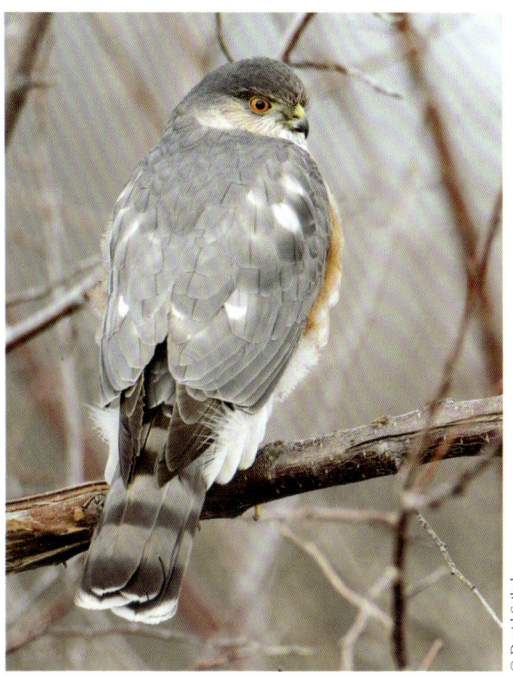

© David Seibel

Habitat and Distribution: Found statewide. Has been recorded in all 105 counties. Usually seen in wooded or brushy areas, both rural and urban.

Seasonal Occurrence: Migrant and winter resident. Present from September through mid-May. Extremely rare as a nesting species in Kansas, but confirmed breeding records exist for three counties and probable breeding for three more,

all in northeastern and south-central Kansas.

Field Notes: Sharp-shinned Hawks are usually seen in brief encounters as they fly through wooded areas in search of small birds, their primary prey. In winter, they frequently learn to hunt around bird feeders.

immature

© Wayne Rhodus

Cooper's Hawk

Accipiter cooperii

Field Identification: This accipiter resembles the Sharp-shinned Hawk but is larger on average. Male Cooper's and female Sharp-shinned nearly overlap in size. The corners of the Cooper's Hawk's tail are rounded. Immature Cooper's show thin vertical streaks on the chest. See the previous Sharp-shinned Hawk account for a more detailed discussion of how to separate these two species. Size: Length 17 inches; wingspan 31 inches.

Habitat and Distribution: Statewide. There are records from all 105 counties. Typically seen in or near wooded and brushy areas but also hunts in open country. Frequently found in cities and towns.

Seasonal Occurrence: Permanent resident. Seen more during migration and in winter, when many move into Kansas from the north. Increasingly common in summer. Confirmed nesting has been recorded in 43 counties spread across most of the state except for the southwestern corner.

Field Notes: An accipiter seen during the summer is almost certainly this species. Once severely decimated by DDT poisoning and illegal shooting, since the 1980s, Cooper's Hawks have enjoyed a steady increase as a nesting species. Cooper's Hawks prey mostly on birds and small mammals. Along with Sharp-shinned Hawks, Cooper's Hawks often capture birds around bird feeders.

immature

Buteo lineatus

Red-shouldered Hawk

Field Identification:
This shy *Buteo* hawk of riparian woodlands is similar in shape to but smaller than the Red-tailed Hawk. Loud, plaintive cries are often the first clue to their presence. Perched birds have checkered black-and-white marks on the wings and back and dense reddish bars on the breast. When they are soaring overhead, look for prominent translucent crescents at the base of the primary feathers in the wings and a black tail with three or four narrow white bands. Juveniles somewhat resemble immature Red-tailed Hawks, but the streaking on the breast and underparts is more extensive, and the tailband pattern resembles that of the adult. Size: Length 17 inches; wingspan 40 inches.

Habitat and Distribution: Eastern and central Kansas, typically along rivers and streams with mature woodlands. Regularly seen in cities and towns. Currently occurs regularly west to Barber, Reno, and Jewell Counties. There are records from 70 counties. Some wander west of the breeding range, especially in the fall.

Seasonal Occurrence: Permanent resident. There are confirmed nesting records from 24 counties west to Dickinson and Sedgwick County. The current breeding range likely extends another 60 miles west of that line.

Field Notes: This hawk species was once on the Kansas list of threatened and endangered species. Today it is steadily increasing in numbers and expanding its range westward along riparian woodland corridors and in urban areas. They hunt from inconspicuous perches but are often seen soaring high above the forest. In early April, the males perform a dramatic courtship flight that includes steep dives and piercing cries. Be cautious of "heard-only" birds, as Blue Jays often expertly mimic this species.

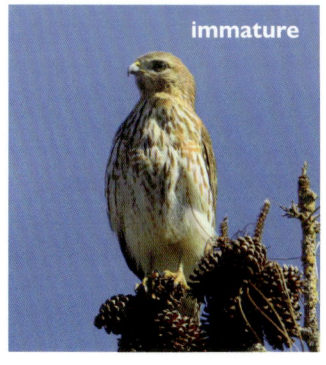

immature

Broad-winged Hawk

Buteo platypterus

Field Identification: This is our smallest *Buteo* hawk. It has a brown back, white breast with reddish bands, and solid reddish throat. The tail has broad black and white bands compared to the narrower tailbands of the Red-shouldered Hawk. In flight, the underwings are mostly white, neatly bordered with black on the wingtips and trailing edges of the wings. The Red-tailed Hawk always has a black bar on the leading edge of the wing, which Broad-winged does not have. There is a rare dark morph. Juveniles are often seen during fall migration. They have the trailing black edge of the wing of the adult, but it is less distinct. The breast on these immature birds is usually heavily streaked, and there is always a heavy malar streak. Size: Length 15 inches; wingspan 34 inches.

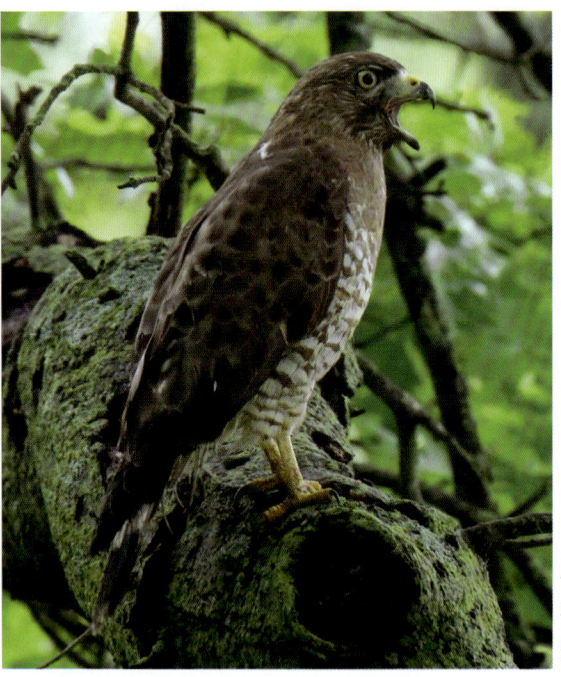

© David Butel

Habitat and Distribution: Deciduous woodlands. Found statewide during migration but more numerous in the eastern half of Kansas. Migrating birds are often seen in cities and towns, especially in western Kansas, where they have fewer woodland options. There are records from 102 counties.

Seasonal Occurrence: Primarily a spring and fall migrant. Rare summer resident. Spring migrants are seen from mid-April through mid-May, fall migrants from mid-August through early October. Confirmed nesting has been reported from six counties in northeastern Kansas within the Kansas River watershed.

There are multiple summer records of Broad-wings from the Cross Timbers region, including apparent nest building and adults with juveniles.

Field Notes: Broad-winged Hawks are often seen in small flocks during the spring migration. They are somewhat tame and more approachable than other raptors. In the fall, most sightings in Kansas are of single birds.

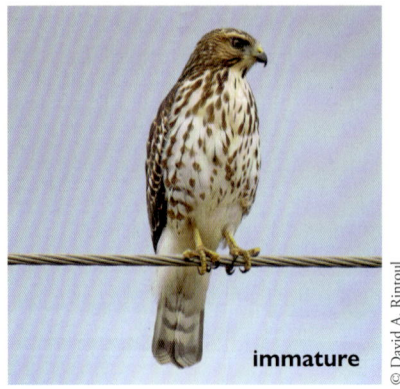

immature

© David A. Rintoul

Buteo swainsoni

Swainson's Hawk

Field Identification: This is a hawk of the open country with several color morphs. Light-morph adults are the most numerous. They are dark brown on the back and white on the underparts, with a reddish chest and white face. The tail has narrow bands with a wider terminal band. Other morphs show rufous on the belly and chest. On soaring birds, the two-toned underwing is white on the leading edge and dark gray on the trailing portion. Juveniles are heavily streaked on the underparts, usually with a vertical streak extending below the eye. Size: Length 19 inches; wingspan 51 inches.

Habitat and Distribution:
Found statewide but more numerous in the western half of Kansas. Usually found near open country. Has been recorded in all 105 counties.

Seasonal Occurrence: Migrant and summer resident. Spring arrival is in early April, fall departure in early October. There are confirmed nesting records from 61 counties, largely in the western half of the state but also from 15 counties in the eastern half, mostly in the northeast.

Field Notes: Swainson's Hawks sometimes follow farm equipment to feed on exposed insects and rodents. The entire population winters in the grasslands of Argentina and nests in western North America, as far north as central Canada. Sometimes Swainson's migrate in large

flocks that can number in the hundreds or even thousands during the fall. These flocks often rest for the night in hayfields or plowed fields. It is an amazing sight to see hundreds of these hawks calmly perched on the ground or on hay bales, awaiting a favorable wind to carry them south.

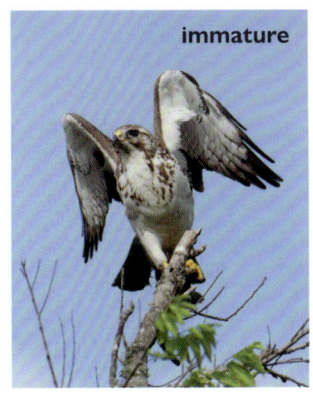

immature

Red-tailed Hawk

Buteo jamaicensis

Eastern Red-tailed Hawk, photo composite

© David A. Rintoul

Field Identification: The Red-tailed Hawk is by far the most widespread and common Kansas hawk. This species is highly variable, more so than any other species of bird that occurs in Kansas. Typical eastern adult birds have a red tail, black terminal tailband, white throat, and white underparts with a belly band of streaks. First-year birds have a brown tail with a series of darker bars. On soaring birds, look first for a prominent dark bar on the leading edge of the underwing, which automatically separates Red-tails from all other buteos. There are several subspecies with dramatically different plumages. The Eastern (*borealis* subspecies) is the most common and represents all breeding birds statewide. Several additional subspecies move into Kansas each winter from north and west of the Great Plains. Western adults have an all-dark head, including the throat; rufous wing linings; and narrow dark tailbands on

the red tail above the wider black terminal tailband. Some Western Red-tailed Hawks (*calarus* subspecies) have all-black bodies but retain the red tail. Harlan's Red-tail winters in Kansas and is the most variable of all the subspecies. No two Harlan's are quite alike. "Typical" dark-morph Harlan's are also mostly black with white streaking on the breast and a mottled white tail that has an irregular black tailband. There is also a light morph of Harlan's. Size: Length 19 inches; wingspan 49 inches.

Habitat and Distribution: Found statewide in almost any habitat. There are records from all counties.

Seasonal Occurrence: Permanent resident. In the winter months, additional birds from the north and west increase the Kansas population. There are confirmed nesting records from 100 counties.

Buteo jamaicensis

Red-tailed Hawk

© David A. Rintoul

Western Red-tailed Hawk

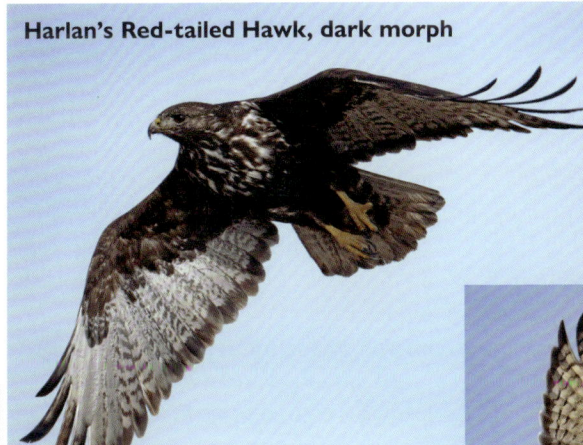

Harlan's Red-tailed Hawk, dark morph

© David A. Rintoul

Harlan's Red-tailed Hawk, light morph

Field Notes: Red-tails are the most common and widespread hawk in Kansas. During winter, there may be one or more per mile in a typical drive through rural areas. Red-tails usually begin nesting in late February and early March. Their large stick nests in tall trees are easy to spot before the trees have leaves.

© David A. Rintoul

Ferruginous Hawk

Buteo regalis

light morph

Field Identification: This is the largest *Buteo* hawk in Kansas. Typical adults are a pale red color on the back, with white head and breast, pale whitish tail, and dark reddish legs that are easily seen on soaring birds. Look for large white patches on the upper wings. There is a rare dark morph that can appear black or deep rufous. Dark-morph individuals retain the white wing-patches and white tail. The Ferruginous Hawk bill is larger than that of other buteos and where the two mandibles of the bill hinge together (referred to as the "gape") has a center line that reaches the center of the eye. This large gape enables the birds to swallow large prey. Juveniles have faint bands on their whitish tails. Size: Length 23 inches; wingspan 56 inches.

Habitat and Distribution: Found in the western half of Kansas in open country, especially near prairie dog towns. Nests in the shortgrass prairies of western Kansas and the chalk bluffs of the Smoky Hill River valley. During winter, a few wander to eastern Kansas. There are records from 84 counties.

Seasonal Occurrence: Permanent resident. Uncommon but widespread during migration and in winter, from late September through early April. Some remain to nest where appropriate habitat is available. There are nesting records from 18 western counties.

Field Notes: This large, beautiful hawk is a

western Kansas specialty. It can be relatively tame outside the nesting season and sometimes will tolerate a close approach. Where prairie dog towns have not been eliminated, these hawks often gather and perch on the ground on prairie dog mounds. The Cimarron National Grassland in southwestern Kansas is a good place to look for these striking raptors during winter.

light morph

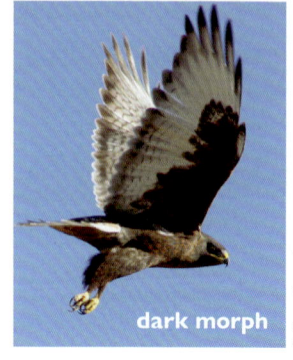

dark morph

© Steve Rottenborn

154

Buteo lagopus

Rough-legged Hawk

light morph

Field Identification: Like most other *Buteos*, the Rough-legged Hawk occurs in light and dark morphs. Soaring light-morph birds from below show a white tail with a broad black band, white underwings, a large black patch at the wrist, and a black belly. Perched light-morph birds show a pale head and a broad black belly band. Dark-morph birds in flight have mostly white flight feathers with darker wing linings. Perched dark-morph birds look uniformly black except for a gray banded tail. Size: Length 21 inches; wingspan 53 inches.

Habitat and Distribution: Statewide in open country. Has been reported from all 105 counties. It strongly prefers grassland habitats over cultivated areas.

Seasonal Occurrence: Present in Kansas from October through April. Most abundant from November through March. Nests on the arctic tundra.

Field Notes: This arctic-nesting hawk moves into Kansas during the winter. It often hovers while foraging. Dark morphs are frequently seen. In grasslands such as the Flint Hills, this species can occasionally outnumber the Red-tailed Hawk during the winter months.

light morph

dark morph

© David A. Rintoul

American Kestrel

Falco sparverius

Field Identification: This small, dove-sized falcon has a double-mustache mark on the face and a rufous back with black bars. The wings are blue on males, rufous on females. Females have coarse reddish-brown streaks on the breast. All falcons have angular pointed wings. This helps distinguish them from accipiters, which have rounded wings. Size: Length 9 inches; wingspan 22 inches.

Habitat and Distribution: Statewide. Has been recorded in all 105 counties. Found in open areas, usually with a few trees nearby.

Seasonal Occurrence: Permanent resident. Most common in summer and during migration. There are confirmed nesting records from 77 counties. Smaller numbers remain in Kansas during winter months.

Field Notes: These small falcons are common along roadsides and are frequently seen perched on power lines or fences. They often hover when hunting. They nest in tree cavities and will readily use nesting boxes when placed in appropriate locations. In September, migrating kestrels moving through Kansas take advantage of abundant grasshoppers to fatten up for the leaner winter months ahead. This is when they are at their highest abundance.

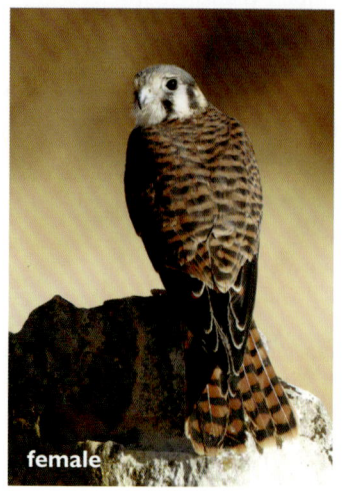

female

Falco columbarius

Merlin

Field Identification: This small falcon is slightly larger than a kestrel. Two subspecies occur in Kansas. All Merlins have a strongly banded tail and a heavily streaked breast. Males are blue on the back and wings, females brown. Males of the Taiga subspecies have a narrow "mustache" marking called the malar streak. Taiga males are dark blue-gray on the back, females dark brown, and the breast streaking is dark. Males of the Prairie subspecies are light powder blue on the back, and females have sandy-brown backs; neither sex has malar streaks; and the breast streaking is lighter brown than that of the Taiga subspecies. Their flight is more powerful and direct than that of the American Kestrel. Size: Length 10 inches; wingspan 24 inches.

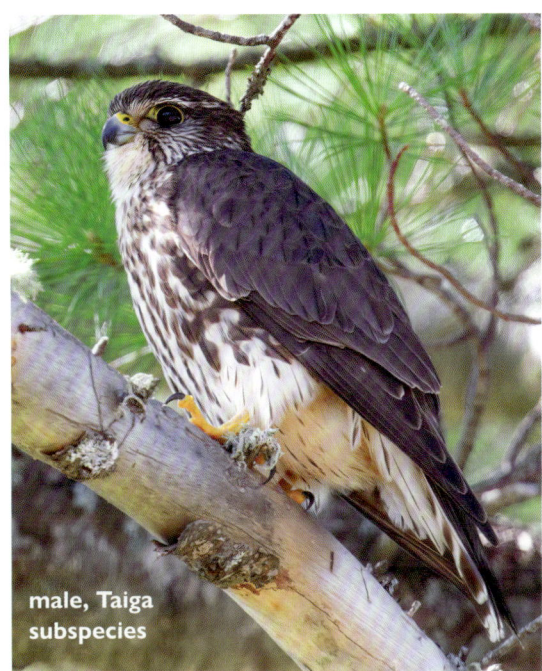

male, Taiga subspecies

© Vic Berardi

Habitat and Distribution: Merlins occur statewide and have been reported from all 105 counties. They are seen in a variety of habitats, including prairies, wetlands, woodlands, and urban areas. Both subspecies occur statewide. In Kansas, the Prairie subspecies is seen more often than the Taiga, especially in central and western Kansas.

Seasonal Occurrence: Migrant and winter resident. Seen in Kansas from early September through late April.

Field Notes: Merlins are typically seen in fast, brief encounters. They are exceptionally aggressive and frequently harass other raptors, including those much larger in size. In winter, Prairie Merlins sometimes form small communal roosts that remain in the same area for much of the season. These roosts have been documented in Cowley and Sedgwick Counties and no doubt occur elsewhere.

female

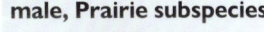

male, Prairie subspecies

© David A. Rintoul

Peregrine Falcon

Falco peregrinus

Field Identification:
Peregrine Falcons are large and powerful, with long, pointed wings and swift flight. Adults are dark bluish gray above, with a prominent "helmet" marking on the head, fine bars on the belly, and a banded tail. Immature birds are similar but browner, with heavy streaking on the breast. Size: Length 16 inches; wingspan 41 inches.

Habitat and Distribution:
Peregrine Falcons have been recorded throughout the state, with records from 95 counties. Migrating birds can be seen almost anywhere and in a variety of habitats. Frequently seen at wetlands, at reservoirs, and along rivers. They also visit larger cities, where they hunt Rock Pigeons and roost atop tall buildings.

prey with a blow from their large feet. They are regularly seen at Quivira NWR and other wetlands, where several individuals at a time can be found at the peak of shorebird migration. When large flocks of shorebirds suddenly take flight, it may be due to a cruising Peregrine.

Seasonal Occurrence: Migrant in spring and fall. Most spring birds are seen in April and May, fall birds in August and September. A few remain during the winter months. Peregrines have successfully nested on special platforms installed on tall buildings in downtown Topeka.

Field Notes: Peregrines usually hunt by diving out of the sky at speeds that can reach almost 200 mph and killing their

Falco mexicanus

Prairie Falcon

Field Identification:
Similar in size and shape to the Peregrine Falcon, the Prairie Falcon is pale sandy brown on the upper body, with a thin mustache mark on the white face. The underparts are white and streaked with brown. On flying birds, look for the black axillaries, or "wing-pits," where the wings meet the body. Size: Length 16 inches; wingspan 30 inches.

Habitat and Distribution:
Found in open country, mostly in the western two-thirds of the state. There are records from 104 counties. Decidedly scarce east of the Flint Hills.

Seasonal Occurrence:
Migrant and winter resident. Most are seen from September through March, but there are records from all months of the year.

Field Notes:
After the nesting season, Prairie Falcons move onto the plains from the western mountains. Their numbers are greatest in winter. They are raptors of the prairies and are most likely to be seen in the western counties. They typically hunt by flying swiftly at low altitude and taking birds they startle into flight. They also eat small mammals. Like Merlins, they show aggression toward Red-tailed hawks and other raptors. Their favorite perch is atop utility poles in open country.

American Barn Owl

Tyto furcata

Field Identification: A large, pale owl of open country, the American Barn Owl has a heart-shaped facial disk with dark eyes and stands upright on long legs. The wings and upperparts are blue and tawny with white speckling. Size: Length 16 inches; wingspan 42 inches.

Habitat and Distribution: Found statewide, with records from 101 counties. Most numerous in the western half of Kansas. It favors areas that are dominated by grasslands but requires agricultural buildings, dense tree rows, or holes in embankments for nest sites and for roosting during the day.

Seasonal Occurrence: Permanent resident. There are confirmed nesting records from 56 counties.

Field Notes: American Barn Owls are most numerous in the Red Hills and Smoky Hills, where there are many abandoned barns and buildings as well as natural crevices for nest and roost sites. The Cimarron National Grassland in Morton County is a reliable place to find them. American Barn Owls lay more eggs in a clutch than most other owls. The average clutch size is 5, and clutches of up to 13 eggs have been documented.

Megascops asio

Eastern Screech-Owl

Field Identification: This small owl has ear tufts, large yellow eyes, and vertical streaks on the breast. Most are gray, but in Kansas, about 10 percent of the population is red. Size: Length 8 inches; wingspan 20 inches.

Habitat and Distribution: Found statewide in wooded areas. There are records from 103 counties. Least common in western Kansas. It is frequently found in cities and towns.

Seasonal Occurrence: Permanent resident. There are confirmed nesting records from 46 counties, predominantly in central and eastern Kansas.

Field Notes: Although many people are unaware of them, these small owls are common across most of Kansas, often in cities and towns. Look for them in areas with mature trees, which provide cavities for nesting and roosting. Listen for their whinnying calls at dusk and after dark. Flocks of songbirds often noisily mob predators. While the predator can be anything from a house cat to a rat snake, Screech-Owls are frequently the object of these attentions. Check carefully when you observe mobbing behavior!

gray morph

red morph

161

Great Horned Owl

Bubo virginianus

Field Identification:
This large, dark owl has widely spaced ear tufts and is barred below with a white bib. Size: Length 22 inches; wingspan 44 inches.

Habitat and Distribution:
Found statewide in a variety of habitats and roosts in trees and abandoned buildings. There are records from all 105 counties.

Seasonal Occurrence:
Permanent resident. There are confirmed nesting records from all 105 counties.

Field Notes: This is the familiar "hoot-owl" of Kansas. Its deep, resonant hooting calls are unlike those of any other Kansas owl. It is often seen at dawn and dusk perched in treetops or on roadside utility poles. In early spring, check trees for large stick nests and look for the telltale "ears" sticking up. These owls do not build nests but use nests originally built by hawks and crows as well as tree cavities. They nest earlier in the year than any other bird species in Kansas. They mate in January and are incubating eggs by February. In winter, crows gather in noisy flocks when they find a Great Horned Owl. These owls are at the top of the food chain and will take any prey they can kill, including mammals as large as cats and skunks and birds as large as crows.

Strix varia

Barred Owl

Field Identification:
This large, earless owl has dark eyes, a heavily streaked breast, a brown back marked with white, and a facial disk with concentric rings that accentuate the round head. Its loud, booming call is often interpreted as *Who-cooks-for-you?* Size: Length 21 inches; wingspan 42 inches.

Habitat and Distribution:
Barred Owls prefer densely wooded areas, especially along rivers and streams. There are currently records from 78 counties. They are found throughout the eastern half but are most numerous in the southeastern third of the state. Their range is gradually expanding westward as woodlands mature along river systems. They are currently found west to Meade and Smith Counties. This westward expansion is expected to continue in future years. Wanderers sometimes appear west of the breeding range.

Seasonal Occurrence: Permanent resident. There are confirmed nesting records from 33 counties, mostly east of the Flint Hills. Most likely to be detected in spring, when courtship is taking place and birds are vocalizing.

Field Notes: Barred Owls make a variety of hooting and growling sounds when aroused or responding to other owls' calls. Sometimes they are seen perched in trees or on utility poles during daylight hours. Barred Owls have adapted well to humans and regularly nest in cities and towns.

Long-eared Owl

Asio otus

Field Identification: This medium-sized owl has long ear tufts and a tawny-orange facial disk. It is sometimes confused with the Great Horned Owl but is smaller and slenderer and has vertical streaks on the chest instead of the horizontal barring found on the Great Horned Owl. Size: Length 16 inches; wingspan 36 inches.

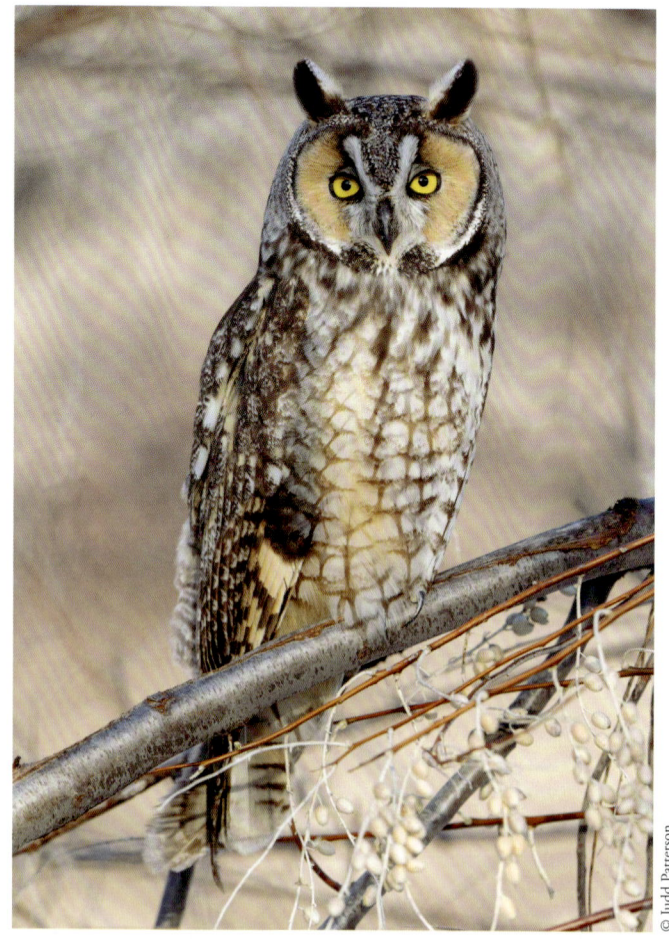

© Judd Patterson

Habitat and Distribution: Found statewide. There are records from 82 counties. It hunts mostly in open country but nests and roosts nearby in dense wooded areas. In winter, it often seeks dense cedar groves for roosting.

Seasonal Occurrence: Present year-round in Kansas, but in varying numbers. Wintering birds sometimes gather in communal roosts near areas with abundant rodents. Migrants are seen between September and April. They seem to be most often detected in April and October. A few remain to nest. There are confirmed nesting records from 22 counties scattered across the state.

Field Notes: In addition to being unpredictable in occurrence, Long-eared Owls are stealthy and often go undetected until they are startled from a concealed perch. The "ears" on this and other species of owls have nothing to do with hearing but are instead aids to concealment. They typically perch near the trunk of a tree and stretch vertically with ears erect, blending with tree branches. While winter roost sites often change from year to year, a few areas such as Cedar Bluff Reservoir and Wilson State Park attract Long-eared Owls nearly every year.

Asio flammeus

Short-eared Owl

Field Identification:
This round-headed owl of open country has short "ears" that are seldom visible. The brown upperparts are spotted with white and streaked below. The black plumage around the eyes is prominent on the pale face. Size: Length 15 inches; wingspan 38 inches.

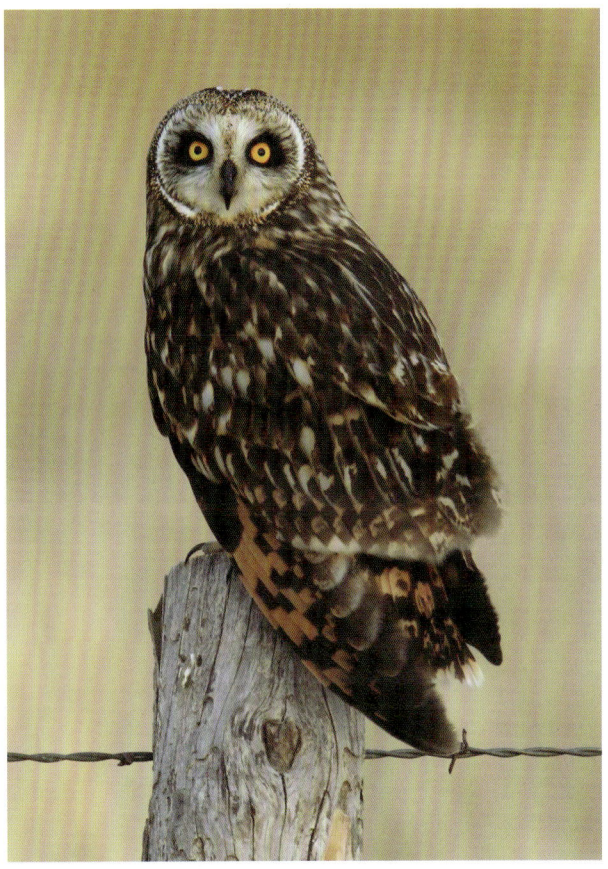

Habitat and Distribution: Found statewide, always in grasslands or marshy areas. Has been recorded in 102 counties.

Seasonal Occurrence:
Present year-round in Kansas. Spring migrants are seen in March and April, fall migrants in October and November. They winter throughout the state, but numbers are erratic from year to year, possibly due to fluctuations in rodent populations. Rare in summer, but there are confirmed nesting records from 17 counties.

mothlike fashion, uttering their peculiar high-pitched barking calls. Marais des Cygnes WA, Quivira NWR, and Slate Creek Wetlands are usually reliable locations for observing these birds during the winter.

Field Notes: Unlike most owls, this species frequently hunts in the hours just before dusk and just after dawn. During the winter, areas with Northern Harriers often have Short-eared Owls. As the harriers end their hunting for the day, the Short-eared Owls begin, patrolling low over the grasslands in an erratic

Snowy Owl

Bubo scandiacus

Field Identification: This large owl is easily identified. The entire plumage is white, with a variable amount of black spots and bars. Younger individuals have the greatest number of dark markings, which gradually disappear as they reach maturity. Some adult males can appear to be nearly pure white. Large yellow eyes. Size: Length 23 inches; wingspan 60 inches.

Habitat and Distribution: Open country and wetlands statewide. There are records from 90 counties.

Seasonal Occurrence: A winter visitor that can be seen between November and March. Most abundant in December and January.

Field Notes: The Snowy Owl is an irruptive species that moves south in some years from its arctic haunts, perhaps in response to food availability. In these invasion years, many are reported across the state. In most years, only a few or none are seen in Kansas. This is a bird with a bit of a mystique, and birders go to great lengths to see them when they are reported. Many Snowy Owls that reach Kansas are emaciated, and care should be taken to not disturb them.

Athene cunicularia

Burrowing Owl

adult with young

Field Identification: This unique prairie owl of western Kansas has long legs; a round, flat head; and a broad white band above widely spaced yellow eyes. It is usually seen standing in prairie dog towns or on top of wooden fence posts. Size: Length 9 inches; wingspan 21 inches.

Habitat and Distribution: Always seen in open country, most often in prairie dog towns. Formerly more common and widespread in Kansas, it is now restricted mostly to the western half of Kansas, where prairie dog towns still exist. There are records from 82 counties, including many eastern counties where this species no longer occurs. During migration, a few turn up east of the breeding range.

Seasonal Occurrence: Migrant and summer resident. Summer residents arrive in late March and depart in October. There are nesting records from 46 counties in the western half of the state. Migrants join the resident population in spring and fall. A few occasionally linger as late as December during mild winters.

Field Notes: Burrowing Owls usually nest in abandoned burrows of prairie dogs and other mammals. Unfortunately, many prairie dog towns in Kansas have been destroyed. Like those of other grassland species, Burrowing Owl populations are declining as habitat is eliminated. They do not prey on prairie dogs but eat insects.

Common Nighthawk

Chordeiles minor

Field Identification: Common Nighthawks are graceful birds of the summer skies. They are mottled gray above and barred on the underparts and tail. They have large eyes and a tiny bill. In flight, notice the long, crooked wings with a prominent white patch near the wingtips. Size: Length 9 inches; wingspan 24 inches.

Habitat and Distribution: Found statewide in grasslands. Commonly seen and heard in the skies above cities and towns. There are records from all 105 counties.

Seasonal Occurrence: Summer resident. Arrives in late April and departs in early October. There are confirmed nesting records from 45 counties. The nests of this species are hard to find. Probable nesting has been recorded in nearly 50 other counties.

Field Notes: Less common than they were formerly, Common Nighthawks are still a frequent sight in the summer skies of Kansas. Watch for them near dusk as they forage overhead for flying insects. In rural areas, they often perch and sleep atop wooden fence posts during the day. When the first cool weather pushes into Kansas during September, large migrating flocks of nighthawks take advantage of the favorable winds ahead of the storm. These low-flying flocks make an impressive sight as they fly acrobatically ahead of the darkening storm clouds.

© David Seibel

Phalaenoptilus nuttalli

Common Poorwill

Field Identification: This is the smallest member of the nightjar family in Kansas. It is mottled gray or brown above, is barred below, and has white tail corners visible in flight. It has a white throat bordered by blackish-brown plumage on the chin and breast. Size: Length 8 inches; wingspan 17 inches.

Habitat and Distribution: Common Poorwills have a patchy distribution because of their specialized habitat, which is characterized by steep slopes in grassland areas that are capped with exposed rocky outcrops. Some of their favored locations are the Arikaree Breaks, Flint Hills, Point of Rocks in Morton County, Red Hills, and bluffs within the Saline, Smoky Hill, and Solomon river systems. Has been reported from 77 counties.

Seasonal Occurrence: Migrant and summer resident. It is present between April and October. There are confirmed breeding records from 17 counties. It is believed that it is more widespread as a breeding species than has been confirmed.

Field Notes: Like other nightjars, Common Poorwills are more often heard than seen. At dusk in hilly country, listen for their simple two-note call. Common Poorwills are often flushed from dirt roads after dark. Look for their eyes reflecting red in your headlight beams. Because it tends to occur in locations that are infrequently visited by birders, and because of its completely nocturnal habits, this is one of the least understood bird species in Kansas.

Chuck-will's-widow

Antrostomus carolinensis

Field Identification: This is the largest nightjar in Kansas and the brownest in color. The male has long white shafts on the outer tail when flushed, while the female has buffy tail corners. It is named for the four-note call, given incessantly after dark. The call is accented on the last syllable. Size: Length 12 inches; wingspan 26 inches.

Habitat and Distribution: Found in much of the eastern half of Kansas. Has been reported from 77 counties, including all counties in and east of the Flint Hills. In southern Kansas, the range extends farther west to Meade and Pawnee Counties. They are always found in wooded areas along streams or on hillsides.

Seasonal Occurrence: Summer resident. Present from mid-April through mid-September. In late summer, they call less frequently. There are confirmed nesting records from 21 counties in the eastern third of the state and farther west in southern Kansas to Clark and Edwards Counties.

Field Notes: Chuck-will's-widows are more numerous and widespread in Kansas than the closely related Eastern Whip-poor-will. At eastern Kansas sites such as Elk City Reservoir, Marais des Cygnes WA, and Toronto Reservoir, their twilight chorus in spring and summer is a signature sound of summer. Most sightings are of birds flushed from the ground in wooded areas. They blend in with leaf litter and remain motionless until you nearly step on them.

Antrostomus vociferus

Eastern Whip-poor-will

© Judd Patterson

Field Identification: This woodland nightjar of eastern Kansas is smaller and grayer than the Chuck-will's-widow. The best field marks on birds flushed into flight are the cleanly defined white tail corners and black upper tail of the male, different from the white shafts on the tail corners of the Chuck-will's-widow. The female is like the female Chuck-will's-widow except for the smaller size. The three-note call has a higher pitch and more rapid cadence than the call of the Chuck-will's-widow, with the strongest accent on the first syllable. Size: Length 10 inches; wingspan 19 inches.

Habitat and Distribution: Found in woodlands of eastern Kansas, mostly east of the Flint Hills. Wanderers regularly appear farther west, most often during spring migration. Has been reported from 61 counties.

Seasonal Occurrence: Migrant and summer resident. Arrives in late April and remains through early fall. After they fall silent in late summer, they are reported only occasionally. There are confirmed nesting records from 17 counties east of the Flint Hills.

Field Notes: One of the best ways to observe this and other nightjars is to slowly drive quiet roads through wooded hills after sundown. Look for their glowing red eyes in the headlights. A cautious approach may allow you to illuminate them with your headlights and look at them through binoculars.

Rock Pigeon

Columba livia

Field Identification: The Rock Pigeon is a large, stocky, and ubiquitous pigeon of urban areas and farmyards. Their color is highly variable. Adults are most commonly gray but can also be white, brown, black, or any combination of these colors. They almost always have a white rump and white wing linings. Size: Length 12 inches; wingspan 28 inches.

Habitat and Distribution: Cities and towns statewide, especially around tall buildings and grain elevators as well as at farms and rural bridges. There are records from all 105 counties.

Seasonal Occurrence: Permanent resident. Confirmed breeding has been recorded in 81 counties.

Field Notes: This species is not native to North America. It was introduced by European immigrants and has since spread across North America. In Kansas, it is common in the larger cities, and even the smallest towns usually have a small resident population of these pigeons. Older farms with wooden barns and outbuildings also attract Rock Pigeons. In remote rural areas, it is not uncommon to find a few pairs nesting under bridges far from the nearest town or farmyard.

Streptopelia decaocto

Eurasian Collared-Dove

Field Identification: This large, pale dove has a sharply defined black collar on the back of the neck and a long, broad tail with large white corners. Dark wingtips are visible on perched birds and flying birds. In flight, it sometimes soars briefly. It commonly will perch on top of utility poles and other high spots. Its loud three-note cooing call is given throughout the day. Size: Length 13 inches; wingspan 22 inches.

Habitat and Distribution: Cities and towns throughout the state. It is increasingly common in rural areas, especially near farmyards. Has been reported from all 105 counties.

Seasonal Occurrence: Permanent resident. Confirmed breeding has been reported from only about 31 counties, but it probably nests in all 105 counties.

Field Notes: This species is native to Eurasia and was introduced to the Bahamas in the 1980s. From there, it spread to Florida and then rapidly expanded its range across the US. It is now widespread and abundant in Kansas towns, especially in western Kansas. It is classified as a game bird by the Kansas Department of Wildlife and Parks.

White-winged Dove

Zenaida asiatica

Field Identification: This is a large, stocky, brownish-gray dove with a relatively short tail. Prominent white wing-patches are seen in flight and are visible as a white crescent on the edge of the folded wing of perched birds. The eyes have a red iris and are surrounded by a powder-blue orbital ring. There is a black stripe on the cheek. The tail is black with a broad white tip. Size: Length 11 inches; wingspan 19 inches.

Habitat and Distribution: Cities and towns statewide. Only occasionally seen in rural habitats. Has been reported in 96 counties. The greatest concentrations are in southwestern Kansas. Least common in the northern counties and east of the Flint Hills.

Seasonal Occurrence: Now considered a permanent resident, but some probably move out of the northern counties in winter. There are confirmed breeding records from 10 counties.

Field Notes: As recently as the 1990s, sightings of this southern dove in Kansas were cause for excitement among birders. It has now expanded its range to include most of the state and is common and well-established in many areas. Roosts of over 100 birds regularly occur in Garden City. It readily comes to bird feeders. Many records from northern and eastern Kansas are of single birds seen at feeders with other dove species.

Zenaida macroura

Mourning Dove

Field Identification: This familiar dove is brownish gray with black spots on the wings and a long, wedge-shaped tail with wide white borders. It flushes abruptly with loud, whistling wings. The song of low cooing notes is given most often at dawn and dusk. Size: Length 12 inches; wingspan 18 inches.

Habitat and Distribution: Statewide in a wide variety of rural and urban habitats. Has been reported in all 105 counties.

Seasonal Occurrence: Permanent resident, but with substantial population movement. Abundant migrant and summer resident. There are confirmed breeding records from all 105 counties. A few remain for the winter, mostly in southern Kansas.

Field Notes: This is one of the most abundant birds in Kansas from April through September. It is also one of the most popular game birds in the state. The hunting season for doves begins around Labor Day and is the traditional start of the hunting season in Kansas. During late summer, large conspicuous flocks begin to gather, peaking in numbers during late August. The first cool fronts of September push more into Kansas from the northern states, and these birds linger until successive cold fronts send most of them farther south.

Yellow-billed Cuckoo

Coccyzus americanus

Field Identification: Long and slender, this bird can be confused with a Mourning Dove or a Brown Thrasher at first glance. It has a brown back and crown and pure white throat and underparts. The undertail has large white spots. Large rufous patches are visible on the wings of flying birds. The long, stout bill is mostly yellow. Size: Length 12 inches; wingspan 18 inches.

Habitat and Distribution: Statewide in brushy wooded areas, often near streams or ponds. Has been reported from all 105 counties but is less common in western Kansas.

Seasonal Occurrence: Summer resident. Arrives in early May and departs by late September. There are confirmed nesting records from 77 counties.

Field Notes: This is a common but secretive bird. Many residents know it as the "rain crow." Its guttural calls are often considered a harbinger of rainfall. Despite the striking vocalizations, it can be difficult to see. Because of its habit of flying low over the ground, it is often killed by vehicles on highways and by flying into windows. Populations of this species are in serious decline in some parts of the US, but Kansas populations appear to be stable.

Coccyzus erythropthalmus

Black-billed Cuckoo

Field Identification: This species resembles the Yellow-billed Cuckoo in many respects. It differs by having a black bill with a red eye-ring, no rufous color in the wings, and much smaller and fainter white spots on the undertail. The voice is a monotone three-note series of *coo-coo-coo* notes. Size: Length 12 inches; wingspan 18 inches.

Habitat and Distribution: Brushy wooded areas. Seen more often in mature woodlands than the Yellow-billed Cuckoo. There are records from 92 counties. Least common in the western counties.

Seasonal Occurrence: Migrant and summer resident. There are confirmed breeding records from 22 counties, mostly in the eastern half of Kansas. Most sightings are from late May and June. Fall migrants depart by late September.

Field Notes: This cuckoo is seen far less often than the Yellow-billed Cuckoo and is more secretive. Kansas is at the southern edge of its breeding range, which lies mostly in the northern states and Canada. It can turn up anywhere in the state during migration.

Greater Roadrunner

Geococcyx californianus

Field Identification: A unique terrestrial bird of rugged country, the Greater Roadrunner is easily identified by its exceptionally long tail and long, pointed bill. The ragged crest feathers are not always raised. It is occasionally seen flying for short distances but prefers to run on its strong legs. The song is a low series of cooing notes, often the first clue to its presence. Size: Length 23 inches; wingspan 22 inches.

Habitat and Distribution: Rolling hills, rugged arroyos, and draws in the south-central counties from Cowley County west to Meade County and north as far as Barton County. Regularly wanders well beyond this area. This species is slowly expanding its range northward in Kansas. Has been observed in 51 counties.

Seasonal Occurrence: Permanent resident. There are confirmed nesting records from 11 south-central counties as far north as Barton County.

Field Notes: Roadrunners are most likely to be seen in Barber, Comanche, and Kiowa Counties, where the rough terrain they prefer is widespread. They are often extremely wary, but at times they can be peculiarly tame and approachable. Little is known about the Kansas population. Deliberate searching for them can be frustrating because they are scarce and elusive and wander widely.

Megaceryle alcyon

Belted Kingfisher

female

Field Identification: This large-headed bird has a prominent ragged crest and big spiked bill. It is blue gray above and white below. The male has one blue band across the breast, the female an additional red band. Size: Length 13 inches; wingspan 20 inches.

Habitat and Distribution: Found statewide at streams, rivers, lakes, and ponds. Nests in burrows excavated in steep banks. They seem to prefer smaller streams with slow-moving water. There are records from all 105 counties.

Seasonal Occurrence: Permanent resident, but abundance varies by season. Most numerous and widespread during spring and fall migration. Many remain to nest. There are confirmed nesting records from 38 counties, mostly in the eastern half but also from counties in the western half of Kansas. There are another 20 counties with

probable breeding. A few remain in winter if there is open water.

Field Notes: Kingfishers are easy to find when present because of their conspicuous perches above water and their frequent loud, rattling calls. They sometimes hover before diving into the water to capture small fish.

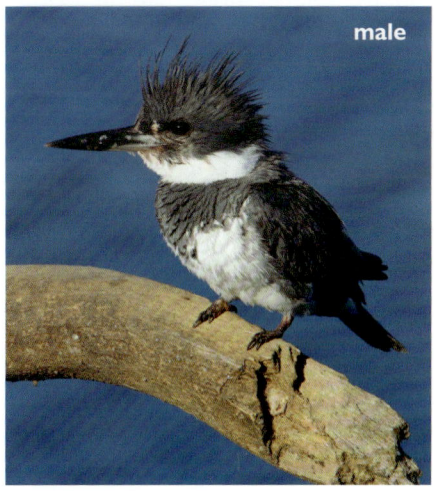

male

179

Red-headed Woodpecker

Melanerpes erythrocephalus

Field Identification: The adult is unmistakable, with a bright red head and a black-and-white pattern on the body. Juveniles have the same color pattern, but the head is streaked brown instead of red. Size: Length 9 inches; wingspan 17 inches.

Habitat and Distribution: Found statewide. Prefers open country with some trees in summer. Its habitat preference changes in the colder months. Those that remain for the winter in the east inhabit oak-hickory woodlands and pecan orchards. There are records from all 105 counties.

can be numerous in oak woodland and pecan groves in eastern Kansas when the acorn or nut crop has been good. Because of their habit of flying low over the ground, they are often hit by cars.

Seasonal Occurrence: This species is found year-round in Kansas, but the distribution changes by season. It is a summer resident throughout the state, with confirmed nesting records from 95 counties. In the winter, it occurs only in eastern and southeastern Kansas. In September, loosely organized flocks can be seen migrating from the western plains eastward to the forested areas where they will spend the winter.

Field Notes: These boldly plumaged birds are conspicuous in summer, especially in open areas of western Kansas, where they subsist on insects and seeds. In winter, they subsist primarily on acorns and nuts, and they

immature

© David Seibel

Melanerpes carolinus

Red-bellied Woodpecker

Field Identification:
This woodpecker has horizontal black and white bars on the back and mostly light underparts. The male has a red cap and hindneck. On the female, only the hindneck is red. Look for the conspicuous white rump and wing-patches on flying birds. Their staccato "laughing" calls are often the first clue to their presence in wooded areas. Size: Length 9 inches; wingspan 16 inches.

Habitat and Distribution: Found statewide but is most numerous in the eastern two-thirds of Kansas. Prefers woodlands along streams, upland forests, and towns with mature trees. Their range is gradually expanding westward. There are records from all 105 counties.

Seasonal Occurrence: Permanent resident. There are confirmed nesting records from 70 counties. These include almost all counties in central and eastern Kansas. In western Kansas, there are nesting records from most counties bordering Nebraska west to the Colorado state line. Most of the counties lacking confirmed breeding are in the western third of the state.

Field Notes: Red-bellied Woodpeckers are named after one of their least noticeable plumage characteristics. The red belly is at best a rosy wash visible only when the bird is in hand. It is a frequent visitor to bird feeders. When present, the rest of the birds at the feeder must wait until the Red-bellied has had its fill of suet, nuts, or sunflower seeds.

female

© Judd Patterson

Yellow-bellied Sapsucker

Sphyrapicus varius

Field Identification: This medium-sized woodpecker is most often seen on pine trees. In all plumages, look for the vertical white mark on the flanks, found on no other Kansas woodpecker. The head is boldly striped in black and white, with a red crown. Males have red throats, females white. The back is barred with white and black. Juveniles are brownish on the head and breast. Size: Length 8 inches; wingspan 16 inches.

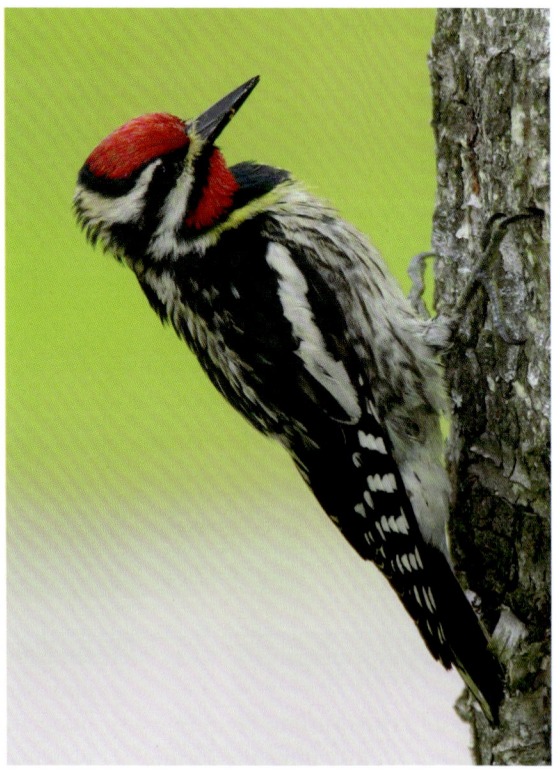

Habitat and Distribution: Found statewide, usually in cemeteries, yards, parks, and other areas with numerous conifers. While it prefers pines, it is frequently also seen on deciduous trees, especially pecans and sweetgums. Has been recorded in 97 counties in Kansas. Sightings are most likely in central and eastern Kansas, fewer in the western third of the state.

Seasonal Occurrence: Winter resident. Arrives in early October and departs in mid-April. Migrants in late fall and early spring supplement the local winter populations.

Field Notes: These solitary woodpeckers are quiet and retiring and are easily overlooked. They hide on the back side of tree trunks and then quietly fly away.

It is helpful to learn the quiet mewing call of this elusive species, as it is often the first clue that one is nearby. Look on pine and sweetgum trees for long, horizontal rows of holes drilled in the bark. Sapsuckers create these and then return for sap and trapped insects.

female

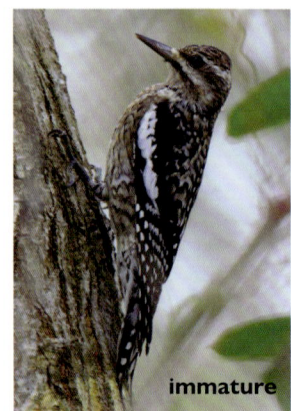

immature

Dryobates pubescens

Downy Woodpecker

Field Identification:
This is the most common and widespread woodpecker in Kansas and also the smallest. The most prominent field mark is the broad vertical white stripe down the middle of the back. The head is striped black and white. The male has a patch of red on the back of its head. It has black wings and white underparts. Its call is a shrill descending whinny or a single *pik* note. Size: Length 7 inches; wingspan 12 inches.

Habitat and Distribution: Found statewide anywhere trees are present. It is abundant in the east, less common in western Kansas. There are records from all 105 counties.

Seasonal Occurrence: Permanent resident. It nests statewide, with confirmed breeding recorded from 87 counties.

Field Notes: Downy Woodpeckers are not afraid of people and are found in nearly every city and town in the state. They are frequent visitors to bird feeders. In addition to foraging for insects on trees, they feed on corn, milo, weeds, and yucca stalks. In winter, smaller woodland birds form mixed-species flocks that forage for food together. These may include Downy Woodpeckers as well as chickadees, kinglets, creepers, nuthatches, titmice, and wrens.

female

183

Hairy Woodpecker

Dryobates villosus

Field Identification: This species is like the Downy Woodpecker but is larger, with a noticeably longer, stouter bill. The outer tail feathers are all white on the Hairy Woodpecker, white with black spots on the Downy Woodpecker. The calls of the Hairy are like those of the Downy but significantly louder and lower-pitched. Those seen in most of Kansas have extensive white spots on the back and wings. A few found in western Kansas are of the Rocky Mountain subspecies and show almost no white spotting on the back. Size: Length 9 inches; wingspan 15 inches.

Habitat and Distribution: Statewide. There are records from all 105 counties. This species prefers larger and more mature tracts of woodlands than the Downy Woodpecker.

Seasonal Occurrence: Permanent resident. There are confirmed nesting records from 52 counties. Many counties in western Kansas lack nesting records, although this species likely nests in many of them.

Field Notes: Hairy Woodpeckers are more selective of habitat than Downy Woodpeckers and have larger feeding and nesting territories. They are common in cities and towns but rarely visit bird feeders. Based on Christmas Bird Count and breeding bird survey data, the Kansas population is roughly 25 percent that of the Downy Woodpecker.

female

Dryocopus pileatus

Pileated Woodpecker

Field Identification: This exceptionally large woodpecker of the eastern woods is mostly black, with a broad white stripe on the neck and a large red crest. Bright white wing linings are visible when in flight. Some of its calls resemble those of the Northern Flicker, but they are much louder and primeval-sounding. The drumming is louder than that of any other woodpecker. Size: Length 17 inches; wingspan 29 inches.

Habitat and Distribution: Found mostly in mature woodlands in the eastern half of Kansas, especially along rivers. Has been reported from 64 counties.

Seasonal Occurrence: Permanent resident. There are confirmed nesting records from 16 counties.

Field Notes: Formerly a rare species in Kansas, the Pileated Woodpecker has substantially increased its range and population over the past several decades. As of 2023, it has been observed as far west as Clark, Ellsworth, Ford, and Jewell Counties as woodlands along the major river systems have matured. Despite its size and distinctive appearance, it is often elusive and difficult to see. When approached, it usually slips around to the back side of the tree before flying quietly away. Sightings are frequently of birds flying above the forest canopy. Their massive excavations for grubs in dead trees are deep, distinctive, and a good clue to their presence.

female

185

Northern Flicker

Colaptes auratus

Yellow-shafted male

© David Seibel

Yellow-shafted female

© Judd Patterson

Field Identification: This large woodpecker has a brown back with narrow black bars, a white breast with round spots, and a black bib marking on the chest. The wing linings are yellow or red depending on the subspecies. Males of the Red-shafted subspecies have a gray face and red mustache mark, while those of the Yellow-shafted subspecies have a brown face and black mustache mark. Females lack mustache marks. Size: Length 12 inches; wingspan 20 inches.

Habitat and Distribution: Statewide in brushy open country, woodlands, and urban areas. There are records from all 105 counties.

Northern Flicker

Colaptes auratus

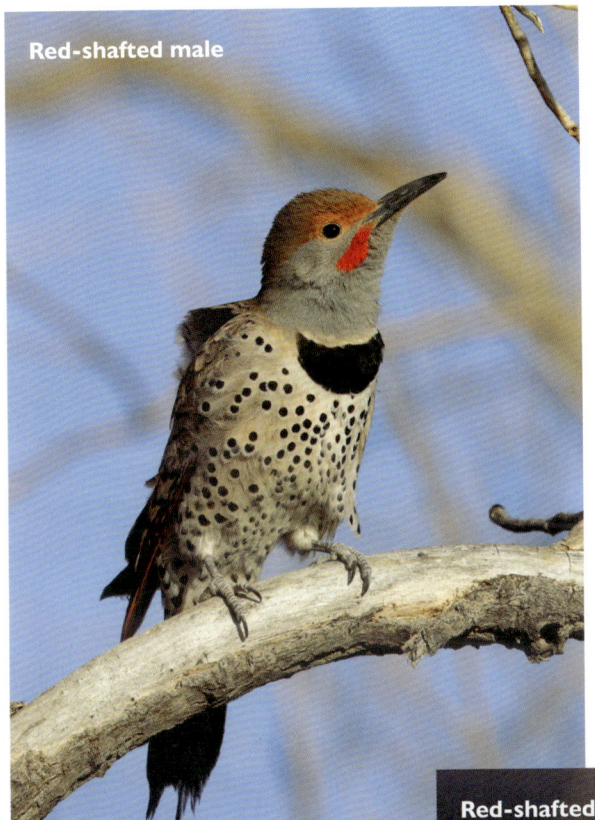

Red-shafted male

Field Notes: Northern Flickers are named for their peculiar calls, which are often interpreted as *flicker flicker flicker flic*. Most flickers seen in the summer in Kansas are of the Yellow-shafted subspecies. In winter, Red-shafted birds move onto the plains from the mountains. The two subspecies hybridize on the Great Plains, and intermediate birds are often observed. Flickers are fond of eating ants and other insects while perched on the ground.

Red-shafted female

Seasonal Occurrence: Present year-round, although populations fluctuate by season. In spring and fall, large flocks of migrating flickers are often observed, especially in March and October. Many remain statewide in winter. Widespread in summer but not as numerous as in winter. There are confirmed nesting records from 81 counties.

Chimney Swift

Chaetura pelagica

Field Identification: These birds have slender, streamlined bodies, with long pointed wings, and unmarked gray plumage, lightest on the throat. Their flight is rapid and acrobatic, alternating gliding with deft wingbeats. Their frequent, chittering calls, given in flight, are a familiar summer sound in Kansas towns with older buildings. Size: Length 5 inches; wingspan 14 inches.

Habitat and Distribution: Found statewide, but more common in the eastern two-thirds of the state. There are records from all 105 counties. Typically seen in or near farmyards, towns, and cities.

Seasonal Occurrence: Migrant and summer resident. Migrants arrive in mid-April and depart in mid-October. Confirmed nesting has been reported from 57 counties. Most of the counties without breeding records are in the west.

Field Notes: Historically, Chimney Swifts have benefited significantly from humans. Brick chimneys on houses, especially those constructed prior to the 1960s, offer a perfect nesting habitat for them. Contemporary chimney construction is usually not suitable for Chimney Swifts. As older chimneys disappear, so do the swifts. During fall migration, Chimney Swifts travel in massive flocks, which roost in tall brick chimneys at schools and industrial facilities. At sunset, they descend into these stacks in impressive tornado-shaped swirling flocks.

© David Seibel

Purple Martin

Progne subis

male and female

Field Identification: The Purple Martin is a large member of the swallow family. The male is entirely iridescent purple. Females and juveniles are dark gray above and mottled white and gray below. Size: Length 8 inches; wingspan 18 inches.

Habitat and Distribution: Found statewide, but rare and local in the west. There are records from 103 counties. In summer, it is seen almost exclusively around man-made martin houses. Foraging and migrating birds are observed in a variety of habitats, typically near water.

Seasonal Occurrence: Migrant and summer resident. Arrives in southern Kansas in mid-March, departs in August and September. There are confirmed breeding records from 80 counties. Most counties lacking breeding records are in the western third of the state.

Field Notes: Many homeowners go to considerable lengths to attract Purple Martins to multichambered martin houses installed in their yards. After nesting, martins begin gathering in large flocks prior to migrating. These flocks often grow to tens of thousands of birds and occur annually in Kansas City, Wichita, and elsewhere. They typically roost together at night, then forage widely during the day, returning to the roost at dusk. These large flocks depart for the tropics with the first cool fronts of September.

Tree Swallow

Tachycineta bicolor

Field Identification: Adult males are iridescent blue green above and white below. Some females are near the males in coloration; others are more brownish above. Size: Length 6 inches; wingspan 15 inches.

Habitat and Distribution: Occurs statewide as a migrant, usually near water. There are records from 104 counties.

Seasonal Occurrence: Migrant and summer resident. There are confirmed nesting records from 48 central and eastern counties and a handful of western counties. Arrives in March and departs in October. Occasionally lingers into late fall when the weather is mild.

Field Notes: Tree Swallows are the most cold-tolerant member of the swallow family and can be seen earlier in the spring and later in the fall than other swallow species. Look for them hawking for insects near the surface of marshes, lakes, and ponds. In recent years, their breeding range has expanded. They nest in cavities of dead trees located in or near bodies of water. They also nest in bluebird nesting boxes erected in many parks and reservoirs.

Stelgidopteryx serripennis # Northern Rough-winged Swallow

Field Identification: This widespread swallow is brownish above and white below. The key mark to separate this species from the somewhat similar Bank Swallow is the smudgy brown throat and breast. The tail is wider and squarer on the end than the notched and thinner tail of the Bank Swallow. Size: Length 5 inches; wingspan 14 inches.

Habitat and Distribution: Statewide, especially in areas located near water, but also in grasslands far from ponds or streams where exposed dirt banks are present. Has been reported from all 105 counties.

Seasonal Occurrence: Migrant and summer resident. Arrives in early April and departs in mid-September. There are confirmed nesting records from 74 counties. Most counties lacking records are in the southwest.

Field Notes: Rough-winged Swallows nest in holes excavated in soft dirt banks. During migration, they can be seen at any body of water, where they forage for insects low over the surface. Most sightings in spring and summer are of pairs or small flocks. In late July, they begin to gather in larger flocks, often with other swallow species. These flocks gather in areas with abundant insects, usually adjacent to wetlands and reservoirs.

Bank Swallow

Riparia riparia

Field Identification: This small swallow is brown above and pure white below with the two colors cleanly separated, as opposed to the smudgier throat and underparts of the Rough-winged Swallow. A sharply defined brown band on the white breast is the most definitive mark. They also have a forked tail. Size: Length 5 inches; wingspan 13 inches.

Habitat and Distribution: Like other swallow species, it is almost always seen near water. Has been reported from 104 counties in the state. Can be seen anywhere during migration.

Seasonal Occurrence: Migrant and summer resident. Arrives in mid-April and departs in mid-September. Confirmed nesting has been recorded in 29 counties, mostly in central and eastern Kansas.

Field Notes: Bank Swallows nest in burrows excavated from earthen banks, often along rivers, lakeshores, or quarries. They nest in colonies of 10 to 100 pairs. These colonies are scattered across Kansas but are not numerous and are widely separated. They begin migrating south in July, gradually assembling into large flocks, often with other swallow species. By early August, flocks in the hundreds or thousands have gathered at wetland sites with abundant insects.

Petrochelidon pyrrhonota

Cliff Swallow

Field Identification: This distinctive swallow has a dark back, square-cut tail, dark reddish throat and cheeks, and white belly. The most prominent markings are the orange-buff rump and the creamy buff forehead. The rump and forehead markings are easily seen on flying birds, quickly separating the Cliff from all other swallow species. Size: Length 5 inches; wingspan 13 inches.

Habitat and Distribution: This swallow has been recorded in all 105 counties. It can be seen anywhere there is water. Abundant at bridges, which host nesting colonies. Most common in the western two-thirds of Kansas, but populations in eastern Kansas are increasing.

Seasonal Occurrence: Migrant and summer resident. Arrives in mid-April and departs in September. Confirmed nesting has been reported in 85 counties.

Field Notes: Cliff Swallows always nest in colonies. The colonies are usually under road bridges with a source of mud nearby. Look for the gourd-shaped mud nests plastered to the undersides of concrete bridges. Bridges that harbor colonies nearly always have birds in the air. Beginning in late July, postbreeding Cliff Swallows gradually congregate in ever larger flocks. These flocks wander in search of abundant insect populations until the first cool fronts of September encourage them to depart southward.

193

Barn Swallow

Hirundo rustica

Field Identification: This common swallow is dark blue above. The underparts are orangish on males and paler on females. The throat is always dark orange. The best field mark is the long, deeply forked tail, found on no other swallow. Size: Length 7 inches; wingspan 15 inches.

Habitat and Distribution: Has been recorded in all 105 counties. Most often seen near barns or other rural buildings. Foraging birds can be seen in a variety of habitats. Barn Swallows are one of the most familiar and widespread summer birds in Kansas.

Seasonal Occurrence: Migrant and summer resident. Arrives in mid-April and departs in mid-October, occasionally lingering to early November. Nesting has been confirmed in 104 counties.

Field Notes: In summer, these birds are a familiar sight, flying low over fields and pastures in acrobatic fashion as they forage for insects. Their mud nests are placed on vertical walls under the eaves of homes or barns and inside older sheds and barns. They frequently nest under bridges. They gather in large flocks with other swallow species in late summer before departing south for the winter.

Archilochus colubris

Ruby-throated Hummingbird

Field Identification: The adult male has a bright ruby-red throat, black mask, black chin, forked black tail, and metallic green back and crown. Females are also metallic green above, with white throat and underparts and a black tail with white tips. Females and immatures usually show a faint buffy wash on the flanks and a variable amount of spotting on the throat. Females and immature birds are seen more frequently than adult males. Size: Length 4 inches; wingspan 4.5 inches.

Habitat and Distribution: In summer, common in wooded areas and suburban yards east of the Flint Hills. Less numerous in central Kansas and rare in the west. During fall migration, it can be seen at hummingbird feeders statewide. There are records from 97 counties. All counties lacking records are in the west.

Seasonal Occurrence: Migrant and summer resident. Arrives in May and departs in early October. Many Ruby-throats seen in Kansas are migrants, and northbound individuals do not linger in spring. Southbound fall migrants begin to arrive in late July and linger into early October. There are confirmed nesting records from 27 central and eastern counties and west to Barber and Rooks Counties. It probably nests in most eastern counties.

Field Notes: This is the only hummingbird found in most of the eastern US. While 10 species of hummingbirds have been documented in Kansas, the overwhelming majority of hummingbirds seen in the state are Ruby-throated. While hummingbirds are frequently seen in wild habitats, perhaps no family of birds is observed more exclusively at feeders. Homeowners who diligently offer appropriate flower plantings and nectar feeders are usually rewarded with hummers, especially during the fall migration.

female

Rufous Hummingbird

Selasphorus rufus

Field Identification:
The adult male is mostly orange with green wings and an orange-red throat. Females and immature birds are seen more often than adult males. They have more green plumage than males but always have at least some orange on their flanks and at the base of the tail feathers. Size: Length 4 inches; wingspan 4.5 inches.

Habitat and Distribution: Almost always seen at feeders. Has been recorded in 55 counties, mostly from the Flint Hills and westward. Migrants can turn up anywhere in Kansas but are most often reported in the western third of the state.

Seasonal Occurrence: Fall migrant. Rufous Hummingbirds are found from July through September. These are postbreeding birds moving out of the Rockies that drift eastward onto the Plains as they migrate south. There are only a handful of spring records.

Field Notes: Dedicated hummingbird enthusiasts always hope for a hummingbird other than the Ruby-throated. This is the most likely of the western hummingbird species to be seen in Kansas and the only one included in this book. Western Kansas towns such as Dodge City, Elkhart, Garden City, Hays, and Larned have earned a reputation for attracting rare hummingbirds to feeders during fall migration. In these western counties, the Rufous Hummingbird is more likely to be seen than the Ruby-throated Hummingbird, which is rare in the west.

female

Myiarchus crinitus

Great Crested Flycatcher

Field Identification: This large flycatcher of woodlands has a gray breast, yellow belly, rusty-red tail, and olive back. The head has an obvious crest. Its call is a loud, rough *wheep*, heard throughout the summer. Other vocalizations have the same tonal quality. Size: Length 9 inches; wingspan 13 inches.

Habitat and Distribution: Found statewide in woodlands, but most common in the eastern two-thirds of Kansas. Prefers woodlands along rivers and streams, but also found in drier upland woods. There are records from all 105 counties. Most records from the westernmost counties are of spring and fall migrants.

Seasonal Occurrence: Summer resident. Arrives in late April and departs in early September. There are confirmed nesting records from 66 counties, mostly in the eastern three-quarters of the state. Most numerous during migration.

Field Notes: In spring and summer, this is one of the most frequently heard woodland birds in central and eastern Kansas. They are heard throughout the day when most other birds are quiet. They aggressively scold intruders of all kinds. Unlike most flycatchers, they nest in tree cavities. After the nesting season, they grow silent and are easy to overlook as they quietly forage in the canopy.

Western Kingbird

Tyrannus verticalis

Field Identification: This large flycatcher of open areas has a gray head, light olive back, pale gray breast, yellow belly, and black tail with clean-cut white edges. The sharp-sounding song and call notes are given incessantly in the nesting season, often even at night. Size: Length 9 inches; wingspan 16 inches.

Habitat and Distribution: Common in the western two-thirds of Kansas in open country and urban areas. Scarcer and local east of the Flint Hills, where they are more often seen in towns than in rural areas. There are records from all 105 counties.

Seasonal Occurrence: Summer resident. There are 95 counties with confirmed nesting records. Arrives in late April and departs in late September.

Field Notes: Western Kingbirds are one of the most common and conspicuous summer birds of central and western Kansas. They are usually seen perched on utility wires or fences, from which they fly out to capture insects. They are often seen in cities and towns.

Tyrannus tyrannus

Eastern Kingbird

Field Identification: Eastern Kingbirds are dark gray above and all white below. The crown and face are darker than the back. The tail has a clean-cut white tip. Size: Length 8 inches; wingspan 15 inches.

Habitat and Distribution: Found statewide. There are records from all 105 counties. Most common in the eastern two-thirds of the state. While both kingbirds are common in Kansas, they occur in subtly different habitats. Western Kingbirds prefer drier open country, whereas Eastern Kingbirds are most often found in slightly moister areas with woody vegetation.

Seasonal Occurrence: Summer resident. There are 95 counties with confirmed nesting records. Arrives in late April and departs in mid-September. Most numerous in May and August, when migrants to and from northern breeding areas are moving though Kansas.

Field Notes: Eastern Kingbirds forage from low perches such as fences and shrubs. All kingbirds are known for their aggressive behavior toward other birds that they perceive as threats, especially crows and hawks. They fearlessly attack these intruders by repeatedly dive-bombing them and have even been seen landing on the backs of flying hawks and vigorously jabbing them with their beaks.

Scissor-tailed Flycatcher

Tyrannus forficatus

Field Identification: The long, streaming tail makes this species unmistakable. Adults are light gray, with salmon-pink flanks and dark wings. Juvenile birds have a shorter tail that is still longer than that of any other flycatcher. The vocalizations resemble those of the Western Kingbird. Size: Length 15 inches; wingspan 15 inches.

Habitat and Distribution: Recorded statewide, with records from all 105 counties. It is most numerous in the eastern and south-central counties. Inhabits open country with scattered trees.

Seasonal Occurrence: Summer resident. There are confirmed nesting records from 73 counties. Most of the counties lacking nesting records are in the northwestern corner and along the Colorado border. Scissor-tails arrive in mid-April, a bit earlier than most other flycatchers. In fall, they depart in late

September and early October. Each year, a few linger into late October.

Field Notes: Scissor-tailed Flycatchers are well-known birds in Kansas. They are in the same genus as the kingbirds, and like them, they perch conspicuously in open country, flying out to capture insects. The long, divided tail allows them exceptional maneuverability in flight. Their nest is often located in an isolated tree in open country. In late September and October, they gather into flocks prior to migrating south. Lucky observers may find flocks of up to several hundred birds lined up on fences or utility wires.

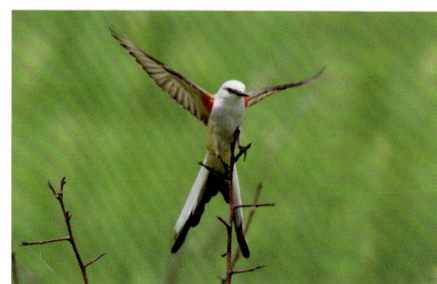

Contopus cooperi

Olive-sided Flycatcher

Field Identification: This large-headed and somewhat short-tailed flycatcher is mostly drab olive gray in color. Dark flanks extend well onto the white breast, creating a "vested" look. There are distinctive small white tufts of feathers on the lower back, which are not always visible. The distinctive three-note call is often interpreted as QUICK *three beers*, trailing off on the last note. Size: Length 7 inches; wingspan 10 inches.

Habitat and Distribution: Found statewide during migration. Has been recorded in 101 counties. In spring, most are seen in central and eastern Kansas. In fall, it is clearly more common in the western counties, where multiple birds can be seen in a single day.

Seasonal Occurrence: Spring migrants are seen almost entirely in May, fall migrants in late August and September.

Field Notes: This neotropical migrant nests in the boreal forests of the Rocky Mountains and Canada and winters in Central and South America. It is seen in Kansas only as a migrant. Olive-sided Flycatchers usually hunt from a high, bare branch, flying out to capture flying insects and returning to the same perch.

Eastern Wood-Pewee

Contopus virens

Field Identification: This flycatcher is closely related to the Olive-sided Flycatcher but is smaller and has obvious wing-bars. The vested look is more muted, and its belly has a yellowish cast in early summer. It is named for its distinctive two-note song, which is both plaintive and lazy in quality. In the westernmost counties, care must be taken to differentiate it from the nearly identical Western Wood-Pewee. Size: Length 6 inches; wingspan 10 inches.

Habitat and distribution: Summer resident in the eastern half of Kansas in mature woodlands adjacent to rivers and streams. Migrants are seen statewide in a variety of habitats with trees. Has been recorded in 99 counties.

Seasonal Occurrence: Migrant and summer resident. Confirmed nesting has been recorded in 39 counties in eastern Kansas west to Barber and Russell Counties, and it undoubtedly nests in most central and eastern counties. Arrives in early May; departs by late September. West of the nesting range it occurs only as a migrant.

Field Notes: During the summer, this is one of the most common bird species of eastern Kansas woodlands. Its song is heard late into the season, even at midday during the hottest weather. Like the Olive-sided, it is fond of high, exposed perches, from which it flies out to capture insects.

Contopus sordidulus

Western Wood-Pewee

Field Identification: Virtually identical to the Eastern Wood-Pewee but typically a few shades darker. The calls of the two species are the most reliable point of differentiation. Western usually has a burry-sounding single call note. However, in fall, most are silent. At exceptionally close range, the base of the lower mandible is dark with less orange on Western, dull orange on Eastern. Western occurs only in the westernmost counties. Size: Length 6 inches; wingspan 10 inches.

Habitat and Distribution: Woodlands along creeks and streams in the westernmost Kansas counties on or near the Colorado border. Often seen in cities and towns in the same part of the state. Migrants have been reported from 32 counties, almost all of which are in the western quarter of the state.

Seasonal Occurrence: Spring and fall migrant in far western counties. Spring migrants are mostly reported in late May and in fall during early September. The first fully confirmed breeding in Kansas was at Historic Lake Scott State Park in the summer of 2021. Singing males in appropriate habitat have been observed during the breeding season in several other counties along the Colorado line, suggesting additional nesting locations.

Field Notes: Telling the two species of wood-Pewees apart is an issue only in the westernmost counties. Eastern Wood-Pewee occurs commonly in the west during migration, and on some days, multiple birds of both species can be seen at favored birding locations, such as the Cimarron National Grassland and Historic Lake Scott State Park. It takes practice to tell the two species apart. Some individuals can't be identified in terms of species.

Eastern Phoebe

Sayornis phoebe

Field Identification:
The Eastern Phoebe is a medium-sized flycatcher lacking wing-bars. Adults are gray above, with a darker face and cap and a white belly; immatures have a yellowish belly. They habitually dip the tail downward while perched. Its song is a series of alternating pairs of burry notes, from which the name is derived. Size: Length 7 inches; wingspan 10.5 inches.

Habitat and Distribution: Statewide along streams and around rural buildings. There are records from all 105 counties.

Seasonal Occurrence:
Migrant and summer resident. Nesting has been confirmed in 84 counties. Absent as a nesting species in the southwestern corner and most of the westernmost counties. Arrives in March and departs in late October. A few sometimes remain into early winter in southern Kansas.

Field Notes: Eastern Phoebes are among the earliest migratory songbirds to arrive each spring. During summer, a stop at almost any rural bridge will produce a pair of these flycatchers. Look under the bridge and you will usually find their nest attached to a vertical surface or a ledge. They are also fond of nesting in barns and other rural buildings.

immature

Sayornis saya

Say's Phoebe

Field Identification: Closely related to the Eastern Phoebe but slightly larger and found in drier western habitats. Mostly light gray, with a darker mask, black tail, and muted rufous-orange on the belly and under the tail. Size: Length 7.5 inches; wingspan 13 inches.

Habitat and Distribution: Prefers arid, open country. There are records from 76 counties. Summer resident in the western half. During migration, it wanders east, and there are records from about half of the eastern counties. Say's Phoebes are usually seen in or near grasslands, conspicuously perched on fences, shrubs, or weed stalks. Also seen around farmyards and other rural buildings.

Seasonal Occurrence: Migrant and summer resident. There are confirmed nesting records from 45 counties in the western half of the state and east to Reno and Jewell Counties. Spring birds arrive in late March and early April, and most depart in late September. A few linger into early winter in the southwest.

Field Notes: Like Eastern Phoebes, they often select bridges or rural structures such as barns or sheds for nest sites. Say's Phoebes are numerous in mid-September, when many southbound migrants can be seen at locations in western Kansas. Say's Phoebe is retreating to the west as Eastern Phoebe expands its range westward, perhaps due to increased woodlands along streams.

Acadian Flycatcher

Empidonax virescens

Field Identification: All *Empidonax* flycatchers are greenish gray above and pale below, with two wing-bars. They habitually flick their wings nervously. The best way to distinguish these similar species is by their songs. The Acadian's song is an emphatic *peet-SUH*. It also shows a distinct eye-ring. Size: Length 6 inches; wingspan 9 inches.

Habitat and Distribution: Found in swampy and moist bottomland forests with dense foliage in the eastern three tiers of counties, extending farther west along the Oklahoma border to Cowley County. Spring migrants are sometimes seen farther west. There are records from 46 counties, mostly in and east of the Flint Hills.

Seasonal Occurrence: Summer resident. Arrives in May, departs in August. Confirmed nesting has been recorded in 11 eastern counties.

Field Notes: Several small flycatchers belonging to the *Empidonax* genus are found in Kansas. Most are seen only in migration. The Acadian Flycatcher is the only *Empidonax* that is a summer resident in most of Kansas. Learning its song is helpful, as it can otherwise go unnoticed in the dense deciduous woodlands it inhabits.

Empidonax traillii

Willow Flycatcher

Field Identification: This *Empidonax* flycatcher is nearly identical to the Alder Flycatcher, so much so that until recently, they were considered one species. The identification of an *Empidonax* lacking an eye-ring, or with only a narrow one, can usually be narrowed down to either Willow or Alder Flycatcher. They are best separated from each other and from other *Empidonax* by voice. The Willow has a sneezy two-note call accented on the first syllable. The single *whit* call note is softer than that of the Least Flycatcher and quite distinct from the sharp call note of the Alder Flycatcher. Size: Length 6 inches; wingspan 8.5 inches.

Habitat and Distribution: Recorded statewide in a variety of wooded and urban habitats during migration. Has been reported from 99 counties. Willow is more likely to be seen in the west than Alder.

Seasonal Occurrence: Spring migrants are seen in May, fall migrants in late August and early September. A small population of Willow Flycatchers remains to nest in five counties in extreme northeastern Kansas along the Missouri and Kansas Rivers.

Field Notes: After the Least Flycatcher, this is the most frequently observed *Empidonax* flycatcher in Kansas. Silent birds are so nearly identical to Alder Flycatchers that they often must be simply noted as Alder/Willow.

Alder Flycatcher

Empidonax alnorum

Field Identification: Any *Empidonax* lacking an eye-ring, or with only a narrow one, is likely either an Alder or Willow Flycatcher. They are best separated from each other and from other *Empidonax* by voice. Alder Flycatcher gives a three-note song accented on the second syllable. The call note of Alder is a sharp *pip* note, quite distinct from the softer *whit* call of Willow Flycatcher. Size: Length 6 inches; wingspan 8.5 inches.

Habitat and Distribution: Can be seen in a variety of wooded and urban habitats. There are records from 75 counties. There are many western counties where Alder has not been reported.

Seasonal Occurrence: Spring migrants are seen in May, fall migrants in late August and early September. There are no nesting records for this species in Kansas.

Field Notes: Singing Alder Flycatchers are often encountered in late May and the first few days of June, after most other migrant passerine species, including other *Empidonax* species have already departed northward.

Empidonax flaviventris

Yellow-bellied Flycatcher

Field Identification: This *Empidonax* is comparatively easy to identify because of the yellow underparts. While other *Empidonax* species sometimes show some yellow below, only this species has a yellow throat, which blends with the yellowish belly. It is the only *Empidonax* with a yellow eye-ring that is likely to be seen in Kansas. Size: Length 5.5 inches; wingspan 8 inches.

Habitat and Distribution: This species is usually seen in the undergrowth of woodlands in eastern Kansas. Has been recorded in 47 counties, mostly east of the Flint Hills but also including about a dozen counties from farther west.

Seasonal Occurrence: Spring migrant during May; fall migrant during August and September.

Field Notes: Yellow-bellied Flycatchers are most likely to be seen in the eastern third of the state. They are usually silent when in Kansas, so it is fortunate that they are marked more distinctly than other members of their genus.

Least Flycatcher

Empidonax minimus

Field Identification: This flycatcher has a slightly shorter tail and wings than other *Empidonax* species. The eye-ring is obvious. It looks big-headed. The song is a dry *che-bek*, and the call note is a dry *pit*. Both vocalizations are heard frequently, simplifying identification. Size: Length 5.25 inches; wingspan 8 inches.

Habitat and Distribution: Found statewide in a variety of habitats. Has been recorded in all 105 counties. Common in central and eastern Kansas, less so in the westernmost counties.

Seasonal Occurrence: Spring migrants are seen from late April through early June. Fall migrants are seen as early as mid-July and continue through mid-September.

Field Notes: This is the most abundant *Empidonax* flycatcher in most of Kansas and the one most likely to be heard singing. Least Flycatchers are most common in early May and early September. On some days in early May, they seem to outnumber all other migratory songbirds.

Empidonax oberholseri

Dusky Flycatcher

Field Identification: The Dusky Flycatcher resembles the Least Flycatcher, but the grayer head slightly contrasts with the olive back, the tail is longer, and it has less of a crested look. Size: Length 6 inches; wingspan 8 inches.

Habitat and Distribution: Found in the western three tiers of counties only. There are records from 18 counties. Seen in woodlands, along streams, and in thickets in towns.

Seasonal Occurrence: Spring migrant in late April and May; fall migrant in August and September.

Field Notes: Identification of *Empidonax* flycatchers becomes even more complex in the western part of the state, where this species, along with three other western *Empidonax* species that were considered too rare to be included in this book, can be encountered. The majority of *Empidonax* seen in the far west are probably this species. Professional ornithologists are usually unwilling to identify *Empidonax* unless they have been captured and identified in hand, but many birders consider it great sport to attempt field identification of these confusingly similar species.

Golden-crowned Kinglet

Regulus satrapa

Field Identification: This tiny winter bird is olive green and gray above, is whitish below, and has black wings with wing-bars. The head is striped black and white with a bright yellow crown, tinged with red on the male. The call is a series of three high-frequency *seep* notes, similar in pitch to the single call note of the Brown Creeper. Always in motion, it flicks its wings frequently. Size: Length 4 inches; wingspan 7 inches.

Habitat and Distribution: Found statewide in towns and wooded areas, especially where there are cedars and pines. There are records from 103 counties. This species is seen more often in central and eastern Kansas than in the west.

Seasonal Occurrence: Migrant and winter resident. Arrives in mid-October and departs in early April. It is present every year, but numbers fluctuate substantially from year to year.

Field Notes: Kinglets are small but fearless and on many occasions can be observed from a distance of a few feet as they investigate the birder who has entered their territory. They often join mixed-species foraging flocks with chickadees, creepers, nuthatches, and titmice. Cemeteries and city parks with abundant cedars and pines are good places to look for these birds. Listen for the high-pitched call notes, often the first clue to their presence.

Corthylio calendula

Ruby-crowned Kinglet

Field Identification: The Ruby-crowned Kinglet is a tiny species with a mostly olive body that is lighter on the underparts. The wings are black with bright white wing-bars. The head is also olive, with a tear-shaped white eye-ring. Males have a red crown that is usually concealed, visible only when highly agitated. Listen for the brief two-note scolding call, which is often heard before they are seen. Size: Length 4.25 inches; wingspan 7.25 inches.

Habitat and Distribution: Found statewide in rural areas, cities, and towns. Like Golden-crowned Kinglets, they are fond of cedars and pines but are equally frequent in brushy deciduous woodland habitats. Has been recorded in all 105 counties.

Seasonal Occurrence: Migrant and winter resident. It was formerly considered rare during winter but is now more regularly observed during the colder months. Spring migrants are found from late March through mid-May, fall migrants from mid-September through late November.

Field Notes: Migrating Ruby-crowned Kinglets arrive earlier in spring and linger longer in the fall than do many other migratory songbirds. In early spring, they sing a loud and melodious song.

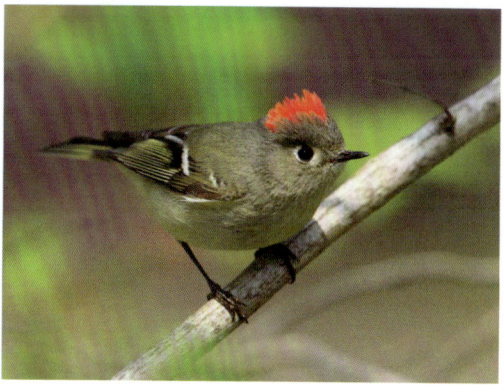

213

White-eyed Vireo

Vireo griseus

Field Identification: This woodland vireo is olive green above, with yellow flanks, a white throat, and two whitish wing-bars. The combination of white iris and yellow "spectacle" plumage around the eye gives an impression of watchful alertness. It is often heard before it is seen. Its vocalizations are peculiar and highly variable, usually beginning and ending with guttural *chek* notes. Size: Length 5 inches; wingspan 7.5 inches.

Habitat and Distribution: Summer residents are found in the eastern fourth of the state. Migrants occur farther west in central and western Kansas. There are records from 64 counties. Inhabits dense edge and brushy understory in riparian woodlands.

Seasonal Occurrence: Migrant and summer resident. In summer, it seems to be most common in the southeast and west to the Cross Timbers region. Confirmed nesting has been reported in nine eastern counties, but the nests are difficult to find, and it likely nests in numerous other eastern counties. It has nested in previous years in woodland timber on the Arkansas River near Derby in Sedgwick County, farther west than other Kansas locations. Arrives in April and departs in September.

Field Notes: Vireos are small songbirds with stouter, thicker bills than warblers. White-eyed Vireos are inquisitive and territorial and usually vocalize when humans approach their brushy haunts.

Vireo flavifrons

Yellow-throated Vireo

Field Identification: This woodland vireo is olive above with white wing-bars, a yellow throat and breast, and yellow spectacles surrounding dark eyes. The song is a series of two- or three-note burry phrases that is somewhat reminiscent of the song of a tanager. Size: Length 5 inches; wingspan 9 inches.

Habitat and Distribution: Mature woodlands east of the Flint Hills. During migration, it occurs farther west. Has been reported in 69 counties, mostly central and eastern but also including records from about 15 western counties.

Seasonal Occurrence: Summer resident. There are confirmed nesting records from 14 eastern counties, many of which are in the northeastern corner of the state. It likely nests in most of the eastern three tiers of counties. Arrives in late April and departs in early September.

Field Notes: In summer, Yellow-throated Vireos are restricted to eastern Kansas, most often in woodlands with mature oaks, hickories, or pecans. The forests around Milford Reservoir in Geary County and in the Cross-Timbers region are their western breeding outposts in Kansas.

Blue-headed Vireo

Vireo solitarius

Field Identification: Look for the slate-gray head, prominent white spectacle markings, olive back, yellow flanks, and two yellow wing-bars on this migratory vireo. The song is melodious and repetitive, like the song of the Red-eyed Vireo but slower and more abbreviated. Size: Length 5.5 inches; wingspan 9 inches.

Habitat and Distribution: Found statewide in woodlands, parks, and cemeteries. There are records from 95 counties. It is most likely to be seen in the central and eastern parts of the state.

Seasonal Occurrence: Spring and fall migrant. Seen in spring during late April and May and in fall from September through late October. It remains in Kansas later into the fall than other vireo species. In some years, a few linger into November.

Field Notes: Like most vireos, this species moves about a bit more deliberately than warblers and other small songbirds. It is encountered in woodlands, towns, and parks during migration, and especially in fall, it often joins foraging flocks of chickadees, warblers, and other species.

Vireo bellii

Bell's Vireo

Field Identification: Bell's Vireo is olive above and yellow below, with a white throat, a whitish line above the eye, and a thin black line through the eye. There is one light wing-bar that fades as the plumage becomes worn. It is usually heard before it is seen. The song is a distinctive gurgling chatter that begins and ends abruptly. Size: Length 5 inches; wingspan 7 inches.

Habitat and Distribution: Found statewide, but most common in the central counties. There are records from 104 counties. Inhabits large, shrubby thickets surrounded by prairies or pastures.

Seasonal Occurrence: Migrant and summer resident. Arrives in late April and departs in late September. Confirmed nesting has been reported from 56 counties, probable nesting from more than 30 other counties. Many counties in the western third of the state lack breeding records. During migration, this vireo can be seen in atypical habitats.

Field Notes: The open-country habitat this vireo prefers is typified by the upland prairies of Sand Hills State Park and Quivira NWR, where large stands of brushy plant species such as sandhill plum, sumac, and rough-leaf dogwood dot the sprawling mixed-grass prairie. In summer months, check roadside shrub thickets. You are more likely to hear than see them. Like several other vireo species, Bell's sings throughout the day even in the hottest weather.

Philadelphia Vireo

Vireo philadelphicus

© Judd Patterson

Field Identification: This is a vireo without wing-bars, almost as plain as the Warbling Vireo. It is separated from the Warbling by the dark line passing through the eye that extends to the base of the bill, a darker cap, and a yellow or yellowish throat. Its underparts are more yellow than those of the Warbling Vireo and are brighter yellow in fall than in spring. Some Warbling Vireos in fall can be almost as yellow on the underparts, so some caution is in order. At all seasons, the dark eye-line is the most reliable field mark. Size: Length 5 inches; wingspan 8 inches.

Habitat and Distribution: Seen in woodlands, towns, and cities during migration. Seen mostly in the eastern third of the state, rarely farther west. There are records from 76 counties. Has been reported from about half of the western counties.

Seasonal Occurrence: Seen only during migration. Nearly all spring records are from May. Fall records are almost entirely from September.

Field Notes: This is the least common migratory vireo in Kansas. It does not sing as frequently as other vireos, and when it does, the song is confusingly similar to that of the more common and widespread Red-eyed Vireo.

Vireo gilvus

Warbling Vireo

Field Identification: A very plain-looking vireo without wing-bars, the Warbling Vireo is pale gray above and white below. In the fall, some birds have a light yellow wash on the flanks and at the base of the undertail. A white line above the eye is the only noticeable field mark. The song is a rapid warbling ending in a single high note. A wheezy scolding call is often heard. Size: Length 5.5 inches; wingspan 8 inches.

Habitat and Distribution: Found statewide in riparian woodlands, forest edges, cities, and towns. While they can be found in a variety of open woodland habitats, they are especially fond of cottonwood trees for both foraging and nesting. There are records from all 105 counties.

Seasonal Occurrence: Migrant and summer resident. Arrives in late April, departs in late September. Confirmed nesting has been documented statewide in 61 counties. Probably nests in all 105 counties. Widespread and common in migration during May and September.

Field Notes: This is the most widespread and numerous vireo species in Kansas. Its song is heard throughout the summer, even on the hottest afternoons. Adults often sing while sitting on the nest, which is a woven cup hanging near the end of a branch in a clump of leaves.

Red-eyed Vireo

Vireo olivaceus

Field Identification: Another vireo lacking wing-bars, this species has a prominent white eye-line passing through the bright red eye and a gray cap bordered with black. Otherwise, it is olive green above and white below. In spring and early summer, the lower belly shows a considerable amount of yellow. Its song is a repeated warbling phrase, somewhat like that of the Robin. Size: Length 6 inches; wingspan 10 inches.

Habitat and Distribution: Mature woodlands as well as cities and towns with mature trees. Has been recorded in 102 counties. Migrating birds can be seen in a variety of locations and habitats statewide.

Seasonal Occurrence: Migrant and summer resident. In summer, breeds in woodlands and towns in the eastern half of Kansas and in mature riparian woodlands farther west. Confirmed nesting has been reported in 23 counties, most of which are in the eastern half of Kansas, especially from the Flint Hills eastward. Like most other vireos, Red-eyed Vireos build nests that are difficult to find, so documenting breeding is challenging. Arrives in late April and departs in early October.

Field Notes: This is one of the most common summer species of the eastern Kansas woodlands. Like the Warbling Vireo, it sings persistently throughout the day well into the late summer and even in hot weather.

Leiothlypis peregrina

Tennessee Warbler

Field Identification: This is one of many wood-warbler species that migrate through Kansas. The gray cap of spring males contrasts with an olive-green back and white breast and throat. There is a dark line through the eye and no markings on the underparts. Spring females lack the contrasting cap, and the underparts are tinged with yellow. Fall birds can be confused with the Orange-crowned Warbler but always have white undertail coverts. Size: Length 5 inches; wingspan 8 inches.

Habitat and Distribution: Common migrant in eastern Kansas, rare in the west. Found in mature woodlands, wooded urban areas, and parks. There are records from 91 counties.

Seasonal Occurrence: Spring migrant from late April through late May; fall migrant from September through early October. It is more numerous in spring than in fall.

Field Notes: Although it is still one of the most common migratory warblers, this species seems to be declining in numbers as habitat loss accelerates in both the tropical wintering areas and the boreal forest nesting grounds. Learning the Tennessee Warbler's song is helpful in locating and identifying this species. Males sing incessantly as they journey north to their nesting grounds. In cities and towns in the eastern part of the state, many singing birds can be heard in May. Tennessee Warblers tend to remain high in the tree canopy.

Orange-crowned Warbler

Leiothlypis celata

Field Identification: This is one of the most nondescript warbler species and is variable in appearance. Males are greenish yellow with blurry streaking on the breast and no wing-bars. Females are similar, but some are grayer in color. It always has yellow undertail coverts, which helps distinguish this species from the Tennessee Warbler in fall. As with all warblers, learning the song and call notes helps to identify this species. The monotone trilling song in spring is often heard. In fall, the distinctive *chip* note is unlike that of any other warbler. Size: Length 5 inches; wingspan 7 inches.

Habitat and Distribution: A common migrant statewide in woodlands, weedy fields, and brush. There are records from all 105 counties.

Seasonal Occurrence: Spring migrant from mid-April through mid-May; fall migrant from early September through early November. A few are seen during the winter months in most years.

Field Notes: This is a common warbler throughout Kansas during both spring and fall migration. In fall, it is fond of weedy fields, especially stands of sunflowers. In spring, it usually arrives 7–14 days earlier than most other warbler species. It is present later in the fall than any other warbler except the Yellow-rumped Warbler.

Leiothlypis ruficapilla

Nashville Warbler

Field Identification: The Nashville Warbler has a gray head, white eye-ring, green back and wings, and yellow throat and underparts; it lacks wing-bars. Fall birds are often duller in color but always have the gray head and white eye-ring. Fall adults retain the yellow throat, but the throat of immature birds is sometimes paler. Size: Length 4.5 inches; wingspan 7.5 inches.

Habitat and Distribution: Found statewide in woodland edges, brushy thickets, and weedy areas. There are records from 103 counties.

Seasonal Occurrence: Spring migrant from mid-April through late May; fall migrant from late August through late October.

Field Notes: This is one of the most common migrant warbler species in Kansas. When identifying warblers, first check for wing-bars. This warbler has no wing-bars and is quickly separated from the rest of the plain-winged species by the gray head and white eye-ring. In the fall, they are often found in small flocks, especially in weedy fields and shrubby habitats, particularly rough-leaf dogwood thickets.

Wilson's Warbler

Cardellina pusilla

Field Identification: The male is olive green above and yellow below, with a black cap and no wing-bars. The female is similar but has an olive-green cap. Females can easily be confused with female Yellow Warblers. Size: Length 5 inches; wingspan 7 inches.

Habitat and Distribution: Found statewide in dense thickets and forest understory. There are records from all 105 counties.

Seasonal Occurrence: Spring migration occurs from late April through late May, with peak numbers in the middle of May. Fall migration occurs from late August through early October, with peak numbers in mid-September.

Field Notes: Wilson's Warblers are an uncommon species during the spring migration. In September, and especially in western Kansas, they are often the most abundant warbler and can seem to be everywhere. Learning their distinctive call note is useful when attempting to separate them from less common species.

female

© Judd Patterson

Geothlypis formosa

Kentucky Warbler

Field Identification: This forest warbler is olive above and entirely yellow below. It has a black mask and a broad yellow eyebrow and lacks wing-bars. Size: Length 5 inches; wingspan 8 inches.

Habitat and Distribution: Found in eastern Kansas woodlands, always with mature trees and dense understory vegetation. Migrants sometimes occur farther west, usually in the spring. There are records from 57 counties.

Seasonal Occurrence: Summer resident. Nesting has been fully confirmed in only 18 counties from the Flint Hills east to the Missouri border, but it likely nests in most of the four eastern tiers of counties. Arrives in late April and departs in early September.

Field Notes: Kentucky Warblers are secretive and well camouflaged. They sing a song resembling that of the Carolina Wren from low perches in dense understory. Learn their song and you may be rewarded with a good look at this elusive warbler. Patience will be required.

female

225

Common Yellowthroat

Geothlypis trichas

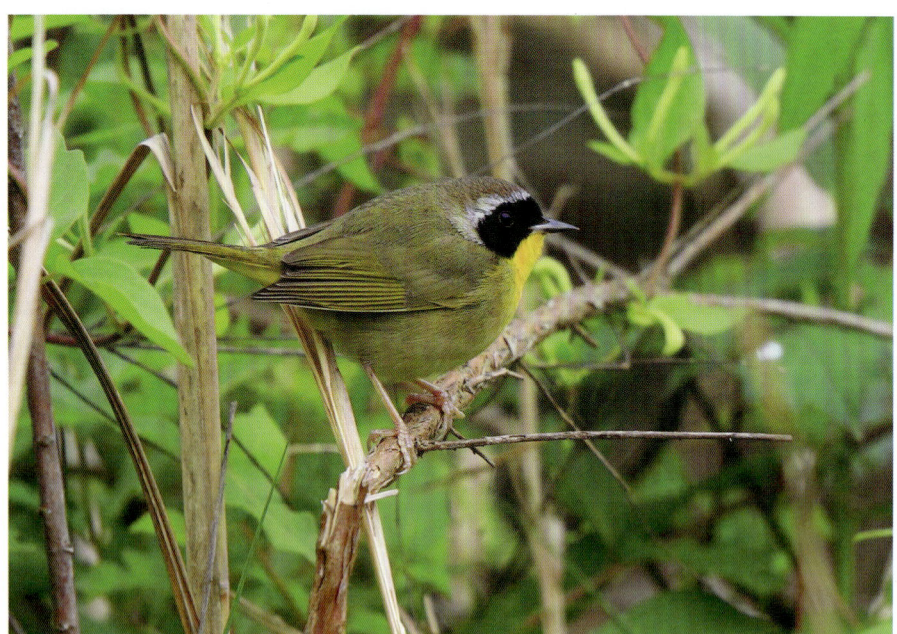

Field Identification: This tiny marsh warbler has an olive back and wings, yellow on the throat and breast, and no wing-bars. The male has a black mask bordered with white. Females have an olive face. The song is usually interpreted as *wichity-wichity-wichity-wich* and is a familiar sound of wetland areas. Size: Length 5 inches; wingspan 7 inches.

Habitat and Distribution: Found statewide in a variety of moist habitats with dense herbaceous weedy cover. There are records from all 105 counties. Wetland areas dominated by cattails usually harbor nesting populations of this warbler. Even small patches of habitat, such as a wet swale in a pasture, are often home to a resident nesting pair.

Seasonal Occurrence: Migrant and summer resident. Arrives in mid-April and departs in mid-October. This is a common nesting species, with confirmed nesting records from about 36 counties and probable nesting in another 40 counties. Most of the counties without nesting records are in western Kansas, where wetland habitats are scarce or absent. Occasionally a few remain into the winter months at wetlands where there is open water. There are records from all months of the year.

Field Notes: During migration, Common Yellowthroats are often seen in atypical habitats such as suburban yards and wooded areas.

female

Setophaga petechia

Yellow Warbler

Field Identification: Male Yellow Warblers are greenish yellow above with yellow wing-bars and yellow below, with bold red streaks on the breast. Females are similar, but the breast is all yellow without streaking. The undertail coverts and large spots on the underside of the tail are yellow. Size: Length 5 inches; wingspan 8 inches.

Habitat and Distribution: Found statewide. There are records from all 105 counties. The nesting population is widespread but localized. It prefers to nest along streams and lakeshores with willow or cottonwood trees. During migration, it occurs statewide in woodlands, cities, and parks.

Seasonal Occurrence: Migrant and summer resident. Arrives in late April and departs in late September. It is abundant during migration. This is often the most abundant warbler species in Kansas during the month of May. There are confirmed nesting records from 27 counties scattered widely across the state.

Field Notes: Only a few species of warbler remain to nest in Kansas; of them, the Yellow Warbler is the most widespread except for the Common Yellowthroat.

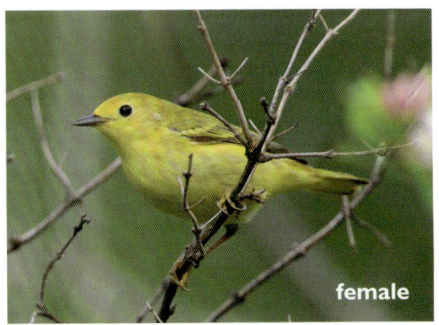

female

© David Seibel

Prothonotary Warbler

Protonotaria citrea

Field Identification: This is an eastern warbler with brilliant gold plumage on the body and head and blue-gray wings with no wing-bars. The dark eyes and bill contrast strongly with the golden plumage. Size: Length 5.5 inches; wingspan 9 inches.

Habitat and Distribution: Found in eastern Kansas around lakes, ponds, and slow-moving streams. Nests in cavities of dead trees standing in or near the water. Nests from the Missouri border west to the Flint Hills. Migrants occur farther west to central and southwestern Kansas, especially in spring. There are records from 64 counties.

Seasonal Occurrence: Summer resident. Arrives in late April and departs in early September. There are confirmed nesting records from 22 eastern counties west to Cowley and Riley Counties.

Field Notes: This beautiful warbler is usually observed on low branches near the water's edge. Listen for the ascending *sweet-sweet-sweet-sweet* call. They are not afraid of humans and can often be seen and heard near popular fishing and camping areas. Look for them at almost any lake or river in eastern Kansas with mature timber. This is the only Kansas warbler species that nests in tree cavities.

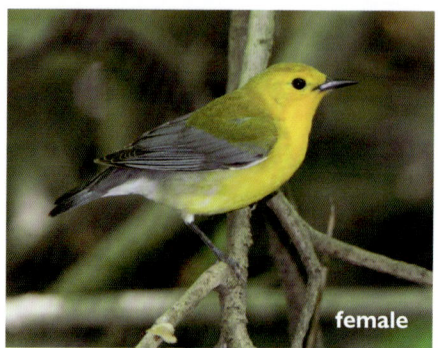

female

Setophaga coronata

Yellow-rumped Warbler

Field Identification: All plumages of this abundant warbler show obvious wing-bars and a bright yellow rump. Spring males have a bright yellow blaze on the flanks and black on the breast. Females have a similar but muted pattern. Winter birds are mostly brownish gray with blurry streaks below, but always look for the yellow rump. The eastern subspecies, called Myrtle Warbler, has a white throat and occurs throughout Kansas. In the westernmost counties, the western subspecies, called Audubon's Warbler, often outnumbers the Myrtle. Western birds have a yellow throat and a single large white wing-bar. Size: Length 5.5 inches; wingspan 9 inches.

Myrtle Warbler, breeding male

Habitat and Distribution: Migrants are found statewide in almost any habitat containing trees or shrubs. There are records from all 105 counties.

Seasonal Occurrence: Spring migrants are seen from early April through mid-May, fall migrants from late September through early November. The fall migration peaks in October, after most other warbler species have already departed south. Some remain through the winter, sometimes in good numbers when there has been a good crop of berries.

Field Notes: This is the most common warbler in Kansas during migration and the one most likely to be seen in the winter months. Dozens or even hundreds can be seen in a single day during migration. They arrive earlier in the spring than any other warbler species.

Myrtle Warbler, nonbreeding

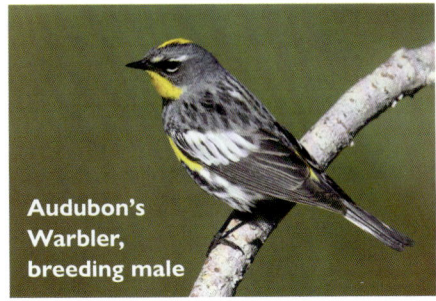

Audubon's Warbler, breeding male

229

Black-throated Green Warbler

Setophaga virens

Field Identification: This boldly marked warbler has a green back and crown, yellow face, blackish wings with bright white wing-bars, and white belly. The throat and breast of the male are black, and the flanks have heavy black streaks. On females, the throat is pale yellow and the black streaks on the flanks are not as heavy. Immatures seen in fall lack the black breast and flank markings. Size: Length 5 inches; wingspan 8 inches.

Habitat and Distribution: Migrant statewide, mostly in wooded areas. There are records from 84 counties. Most likely to be found in central and eastern Kansas. Less common in the western half, where about 20 counties lack any record of this species.

Seasonal Occurrence: Spring migrants are seen from late April through late May, with most records in the first two weeks of May. Fall migration is more extended, and migrants are seen as early as late July, continuing through mid-October. The fall peak is usually in the middle of September.

Field Notes: Black-throated Green Warblers are annual migrants but occur at a low density. Their song is a series of buzzy notes that is easily distinguished from the songs of more frequently encountered migratory warblers. Unlike many of the migratory warbler species that occur in Kansas, they are seen as often in the fall as in the spring.

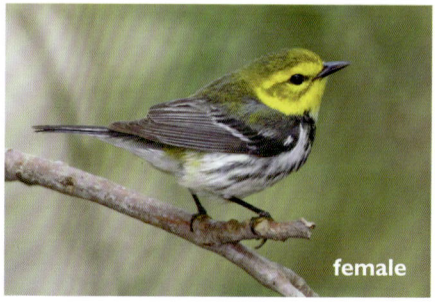

female

Setophaga americana

Northern Parula

Field Identification:
This is a tiny, short-tailed warbler of eastern woodlands with blue head and wings, olive back, and two bright wing-bars. The male has a yellow throat and breast with a rufous breast band. Size: Length 4.5 inches; wingspan 7 inches.

Habitat and Distribution: Inhabits mature riparian forests in eastern Kansas. They strongly prefer woodlands where sycamore trees are present. Migrants west of the breeding range, including many western counties, are seen in a variety of habitats. There are reports from 89 counties.

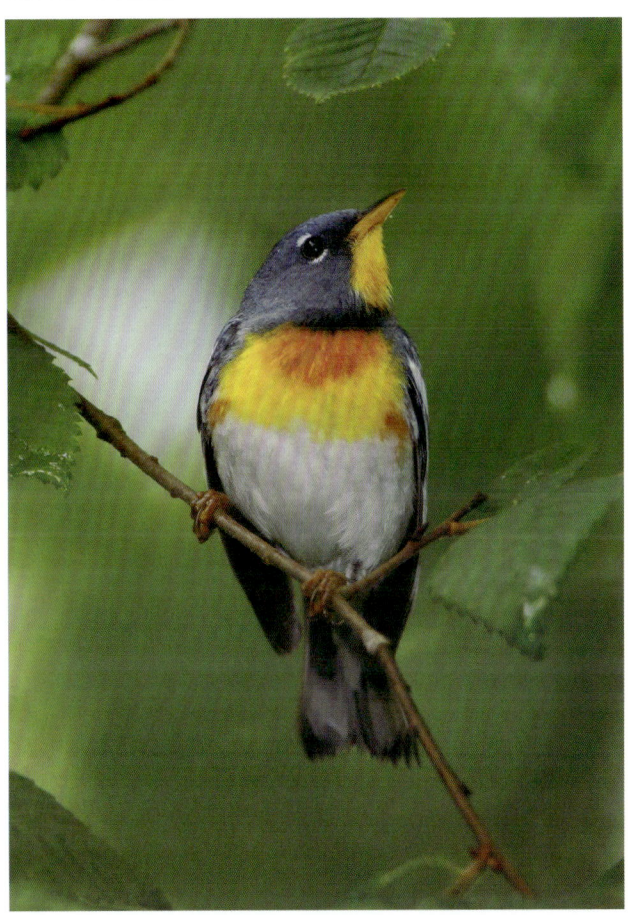

Seasonal Occurrence: Migrant and summer resident. Confirmed breeding has been reported from 24 counties from the Flint Hills eastward. Probably nests in all counties in and east of the Flint Hills. Spring arrival is in early April, and migration continues through May. Fall migrants are seen from August through late September.

Field Notes: Silent Northern Parulas can be difficult to spot as they actively forage high in the forest canopy. Fortunately, the ascending buzzy song is frequently heard in summer along rivers and streams of eastern Kansas, making them easier to find. Any eastern Kansas riparian woodland with sycamores is usually inhabited by this species.

female

© David Seibel

Yellow-throated Warbler

Setophaga dominica

© David Butel

Field Identification: This beautiful warbler has a blue-gray back and wings, white wing-bars, yellow throat, black-and-white facial pattern, white belly, and black streaks on the flanks. The bill is longer than that of most other warblers. Size: Length 5.5 inches; wingspan 8 inches.

Habitat and Distribution: In summer, it is restricted to mature riparian forests with abundant sycamore trees located in the southeastern and easternmost counties. There are records from 50 counties. Migrants have been reported in about 25 counties west of the breeding range and have been seen in a variety of habitats, including city parks and yards.

Seasonal Occurrence: Migrant and summer resident. Breeding individuals arrive in early April and are seldom seen after July. Nesting has been confirmed in only four far-eastern counties, but it likely nests in several other counties in the east. Migrants west of the breeding range are seen in spring from mid-April through mid-May. In fall migration, it is not seen as often. Most fall records are from September.

Field Notes: Yellow-throated Warblers have specialized habitat requirements that are met at only a few locations in Kansas. They are seldom found far from sycamore trees. Marais des Cygnes WA, Schermerhorn Park in Galena, and areas along the Caney River in Chautauqua County are places to look for them. Their song is a series of clear, descending notes.

Mniotilta varia

Black-and-white Warbler

Field Identification: The plumage of the male consists of black and white stripes, with considerable black on the face and throat. Females are similar, but the face and throat are white. It often forages on tree trunks and large branches like a nuthatch. Size: Length 5 inches; wingspan 8 inches.

Habitat and Distribution: Migrants are found statewide in a variety of wooded areas. It is a summer resident east of the Flint Hills in mature oak woodlands, especially those with rocky slopes. There are records from 99 counties.

Seasonal Occurrence: Migrant and summer resident. Spring migrants are seen from early April through late May, fall migrants from late August through early October. Present throughout the summer months in eastern Kansas. Nesting has been confirmed in 12 counties located in and east of the Flint

Hills. Probable nesting has been reported in about 10 additional eastern counties.

Field Notes: Black-and-white Warblers arrive in Kansas earlier in the spring than most other warblers. Their song is a series of repetitious high-pitched double notes, often compared to the cadence of a sewing machine. They are one of the warbler species more likely to be encountered in eastern Kansas during migration.

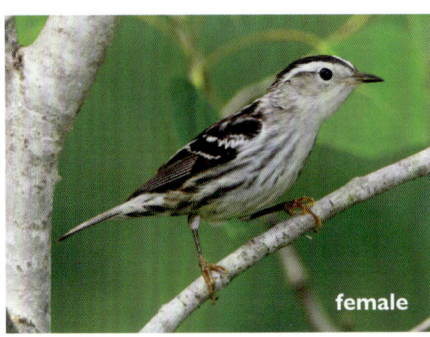

female

American Redstart

Setophaga ruticilla

Field Identification: Adult males are jet black, with a white belly and an orange pattern on tail, wings, and flanks. Females have a similar pattern but are gray and yellow instead of black and orange. Size: Length 5 inches; wingspan 8 inches.

Habitat and Distribution: Migrants are found statewide in a variety of wooded and urban habitats. Summer residents are found in mature woodlands on the eastern edge of the state. Has been reported from 104 counties.

Seasonal Occurrence: Migrant and summer resident. Spring migrants are usually found in the first three weeks of May, fall migrants from late August through late September. Nesting has been confirmed in three counties, and it is a probable breeding species in five more, nearly all located on the Missouri border.

Field Notes: American Redstarts are vocal and active, often fanning their tails as they flit through the understory. This warbler is equally common in fall and in spring. During fall migration, look for them along quiet streams with brushy banks.

female

Seirus aurocapilla

Ovenbird

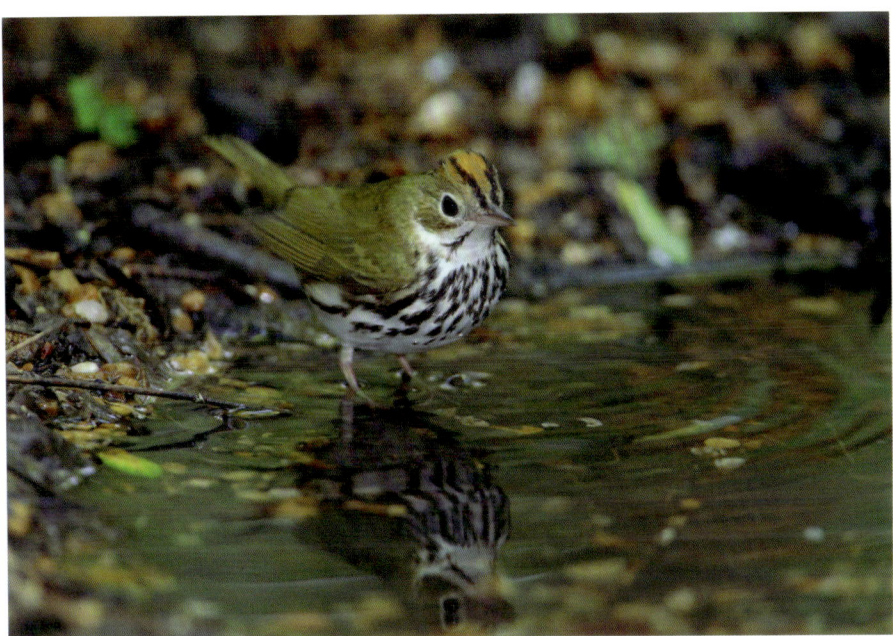

Field Identification: This large, ground-dwelling warbler has olive upperparts, white underparts patterned with bold black streaking, an orange crown bordered with black, and a large, dark eye surrounded by a broad eye-ring. Size: Length 6 inches; wingspan 9.5 inches.

Habitat and Distribution: Occurs statewide during migration, most commonly in the east. Migrants are seen in woodlands and parks where there is brushy understory. There are records from 93 counties.

Seasonal Occurrence: Spring migrants are seen in late April and May, fall migrants from mid-August through September. In summer, a small population remains to nest from Douglas and Miami Counties northward. Nesting has been confirmed in five northeastern counties in the Glaciated Region, where there is extensive oak woodland.

Field Notes: The Ovenbird is usually heard before it is seen. During the spring migration, its loud *teacher-teacher-teacher* call can be heard in forest undergrowth in eastern Kansas. It spends most of its time foraging on the ground under bushes. A "spishing" sound will often entice it to hop into view.

Louisiana Waterthrush

Parkesia motacilla

Field Identification: This large member of the warbler family has a brownish back and crown, white breast with bold black streaking, and a bold white eyebrow mark that widen on the back of the head. There are no streaks on the white throat. It walks with a unique teetering gait on stream banks. A loud, wet *smack* note is often the first indication that one is nearby. Size: Length 6 inches; wingspan 9.5 inches.

Habitat and Distribution: Found in eastern Kansas woodlands along creeks with dense mature timber, especially favoring rocky, steep-banked streams with abundant brush and exposed roots. Found from the Flint Hills eastward. There appears to be a small and poorly understood breeding population in the Red Hills region in Barber, Comanche, and Kiowa Counties. Stray migrants have been reported from a handful of

counties west of the breeding range. There are records from 65 counties.

Seasonal Occurrence: Migrant and summer resident. Breeding birds arrive in early April and begin to depart southward in July. A few fall migrants continue to be seen through early September. There are confirmed breeding records from 24 counties. Probably nests in most counties from the Flint Hills eastward.

Field Notes: The ethereal song of the Louisiana Waterthrush is often heard in early April along the wooded streams of eastern Kansas. In spring and summer, stop at any bridge with timber, even at the smallest creeks, and listen for the song or the loud call notes. If you watch closely along the banks, you might spot one as it teeters along the water's edge or a log in the water.

Parkesia noveboracensis

Northern Waterthrush

Field Identification: This species strongly resembles the closely related Louisiana Waterthrush. However, the narrower eyebrow mark is not as long and is of even width. The white throat has fine streaking. The underparts usually have a yellowish tinge. The call note has a more metallic quality than that of the Louisiana. The songs of the two species are quite different. Size: Length 6 inches; wingspan 9.5 inches.

Habitat and Distribution: This widespread migrant has been reported from 95 counties and occurs statewide during migration.

Seasonal Occurrence: Spring and fall migrant. Spring migrants arrive in the last week of April and depart by late May. Fall migrants occur from August through mid-September. Generally, migrants are seen several weeks later in spring and later in the fall than Louisiana Waterthrush.

Field Notes: The Northern Waterthrush is seen over a much broader area and in a wider variety of habitats than the Louisiana Waterthrush. In western Kansas, where water is scarce, the banks of any body of water or moist woodland habitat will suffice. It is often observed in city parks and suburban yards.

Other Migratory Warblers

Worm-eating Warbler

Golden-winged Warbler

Blue-winged Warbler

Mourning Warbler

Forty-three warbler species have been observed in Kansas at least once. Seeking these elusive and colorful migrants during spring and fall migration is one of the most exciting aspects of birding for many. Kansas lies on the western edge of the flyway for many of these beautiful birds. Consequently, the greatest numbers of warblers are generally observed from the Flint Hills eastward, especially at sites near the Missouri border, such as Atchison State Fishing Lake (SFL), Fitch Natural History Reservation near Lawrence, Marais des Cygnes WA, and Wyandotte County Lake Park. In urban areas, priceless preserved fragments of woodland habitat often attract these birds. Examples include Felker Park in Topeka, Oak Park in Wichita, and the Overland Park Arboretum in suburban Kansas City. Look for eastern species such as Blackburnian, Chestnut-sided, Golden-winged, Hooded, Magnolia, and Palm Warblers at these locations. While most of these warblers inhabit the woodland canopy, other species such as Canada and Mourning Warblers prefer dense thickets in the understory.

In western Kansas,

Other Migratory Warblers

plantings of trees and shrubs in cities and towns on the otherwise treeless High Plains serve as "migrant traps" in both spring and fall. Cemeteries such as those in Elkhart and Garden City are well-known locations for finding warblers. Small towns such as Goodland and Tribune have produced numerous records of rare migrant warblers. Riparian woodlands along major river systems of western Kansas, most notably the Arkansas, Cimarron, and Republican Rivers, also attract warblers. Birders visit these western destinations hoping for western species, including Townsend's and Black-throated Gray Warblers.

Spring migration offers the best opportunity for seeing warblers, especially during the first two weeks of May. In the spring, males are in full song and at the peak of their colorful breeding plumage. The northward migration is often dramatic as these migrants are urgently drawn to their boreal nesting grounds. Warblers and other migrants sometimes arrive rather abruptly in large numbers. The presence of a low-barometric-pressure weather system sometimes triggers a "fallout" as warblers and other migrants

Hooded Warbler

Cape May Warbler

Magnolia Warbler

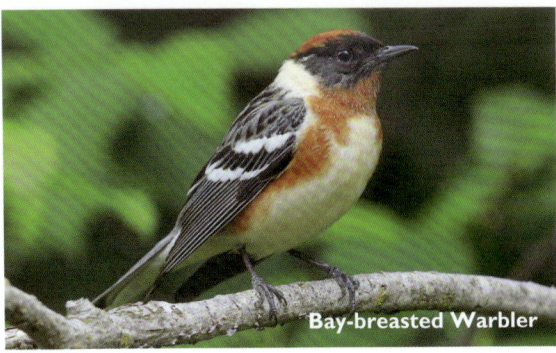
Bay-breasted Warbler

Other Migratory Warblers

Blackburnian Warbler

Chestnut-sided Warbler

Blackpoll Warbler

Palm Warbler

pause to await more favorable weather conditions before continuing north. When these conditions are present, birders visit local hotspots in hopes of finding a variety of warblers.

The southward fall warbler migration reaches its peak during the last week of August and the first two weeks of September. Seeking warblers at this season is often rewarding, but the large fallouts do not occur as often as in the spring migration and are often associated with cool fronts from the north. In the fall, most warblers are in less colorful plumage and are silent except for their brief *chip* notes. In general, fewer birders venture into the field during the fall migration, but a good variety of warblers can be observed by those who take the time to look for them. Fall is a productive time of year to seek warblers in western Kansas. In contrast to warbler dynamics in eastern Kansas, western warbler species tend to be more numerous in fall than in spring.

Many warblers occur in Kansas only as migrants, while a number of others remain to nest during the summer months. Migratory warbler species arrive and depart on different

Other Migratory Warblers

schedules, although these schedules frequently overlap significantly. In the Birding Resources chapter at the end of this book is a useful bar graph, compiled by Chuck Otte, showing the abundance, distribution, and seasonal occurrence of all Kansas warbler species. Many of them are discussed in this book, but some species shown in the graph are considered too rare to be included here.

Pine Warbler

Black-throated Gray Warbler

Townsend's Warbler

Canada Warbler

Carolina Chickadee

Poecile carolinensis

Field Identification: This is the chickadee of the southern counties. It differs from the Black-capped Chickadee in several subtle plumage characteristics. The wings on Carolinas lack the obvious white feather edges of the Black-capped, and the flanks are usually grayer and the black bib slightly smaller. The song is a four-note call usually written as *fee-bee—fee bay*, with the last two notes higher in pitch. Its familiar buzzy *chick-a-dee-dee-dee* call is faster and higher-pitched than the call of the Black-capped. Size: Length 4.75 inches; wingspan 7.5 inches.

Habitat and Distribution: Found in wooded areas south and east of a line between Meade County and Bourbon County. Tends to inhabit moister woodlands than those preferred by the Black-capped Chickadee. Has been recorded in 29 counties.

Seasonal Occurrence: Permanent resident. There are confirmed nesting records from 17 counties in southeastern and southern Kansas.

Field Notes: Differentiating Carolina and Black-capped Chickadees is one of the challenges of birding in southern Kansas. The zone of overlap is narrow, and hybrids have been documented in the contact zone. Carolina Chickadee may slowly be expanding its range to the north. Chickadees in the southernmost tier of counties in Kansas can be confidently identified as Carolinas. The isolated woodlands at Meade State Park are the western extreme for Carolina Chickadees in Kansas. Look for their nest holes in trees with softer dead wood, usually within 10 feet of the ground.

Poecile atricapillus

Black-capped Chickadee

Field Identification: This species is closely related to the Carolina Chickadee and almost identical in appearance. It is slightly larger, shows distinct white edges to the wing feathers, and has a slightly more extensive bib with a more ragged edge. Its song is also different: two whistled *fee-bee* notes. *Chick-a-dee-dee-dee* calls are huskier and deeper. Size: Length 5.25 inches; wingspan 8 inches.

Habitat and Distribution: Common in the eastern two-thirds of the state except in the southernmost counties, where it is replaced by the Carolina Chickadee. There are records from 104 counties, but it is mostly absent from the southern two tiers of counties on the Oklahoma border. In most years, it is absent from much of the western Kansas High Plains, especially south of the Smoky Hill River.

Seasonal Occurrence: Permanent resident. Nesting has been confirmed in 76 counties.

Field Notes: This species is always an entertaining visitor to the bird feeder, where it snatches a seed and retreats to a nearby tree branch to peck it open and devour it. Calling chickadees are often the most visible and vocal members of the mixed-species flocks of songbirds that forage together during migration and through the winter.

Tufted Titmouse

Baeolophus bicolor

Field Identification: This crested relative of chickadees is gray above and white below, with orangish flanks. The dark eye on the pale face creates a somewhat whimsical appearance. Its loud, clear whistled song is a series of two-note phrases; its calls are buzzy scolding notes. Size: Length 6.5 inches; wingspan 10 inches.

Habitat and Distribution: Found in woodlands and towns in the eastern half of Kansas west to Comanche, Rice, and Republic Counties. Most common east of the Flint Hills. Some occasionally wander a bit west of the breeding range. There are records from 72 counties.

Seasonal Occurrence: Permanent resident. Nesting has been confirmed in 47 counties.

Field Notes: Titmice are active, vocal birds, often joining chickadees and other woodland species in loosely organized foraging flocks. They have adapted well to humans and are often observed at bird feeders. Titmice are one of the "traffic cop" species of the woodlands and are often observed scolding predators or intruders.

Polioptila caerulea

Blue-gray Gnatcatcher

Field Identification: This small songbird is uniformly blue gray, darker on the back. Field marks include a long black tail with white edges, a black forehead, and a distinct white eye-ring. Breeding males have a dark line from the base of the bill to above the eye. It gives a variety of wheezy songs and scolding calls. Size: Length 4.5 inches; wingspan 6 inches.

Habitat and Distribution: Nests from the Flint Hills eastward in northern Kansas and westward to the Red Hills in southern Kansas. It is a common migrant across the rest of the state, and there are records from all 105 counties. It prefers wooded areas for nesting but can be seen in a variety of habitats during migration.

Seasonal Occurrence: Migrant and summer resident. Arrival in spring begins as early as late March. Spring migration peaks in late April and early May. Fall migrant in August and September.

Common in summer in the eastern half of the state. There are confirmed nesting records from 42 counties located east of a line from Comanche County to Smith County.

Field Notes: Gnatcatchers are vocal, active residents of eastern woodlands. After the young have fledged, they remain with the parents for several weeks. These noisy family groups are often encountered in July and August throughout eastern and south-central Kansas. This species is not shy and can be attracted by making squeaking or "spishing" sounds.

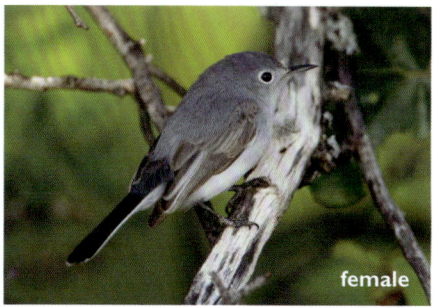
female

Red-breasted Nuthatch

Sitta canadensis

Field Identification: This small nuthatch has a blue back and orangish breast. The head is striped with black and white. Its voice is a monotone series of high-pitched nasal notes. Size: Length 4.5 inches; wingspan 8.5 inches.

Habitat and Distribution: Found statewide, almost always in or near pines or other coniferous trees, most often in cities and towns. There are records from all 105 counties.

Seasonal Occurrence: Mostly a winter resident, arriving as early as mid-August and departing in early May. In a few years, some have remained for the summer and have nested in Kansas towns including Garden City, Junction City, and Wichita. There are confirmed nesting records from four counties.

Field Notes: Red-breasted Nuthatches are one of several birds of the northern forests that are classified as "irruptive," meaning that they move into Kansas in large numbers in some years, appear in low or moderate numbers in other years, and are nearly or completely absent in others. These movements are responses to fluctuations in their food supply, nesting success, or both. In the fall, they can be observed gleaning seeds from pine cones and caching them in crevices of trees for consumption later in the winter. They will also do this with sunflower seeds and peanuts from bird feeders.

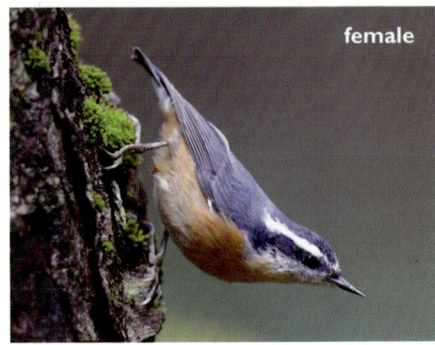

female

Sitta carolinensis

White-breasted Nuthatch

Field Identification: This larger nuthatch has a bluish-gray back, black cap, white face and underparts, and rusty patch under the tail. Vocalizations are nasal notes, lower-pitched than those of the Red-breasted Nuthatch. Size: Length 6 inches; wingspan 11 inches.

Habitat and Distribution: Found statewide, but more common in the east. Inhabits woodlands and towns with mature trees. There are records from all 105 counties.

Seasonal Occurrence: Migrant and permanent resident. In far western Kansas, it is primarily a migrant and wintering species, but it does nest in a few western areas with sufficient riparian woodlands in Cheyenne, Decatur, Gove, and Seward Counties. There are nesting records from 60 counties, mostly in the eastern three-quarters of the state.

Field Notes: White-breasted Nuthatches creep downward headfirst on tree trunks and branches, foraging for insects and grubs in tree-bark crevices. They often join chickadees, creepers, woodpeckers, kinglets, and titmice in foraging flocks during the winter. They are a frequent visitor to bird feeders, where they consume sunflower seeds and suet. They remain vocal during the winter, and their nasal chatter is one of the most frequently heard winter sounds of the woodlands.

Brown Creeper

Certhia americana

Field Identification: Creepers have a brown back with white markings; white underparts; and a thin, decurved bill. They creep upward on tree trunks on their short legs. Their call is a high-pitched single *seep* note. Size: Length 5 inches; wingspan 8 inches.

Habitat and Distribution: Statewide in wooded areas and towns. Less common in the west.

Seasonal Occurrence: Winter resident. Arrives in October and departs in April. There are records from 101 counties.

Field Notes: Brown Creepers fly to the base of a tree and slowly work their way up the trunk, foraging for insects and larvae. When near the top, they fly down to the base of another tree and start over. Their high-frequency call note is often the first clue to their presence. Once heard, they can be located by carefully searching tree trunks and large branches. Creepers are not shy, relying on their camouflage plumage to remain undetected. Their call note is almost identical in pitch to that of the Golden-crowned Kinglet, but the Brown Creeper gives a single long note, while the kinglet's call is three shorter notes given in a series. These two species utilize similar habitats during the winter, and both join mixed-species foraging flocks.

Salpinctes obsoletus

Rock Wren

Field Identification: This is a large wren with a brownish-gray back and crown that are speckled with white. It has whitish underparts, buffy flanks, and long legs. The call heard most often is a sharp *tick-eer*. Its songs are variable and complex. Size: Length 6 inches; wingspan 9 inches.

Habitat and Distribution: Found in the western half of Kansas, but only in areas with rock outcrops on steep slopes. Regularly wanders during migration to eastern Kansas, where it has been seen in about 25 counties. There are records from 69 counties.

Seasonal Occurrence: Migrant and summer resident. Arrives in mid-April and departs in late October. There are confirmed nesting records from 20 counties. A few remain into early winter in some years.

Field Notes: Rock Wrens are easy to find where their preferred habitat exists, especially in the Red Hills, the Smoky Hills, and the Arikaree Breaks in Cheyenne County. Good places to look for them on public lands include Clark SFL, Historic Lake Scott State Park, Cedar Bluff Reservoir, and Point of Rocks in the Cimarron National Grassland. Their songs and calls carry well, partially because of the favorable acoustics of the rocky cliffs they inhabit. They are adept at slipping into rock crevices and reappearing a few feet away.

Carolina Wren

Thryothorus ludovicianus

Field Identification: A wren of the eastern woodlands, the Carolina Wren has a bright reddish-brown back and crown, orange-buff underparts, and a bright white eyebrow. The loud, ringing song is somewhat like that of the Cardinal and can be heard throughout the year, as can its distinctive staccato call notes. Size: Length 5.5 inches; wingspan 7.5 inches.

Habitat and Distribution: Found in brushy woodlands in the eastern half of Kansas. Also inhabits cities and towns. Their range is expanding to the west, and a few have reached the Colorado border counties. There are records from 99 counties.

Seasonal Occurrence: Permanent resident. There are confirmed nesting records from 47 counties in the eastern half of the state. In fall and winter, Carolina Wrens disperse westward and have been seen in most counties in the western half.

Field Notes: Carolina Wrens are common in cities and towns and are frequently seen at bird feeders, especially mealworm and suet feeders. They often select nest sites in sheds, hanging baskets, and other man-made structures. A nest with young birds was found by one of the authors in the back of a rural mailbox. The mail carrier was aware of the nest and was careful not to disturb it when delivering the mail. In the past, Carolina Wrens have suffered significant population losses during cold, snowy winters.

Thryomanes bewickii

Bewick's Wren

Field Identification: The tail of this wren is exceptionally long. It has a brown back, wings, and crown; a distinct white eyebrow; and all-white underparts. The Bewick's can be confused with the Carolina Wren at first glance, but the long tail and all-white underparts make identification straightforward. Size: Length 5 inches; wingspan 7 inches.

Habitat and Distribution: Occurs across most of the state except in the northwestern counties and the High Plains of western Kansas. Prefers brushy habitats in or near prairies, woodland edges, rural buildings, and towns. They nest in the southeastern half of the state, south and east of a line through Seward, Ford, Saline, and Marshall Counties. They are most numerous as a nesting species in the Flint Hills and Red Hills regions. There are records from 89 counties.

Seasonal Occurrence: Migrant and permanent resident. Many depart during the winter months, especially in northern Kansas. There are confirmed nesting records from 38 counties.

Field Notes: Bewick's can be found widely in the state during migration. Their song is complex and melodious, reminiscent of the Song Sparrow, and is usually given from a high, exposed perch. Northern House Wrens compete with this species for cavity nesting sites and will sometimes destroy its nests and eggs.

Northern House Wren

Troglodytes aedon

Field Identification: This wren is mottled brown above and gray and plain below, lacking prominent markings. Its gurgling song and excited scolding calls are variable. Size: Length 4.5 inches; wingspan 6 inches.

Habitat and Distribution: Found statewide, often near human habitations but also in rural areas with brushy woodland edges. There are records from all 105 counties.

Seasonal Occurrence: Migrant and summer resident. Arrives in mid-April and departs in late September. Spring migrants peak in mid-May. Fall migrants peak in late September and early October. Confirmed nesting has been recorded in 93 counties.

Field Notes: This wren often nests in boxes placed in suburban yards. In the summer, it is widespread and vocal. Its rich song is heard throughout the day, even in August, when most other birds have ceased singing. Kansas lies at the southern edge of the nesting range for Northern House Wrens, and the species seems to be slowly shifting its range north as the climate becomes warmer. During fall migration, these wrens are furtive, almost entirely silent, and can be difficult to observe.

Troglodytes hiemalis

Winter Wren

Field Identification: This is the smallest wren in Kansas. It resembles the Northern House Wren, but the tail is much shorter, the plumage is darker brown, and the barring on the flanks and underparts is more pronounced. A two-note call is often the first indication of its presence. Size: Length 4 inches; wingspan 5.5 inches.

Habitat and Distribution: Found statewide in wooded areas with dense brush. Preferred habitat is along small streams, where it lurks along the banks under exposed tree roots and logjams. There are records from 94 counties. There are 11 western counties where it has never been recorded.

Seasonal Occurrence: Winter resident. Arrives in mid-October and departs northward in mid-March. A few linger in spring until mid-April.

Field Notes: Winter Wrens often go overlooked, lurking undetected in the dense streamside tangles and brush piles they inhabit. Sometimes they are disturbed by intruders and rapidly flit in and out of view while vigorously scolding. Shortly after their arrival in early fall and just prior to their departure in spring, you might be lucky enough to hear their complex song, which can have up to 36 notes per second.

Sedge Wren

Cistothorus stellaris

Field Identification: This prairie wren is light buffy overall, with strong black streaking on the back, a rich buffy wash below, and barred wings. The crown is streaked. The wet-sounding *chap* note is often heard. Size: Length 4.5 inches; wingspan 5.5 inches.

Habitat and Distribution: Migrants are found in wet meadows, tallgrass prairies, and densely vegetated wetland areas in central and eastern Kansas. In northeastern Kansas and the Flint Hills, they nest in prairies with tall, dense grass. Has been reported in 74 counties. There are only a handful of records from western Kansas.

Seasonal Occurrence: Migrant and late summer resident. Spring migrants are usually seen in late April and early May. Fall migrants are observed from late September through early November. There are confirmed nesting records

from seven counties and probable nesting records from another 12. Breeding records range from late July through September, mostly in prairie habitats of eastern and northeastern Kansas but also at Quivira NWR and Cheyenne Bottoms.

Field Notes: Sedge Wrens are a mysterious species. Much remains to be learned about them in Kansas, especially their breeding biology. The late-summer nesting birds in northeastern Kansas are thought to be birds that have nested earlier in the summer in the northern states and are making a second nesting attempt. The best chance of seeing this species is in the late summer and fall, when birds can be found in the dense grassy habitats at sites such as Baker Wetlands, Neosho WA, and Slate Creek WA.

Cistothorus palustris

Marsh Wren

Field Identification: This wren of cattail marshes has rich reddish-brown wings, flanks, and rump; white streaks on the black upper back; and a dark, unstreaked crown. The belly is reddish brown, contrasting with the white throat. The white eye-line is brighter and longer than that of the Sedge Wren. The call note is a dry *chip* sound. Size: Length 5 inches; wingspan 6 inches.

Habitat and Distribution: Found statewide and strongly associated with cattail marshes, but migrants are sometimes found in other habitats with dense vegetation. There are records from 93 counties.

Seasonal Occurrence: Migrant; also a rare summer and rare winter resident. Spring migrants are seen in April and May, fall migrants from August through early November. Occasionally nests in Kansas. There are confirmed nesting records from four widely scattered counties. A few remain through the winter months.

Field Notes: Marsh Wrens are more widespread and common migrants in Kansas than Sedge Wrens. The habitats of the two species overlap during migration, but Marsh Wrens are more likely to be associated with cattails and bulrushes. In winter, they are found in areas with dense cattails, especially where open water persists, such as artesian springs or dam outlets. At all seasons, one must be patient to get a good look, as they prefer to remain concealed deep in the cattails.

Eastern Bluebird

Sialia sialis

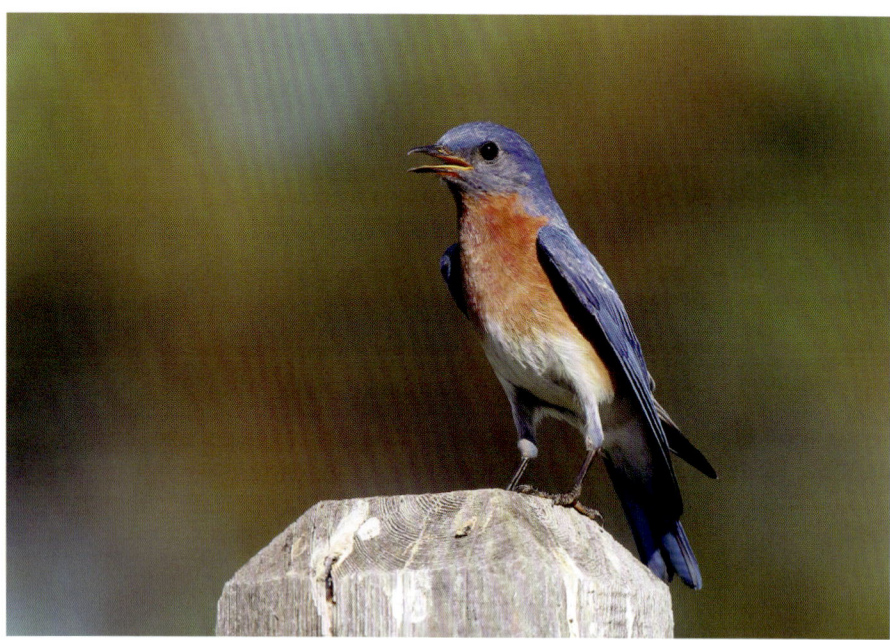

Field Identification: This colorful songbird species is known to many. The male is deep blue above, orange and white below. The female has a similar pattern but is softer gray above and paler orange below. Size: Length 7 inches; wingspan 13 inches.

Habitat and Distribution: This species is found statewide in woodland edges and open woodlands with standing dead trees. In the east, they are widespread and numerous, becoming less so in western Kansas, where their preferred habitat is scarce. They are usually found in the west along rivers and streams with good stands of cottonwood trees. There are records from all 105 counties.

Seasonal Occurrence: Permanent resident in the eastern two-thirds of the state. Found in the west in summer, but most western populations depart during the winter. There are confirmed nesting records from 86 counties.

Field Notes: Eastern Bluebirds are a familiar bird of rural areas. Because of competition with Starlings and House Sparrows for nest holes, populations of this species were severely reduced in the mid-twentieth century. A national effort promoting nest boxes along "Bluebird trails" has helped to increase Eastern Bluebird populations.

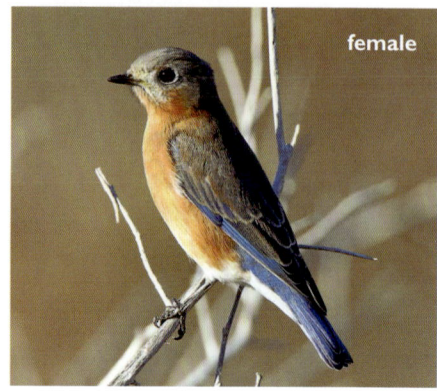

female

Sialia currucoides

Mountain Bluebird

Field Identification:
The male Mountain Bluebird is almost entirely sky blue. The female is light gray above and below, with sky-blue wings and tail. In contrast, all Eastern Bluebirds have orange breasts, simplifying identification of both species. Mountain Bluebirds also have noticeably longer wings and tail. Size: Length 7.25 inches; wingspan 14 inches.

Habitat and Distribution: Typically found in areas with numerous cedar groves in western Kansas. They can be seen across much of western Kansas. In migration, they are sometimes seen in grasslands and other open habitats. They reach their greatest abundance in the Red Hills region of Barber, Comanche, and Kiowa Counties. There are records from 83 counties. A few wander to eastern Kansas in some years.

Seasonal Occurrence: Migrant and winter resident. Arrives in late October and departs in late March. Surprisingly, this high-altitude-breeding species has been confirmed to nest in four Kansas counties.

Field Notes: Mountain Bluebirds descend onto the plains from the Rocky Mountains in winter and congregate in areas where cedar berries are abundant. Because the cedar-berry crop can vary

significantly from year to year, so do populations of these beautiful birds. During major invasion years, hundreds or even thousands can be seen on a single day's drive through the Red Hills. In other years, they are nearly absent.

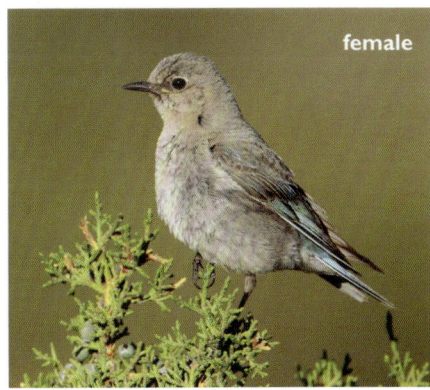

female

Swainson's Thrush

Catharus ustulatus

Field Identification: This migratory thrush is mostly brownish gray above. The underparts are white, with heavy spotting on the breast. Buffy markings on the face separate this species from other thrushes. The song is heard often in the spring and is a series of high-pitched, ascending flutelike notes. Size: Length 7 inches; wingspan 12 inches.

Habitat and Distribution: Found statewide in wooded areas. Frequently seen in parks and yards with trees and shrubs. There are records from all 105 counties.

Seasonal Occurrence: Migrant. Spring migrants are found from late April through early June, fall migrants from September to October.

Field Notes: This is the most common migratory thrush in Kansas and can usually be observed during a morning walk in the woods in May. Swainson's typically flush from the undergrowth as they are approached, then perch briefly on a low branch. In the spring, they linger in Kansas longer than most other migratory songbirds, and it is not uncommon to see small flocks devouring ripe mulberries in early June. During the fall migration, they are silent and retiring and are seen less frequently.

Catharus minimus

Gray-cheeked Thrush

Field Identification: This thrush is more grayish overall than the Swainson's Thrush. The best field mark is the gray face, lacking the buffy eye-ring and facial coloration of Swainson's. The song has the same ethereal quality as that of Swainson's, but the notes are descending instead of ascending. Size: Length 7 inches; wingspan 12 inches.

Habitat and Distribution: Woodland habitats statewide. Often seen in cities and towns. Has been reported from 83 counties. Most likely to be seen in eastern Kansas, but there are records from most of the western counties as well.

Seasonal Occurrence: Spring and fall migrant. Spring birds are seen almost exclusively in the first three weeks of May, fall migrants in mid-September and early October.

Field Notes: In most of Kansas, this species is seen far less often than Swainson's. A good look at the face is required to separate the two species. This is one of the longest-distance migrant songbirds in the world. From the winter range in Brazil, it migrates to breeding grounds in northwestern Alaska and Siberia.

Veery

Catharus fuscescens

Field Identification: This thrush is a rich reddish brown above. The throat and breast have a light buffy wash with sparse and poorly defined reddish spots. The face is pale, making the dark eye prominent. Size: Length 7 inches; wingspan 12 inches.

Habitat and Distribution: Woodland habitats statewide. Has been reported from 64 counties. It is most likely to be seen east of the Flint Hills, but there are records from many central and western counties.

Seasonal Occurrence: Spring and fall migrant. Spring birds are seen almost exclusively in the first three weeks of May, fall migrants in September. Reported much less often in fall.

Field Notes: This is the least common of the migratory thrushes in Kansas, so it is always rewarding when one is spotted. Learning to recognize the calls and songs of the several thrush species is useful. The song of the Veery is perhaps the most beautiful of all the thrushes but is only occasionally heard in Kansas.

Catharus guttatus

Hermit Thrush

Field Identification: The Hermit Thrush is identified by the rusty tail, which contrasts with the brownish upperparts. It also has a complete white eye-ring and a habit of raising and slowly lowering the tail. In the far western counties, birds of several western subspecies, which are similar but lighter gray above, are sometimes seen. These western birds do not habitually raise and lower their tails. The low *chup* call note is a clue to its presence. Size: Length 7 inches; wingspan 12 inches.

Habitat and Distribution: Statewide in wooded areas with brushy undergrowth or dense vines. Has been recorded in 99 counties.

Seasonal Occurrence: Migrant and winter resident. Spring migrants are seen from late March through early May, fall migrants from late September through late November. Most spend the winter south of Kansas, but small numbers remain for the winter in eastern and southern Kansas.

Field Notes: In winter, it prefers areas with honeysuckle and other "evergreen" vines, usually in woodlands with abundant coralberry or other berry bushes. They often share winter habitat with Fox and White-throated Sparrows.

Wood Thrush

Hylocichla mustelina

Field Identification: This is a large thrush with bright reddish-brown upperparts and white underparts with heavy black spots. Its ethereal, flutelike song is often heard at twilight. Size: Length 8 inches; wingspan 13 inches.

Habitat and Distribution: Found in large tracts of mature woodlands in eastern Kansas, mostly east of the Flint Hills. Spring migrants are occasionally seen farther west. There are records from 76 counties.

Seasonal Occurrence: Summer resident. There are confirmed nesting records from 22 counties. Breeding records are mostly from eastern Kansas and west to Sedgwick, Saline, and Cloud Counties as well as several counties farther west. There is even an old breeding record from the woodlands along Sappa Creek in Decatur County. Arrives in late April, and most have departed by early August. A few migrants are seen later into the fall during September and October.

Field Notes: The hauntingly beautiful song of the Wood Thrush is often heard before the bird is seen. Because of habitat loss both on the nesting grounds in North America and the wintering grounds in Central and South America, Wood Thrushes are far less numerous and have a more restricted range in Kansas than they did a hundred years ago. Discovering one today is a good find, although they are still locally common east of the Flint Hills in areas where large enough tracts of mature forest still exist.

Toxostoma rufum

Brown Thrasher

Field Identification: Brown Thrashers are rufous brown above and white with long black streaks below. There are two white wing-bars. The yellow eyes are prominent. They have a long tail and a long, slightly curved beak. The Brown Thrasher song is given in phrases that mimic the songs of other birds, each repeated two or three times. Size: Length 11.5 inches; wingspan 13 inches.

Habitat and Distribution: Found statewide in brushy and edge habitats. There are records from all 105 counties. Frequently observed in urban parks and yards, where they often remain to nest.

Seasonal Occurrence: Migrant and summer resident. The nests are easy to find, and it is a breeding species in 102 counties. Frequently nests in suburban yards if there is suitable shrubbery available. Arrives in April and departs in October. A few individuals remain for the winter in most years, especially in southern Kansas.

Field Notes: In the spring and early summer, Brown Thrashers sing persistently from high, exposed perches, making them easy to observe. Many Brown Thrashers migrate through Kansas in spring and fall on their way to and from nesting grounds in the northern states and Canada. These migrants often travel in loosely cohesive flocks, and large numbers can be seen in May and September.

Curve-billed Thrasher

Toxostoma curvirostre

Field Identification: Uniformly grayish brown above. Underparts are white with large, blurry spots, especially on the breast. The long, decurved bill and bright yellow eyes are prominent. The tail is long with white tips that can be seen on flying birds. The typical call is a loud two- or three-note *whit-wheet*. Size: Length 11 inches; wingspan 13 inches.

Habitat and Distribution: Found in semiarid grasslands and sand sage prairies in the southwestern counties. Also frequents juniper and cedar plantings. It is not shy and is frequently seen in towns such as Elkhart and Hugoton. There are records from 29 counties. They are especially fond of cholla cactus and brush piles. Even a single isolated cholla in southwestern Kansas is worth checking.

Field Notes: This is one of several birds of the southwestern deserts whose range extends into southwestern Kansas. This species is often high on the target list for birders visiting southwestern Kansas. Look for them around Elkhart and Hugoton. In recent years, the cemetery at Hugoton has been a reliable location, as has the River Road in Kearney and Hamilton Counties.

Seasonal Occurrence: It is seen year-round, but populations seem to fluctuate annually. There are confirmed breeding records from six southwestern counties northeast to Finney and Gray Counties. It probably nests at least occasionally in most of the southwestern counties. In fall and winter, it occasionally wanders north and east beyond its typical range.

on nest in cholla

© Pete Janzen

Turdus migratorius

American Robin

© Judd Patterson

Field Identification: The adult is orange below and gray above, with a black head and a bright yellow bill. There are white markings around the eyes. The throat is white with black streaks. Juveniles are scaly above and spotted below, retaining this appearance until September. Size: Length 10 inches; wingspan 17 inches.

Habitat and Distribution: Found statewide in towns, cities, and a variety of rural habitats. There are records from all 105 counties.

Seasonal Occurrence: Present throughout the year, although considerable seasonal population turnover occurs. Nesting has been confirmed in all 105 counties.

Field Notes: American Robins are familiar to most Kansans. Even on the mostly treeless High Plains, they can be found in nearly every town and farmyard in the summer. Those seen in Kansas during the winter are birds that nest in Canada and the northern US. These wintering robins congregate in large flocks in areas with abundant cedars or fruit trees, remaining until the food has been exhausted. Many found in Kansas during the summer migrate south for the winter, returning in the spring. During the spring, the chorus of singing robins begins well before dawn.

immature

Gray Catbird

Dumetella carolinensis

Field Identification: One of the mimic thrushes, Gray Catbirds are slate gray except for black cap, black tail, and rufous undertail coverts. They are named for their mewing call note. The song is long and complex, usually with some mewing or squeaking notes. Catbirds do not repeat phrases as mockingbirds and thrashers do. Size: Length 8.5 inches; wingspan 11 inches.

Habitat and Distribution: Found statewide, but rare in the western third. Inhabits densely brushy habitats. There are records from all 105 counties.

Seasonal Occurrence: Migrant and summer resident. Spring migrants arrive in late April and are most abundant in mid-May. Many remain to nest. Nesting has been confirmed in 66 counties, mostly in central and eastern Kansas. Only a few counties in western Kansas have nesting records. Fall migrants arrive in September and are seen through mid-October.

Field Notes: Catbirds are more secretive than the closely related thrashers and mockingbirds. They only reluctantly leave the security of dense thickets. Even when males are in full song, they usually sing from a concealed perch. However, Gray Catbirds are innately curious, and a patient birder can usually get a good look as they furtively approach to investigate the intruder.

Myadestes townsendi

Townsend's Solitaire

Field Identification: This is a unique thrush of the west. Its entire body is gray except for buffy wing markings, a white eye-ring, and white edges on the long tail. The broad buffy wing-stripe is prominent on flying birds. They can be confused with the Northern Mockingbird at first glance, but mockingbirds are pale white below, with bright white wingpatches, and lack an eye-ring. It is often seen perched high on the top of a cedar tree. Size: Length 8.5 inches; wingspan 14.5 inches.

Habitat and Distribution: Found in stands of pine and cedar trees in western Kansas. Most records are from west of the Flint Hills, but it occasionally wanders farther east. There are records from 87 counties, including about 20 eastern counties.

Seasonal Occurrence: Winter resident. Most are seen from early October through early February. A few remain in spring as late as May.

Field Notes: Townsend's Solitaires nest in the pine forests of the Rocky Mountains, and like the Mountain Bluebird, they move onto the plains during the winter. True to their name, they are usually seen singly. Older cemeteries with abundant mature cedars and pines are good places to look for them. Sometimes they give a single bell-like call note. They are retiring in nature and therefore easy to overlook. Reliable locations to look for them include parks and cemeteries in Colby, Garden City, Hays, and Jetmore.

Northern Mockingbird

Mimus polyglottos

Field Identification: This mimic thrush is flashy and vocal. It has a long and slender appearance. The plumage is gray above and white below. The wings are black with large white wing-patches that are conspicuous in flight. The tail is black with white edges. Size: Length 10 inches; wingspan 14 inches.

Habitat and Distribution: Found statewide in edge habitats and rural areas with scattered shrubs or thickets. There are records from all 105 counties. Often observed near human dwellings.

Seasonal Occurrence: Migrant and permanent resident. Common from early spring through late fall. Widespread as a nesting species. There are 81 counties with confirmed nesting records. The greatest numbers are present in Kansas from April through September. Most depart south in late fall, but some remain throughout the winter, especially in the east and south.

Field Notes: Northern Mockingbirds are conspicuous in summer. They loudly mimic the songs of other birds in repetitive fashion from high, exposed perches. When they fly, the white wing-patches are obvious. At first glance, Northern Mockingbirds can be confused with either of the two species of shrikes that occur in Kansas. Shrikes can be differentiated by their faster wingbeats and short, hooked beaks.

Lanius ludovicianus

Loggerhead Shrike

Field Identification: Loggerhead Shrikes have a black mask and wings, gray back, and white breast. Distinctive white wing-patches are seen as the bird flies away with choppy, rapid wingbeats. Juveniles seen in late summer are similar but are brownish, with fine barring on the breast. The beak is hooked on the upper mandible like that of a hawk. Size: Length 9 inches; wingspan 12 inches.

Habitat and Distribution: Found in open country, especially prairies and pastures with scattered small trees. Often seen perched on utility wires and fences. There are records from all 105 counties.

Seasonal Occurrence: Permanent resident, but populations shift according to the season. Summer resident statewide, with confirmed breeding records from 96 counties. Smaller numbers are present through the winter in most of the state. Many move through Kansas to and from breeding grounds north of Kansas, so numbers are typically highest during spring migration in March and April and fall migration in September and October.

Field Notes: Shrikes prey on insects, small birds, and rodents. They often impale their prey on long thorns, earning them the name "butcher bird." In recent years, their numbers have dropped significantly, both nationally and in Kansas.

Northern Shrike

Lanius borealis

Field Identification: This shrike is similar in appearance to the Loggerhead Shrike but is larger and paler, with a longer bill. The black mask is narrower than that of the Loggerhead, and the black color does not meet above the bill. Birds in their first winter plumage are brownish in color, have a less distinct mask, and are finely barred below. Size: Length 10 inches; wingspan 15 inches.

Habitat and Distribution: Found in open country in northern and western Kansas, roughly north and west of a line from Barber County to Marshall County. Rare but regular east of this line. There are records from 98 counties.

Seasonal Occurrence: Winter resident. Arrives in October and departs in March. Most common between mid-November and late January. Its numbers in Kansas fluctuate from year to year.

Field Notes: This shrike breeds in Canada and Alaska and moves into the lower 48 states during winter. Any shrike seen in northern or western Kansas during the winter is likely to be this species. All shrikes seen in the winter should be carefully identified.

Bombycilla cedrorum

Cedar Waxwing

Field Identification: The swept-back crest, black mask bordered with white, and broad yellow band at the tip of the tail are the most distinctive markings of this species. Otherwise, it has a gray back and wings, warm brown head and chest, and yellow wash on the lower belly. Size: Length 7 inches; wingspan 12 inches.

Habitat and Distribution: Found statewide in open woodlands, cities, and parks where berries are present. Often seen in stands of Eastern Red Cedar in years when the berry crop has been good. There are records from all 105 counties.

Seasonal Occurrence: Winter resident, migrant, and rare summer resident. Winter residents arrive in October and depart in early June. In most years, the greatest numbers are northbound flocks seen in May. A few remain to nest, and there are confirmed nesting records from 28 counties, mostly in the eastern half but also from Scott and Norton Counties in the west.

Field Notes: Flocks of Cedar Waxwings roam widely during the winter, feeding on berries of cedars and other trees, consuming fruit as large as cherries and crabapples. When food is plentiful, waxwings are often abundant. In spring, they also feed on tree buds. Large numbers sometimes remain through May and early June, taking advantage of the mulberry crop. Their calls are thin, high-pitched notes and are given frequently, especially when the flocks are preparing to take flight or land.

Yellow-breasted Chat

Icteria virens

Field Identification: This unique bird of brushy habitats was formerly classified as a species of warbler but has now been placed in its own family. It is olive green above, with a yellow breast, white belly, prominent white spectacle markings, and a long tail. The song and calls are a varied and often bizarre combination of sounds with long pauses in between phrases. Size: Length 8 inches; wingspan 10 inches.

Habitat and Distribution: Found statewide, with records from 103 counties. It occurs in locations where there are extensive stands of brushy shrubs, such as Sandhill Plum with adjacent prairie habitat. Migrants are seen statewide, sometimes in unexpected habitats such as cemeteries and parks. Breeding birds in summer have a patchy, localized distribution.

Seasonal Occurrence: Migrant and summer resident. There are confirmed nesting records from 19 counties and probable nesting from 12 additional counties. Arrives in spring in early May. The greatest number are seen in June and July, when males are singing. Fall migrants are seen through mid-September.

Field Notes: Yellow-breasted Chats are a unique species with a lot of personality. Their songs are a mix of scolds, hollow-sounding single notes, mimicked songs of other species, and a variety of other unusual sounds, often given from a concealed perch. In spring, the male gives a unique display flight, flying high and then descending in a fluttering spiral while singing vigorously. Chats are now absent from many areas where they were formerly found in Kansas. Public lands that support these birds include Elk City Reservoir, Norton WA, Historic Lake Scott State Park, Sand Hills State Park, and Wilson SFL.

Ground Dwellers and Sparrows

Most of the birds in this chapter are sparrows. Some sparrow species are winter residents, some are summer residents, and others occur in Kansas only as migrants in spring and fall. Each species arrives and departs at different times of the year, although there is considerable overlap. In the Birding Resources chapter at the end of this book, there is a detailed bar graph, compiled by Chuck Otte, showing the abundance, distribution, and seasonal occurrence of all Kansas sparrow species. The bar graph is a handy quick reference when you are trying to narrow down the possibilities for sparrows seen at a given time of the year or in a certain part of the state. Please note that a few species shown in the graph are considered too rare to be included in this book.

Grasshopper Sparrow

Horned Lark

Eremophila alpestris

Field Identification: This common ground-dwelling bird of agricultural areas is sandy brown above and white below, with a black-and-yellow pattern on the face and breast. Several subspecies of Horned Lark can be seen in Kansas. On some, the black-and-yellow pattern on the head is bold and colorful, and on others, this pattern is pale and subdued. When the bird is flushed into flight, look for the black tail with thin white edges. Size: Length 7 inches; wingspan 12 inches.

Habitat and Distribution: Statewide in open areas, especially short prairies, plowed fields, and winter wheat. Has been reported from all 105 counties.

Seasonal Occurrence: Permanent resident. Breeding has been confirmed in 73 counties, and it probably nests in all counties at least occasionally. During winter, huge numbers move into Kansas from breeding grounds farther north.

Field Notes: Horned Larks are often seen when they are flushed from the side of the road by approaching vehicles. In winter months, flocks of Horned Larks in the tens of thousands of birds are often reported, especially on the vast expanses of winter-wheat fields in central and western Kansas. These impressive flocks often include huge numbers of Lapland Longspurs.

nonbreeding

Anthus rubescens

American Pipit

Field Identification: This long and slender ground-dwelling bird of open country is light bluish gray above and pale yellow with streaks below and has a thin, pointed bill. When foraging, it walks casually with head held high. In flight, look for the white outer tail feathers. Size: Length 6.5 inches; wingspan 10.5 inches.

Habitat and Distribution: Found statewide in plowed fields and prairies and along shorelines. There are records from all 105 counties.

Seasonal Occurrence: Spring migrant during March and April; fall migrant from late September through November. A few occasionally remain in Kansas through the winter.

Field Notes: Hardier than many migratory songbirds, American Pipits are often seen in early March and late October. Migrating flocks of hundreds of pipits are sometimes seen in muddy plowed fields. They forage on the ground and are often encountered on muddy shorelines at wetlands and reservoirs and in grassland habitats. Learn the two-note flight call of this species, as they are often heard as they fly overhead. They are never seen perching on fences or branches.

Sprague's Pipit

Anthus spragueii

Field Identification: Sprague's is related to the American Pipit but prefers grassland habitat. It is brown and pale overall, with a streaked back, short tail, pale pink legs, and mostly unmarked face. The large, dark eye is prominent on the pale face. Size: Length 6.5 inches; wingspan 10 inches.

Habitat and Distribution: Always found in prairies with short grass. Migrant throughout Kansas but most often reported from the central part of the state. There are records from 71 counties.

Seasonal Occurrence: Spring migrant from late March through late April with a peak in mid-April; fall migrant from late September through late October with a peak in mid-October.

Field Notes: Sprague's Pipits nest on the northern prairies of the US and Canada and winter in southern Texas and Mexico. They are seen in Kansas only during migration and can be common in proper habitat, although they are seldom observed. Hiking through mowed tallgrass or shortgrass prairies in April and October may flush several of these birds. They do not flush until they are nearly underfoot. As they take flight, they utter a loud, distinctive squeaking call. They typically fly for a short distance and then abruptly plunge back to the ground. Finding one perched on the ground is difficult and usually requires great patience.

Calcarius lapponicus

Lapland Longspur

Field Identification:
Longspurs are a unique family of birds that inhabit grasslands and open fields. Of the four longspur species found in Kansas, this is the most abundant by far. It is unmistakable in summer, but in Kansas, it is usually seen in its winter plumage. Unlike other species of longspur in winter plumage, it has bold streaking on the flanks and rufous coloring on the wings. Winter males have a black patch on the breast; females have black breast streaks. Also note the strong black facial markings and the clean-cut white outer tail feathers. The unique rattling calls are given frequently both in flight and on the ground.
Size: Length 6.25 inches; wingspan 11.5 inches.

Habitat and Distribution: Statewide in open country, especially winter-wheat fields and tilled fields. Most abundant in western and central Kansas but also fairly common throughout the east. There are records from 104 counties.

Seasonal Occurrence: Winter resident. Arrives in November and departs in late February and early March.

Field Notes: The vast expanses of central and western Kansas wheat country can seem incredibly bleak and lifeless in the winter, but flocks of Lapland Longspurs, often in association with Horned Larks, can number in the hundreds or thousands in this habitat. These flocks restlessly swirl over the fields, noisily rattling as they fly. After major snowstorms, Lapland Longspurs are often found along roadsides and can be observed at close range.

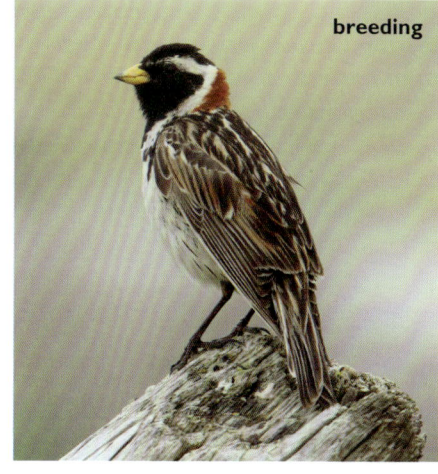

breeding

Smith's Longspur

Calcarius pictus

Field Identification: In winter plumage, this longspur of the eastern prairies has buffy underparts with thin streaks on the breast and flanks. It shows prominent white patches on the wings when in flight, unique among longspurs. Smith's have about twice as much white on the outer tail feathers as Laplands. Males begin to attain breeding plumage in late February and March and are mostly orangish with a strong black-and-white pattern on the head. Their rattling calls are harder- and drier-sounding than those of the Lapland. Size: Length 6.25 inches; wingspan 11.5 inches.

Habitat and Distribution: Grasslands of central and eastern Kansas, especially in and east of the Flint Hills. There are records from 59 counties.

Seasonal Occurrence: Migrant and rare winter resident. Arrives in October and departs in April. Most numerous in March and April, when northbound migrants are moving through Kansas.

Field Notes: Smith's Longspurs are most often observed at mowed prairies in the Flint Hills, especially in the southern half of Kansas. They are sometimes observed in flocks flying overhead in this habitat. When on the ground, their camouflage is superb until they flush into flight a few feet ahead of approaching observers. During migration, they are fairly common in the southern Flint Hills, but this region is almost entirely privately owned, and access is difficult. The best-known place to observe them is the area of mowed fields located just south of Lyon County SFL. This area is privately owned, but the birds can often be seen in flight from the entrance road to the lake.

Rhynchophanes mccownii

Thick-billed Longspur

Field Identification: This is a short-tailed longspur of open country. In all plumages, look for the mostly white tail with black tips and the stouter bill structure. Breeding males are largely grayish white with a black crown, black crescent on the breast, dark gray bill, and rust-red shoulder patch. Winter males have pink bills, are mostly light brown with white underparts, and usually retain some black on the breast. Females have pink bills and streaked crowns and otherwise are pale brown above and white below. Size: Length 6 inches; wingspan 11 inches.

Habitat and Distribution: Shortgrass prairies and cultivated fields on the High Plains. There are records from 44 counties, almost all in the western half of the state.

Seasonal Occurrence: Spring and fall migrant, rare in winter in the southwest. Spring migrants are seen in March and April, fall migrants in October and November. A few linger in the southwestern border counties in the winter months.

Field Notes: Unfamiliar to most Kansas birders, this longspur is likely more common than the relatively small number of records indicate. Look for them in early spring and late fall, often in association with Chestnut-collared Longspurs and Horned Larks. Most of them winter south of Kansas, but in most years, they are recorded on the Cimarron National Grassland Christmas Bird Count.

Chestnut-collared Longspur

Calcarius ornatus

Field Identification: The breeding-plumaged male has black underparts, a yellowish throat, and a chestnut-red nape. Winter birds seen in flight have a mostly white tail with a black triangle at the tip and are paler and less strongly marked than Lapland Longspurs. The winter male has light black bars on the breast, and the female has faint streaking on the flanks. Size: Length 6 inches; wingspan 10.5 inches.

breeding male

Habitat and Distribution: Found in semiarid grasslands and farmlands in central and western Kansas. It is rare in the east. There are records from 70 counties.

Seasonal Occurrence: Migrant and winter resident. Fall migrants are seen from early October through late November. Spring migrants are seen from early March through mid-April. A few are reported in the winter months in some years.

Field Notes: This grassland bird is unfamiliar to many birders. Chestnut-collared Longspurs are seen most often in the early spring, especially when cold fronts delay their migration. Rural roads in Pawnee County have been a reliable location for finding these birds in March, but they are possible anywhere in the western half of the state. Some remain for the winter in the southwest, and they are usually recorded on the Cimarron National Grassland Christmas Bird Count.

nonbreeding male

Pipilo maculatus

Spotted Towhee

Field Identification: The male has an all-black head, black back and wings with numerous white spots, bright red eyes, broad rufous flanks, and a white belly. The female is nearly identical, but the head is soft gray brown instead of black. White corners on the tail are prominent on flying birds. The call note is a harsh mewing sound and is often the first clue to its presence. Size: Length 8.5 inches; wingspan 10.5 inches.

Habitat and Distribution: Found statewide, but most common west of the Flint Hills. Prefers brushy thickets and woodland edges. There are records from all 105 counties.

Seasonal Occurrence: Migrant and winter resident. A small summer nesting population is found along the Nebraska border in the western half of Kansas, where there are confirmed breeding records from 14 counties. Fall migrants arrive in late September and are most abundant in October. Many remain for the winter months. Spring migrants are numerous in April and May.

Field Notes: Spotted Towhees are generally seen singly or in small groups in dense brush, especially berry thickets. They are good at remaining concealed but are also inquisitive, and a few squeaking or "spishing" noises may draw them into the open. When they fly, the black-and-white pattern on the back is prominent.

female

Eastern Towhee

Pipilo erythrophthalmus

Field Identification: The Eastern Towhee is closely related to the Spotted Towhee and looks like it in many respects. Males have an all-black head and back with a single white wing-patch. Females have the same pattern, but the black plumage is replaced with warm brown. The distinctive *chewink* call is unlike the raspy call of the Spotted Towhee. Size: Length 8.5 inches; wingspan 10.5 inches.

Habitat and Distribution: Found mostly from the Flint Hills eastward in brushy areas. Sometimes seen farther west during migration. There are records from 81 counties.

Seasonal Occurrence: Present all year. Summer residents arrive in mid-March and remain through September. Additional migratory individuals are seen in April and October. Some remain through the winter months. Confirmed nesting has been reported from 20 counties from the Flint Hills east to the border and 12 additional eastern counties.

Field Notes: In spring and summer, Eastern Towhees sing their loud *drink-your-tea* song from high, exposed perches on brushy hillsides of eastern and northeastern Kansas. During winter, they are found in similar habitats but are less conspicuous. Eastern and Spotted Towhees were formerly considered to be the same species, then called the Rufous-sided Towhee. Ornithologists and birders are still defining the range limits of these two species in Kansas.

female

© Lillis Boyer

282

Peucaea cassinii

Cassin's Sparrow

Field Identification: This western Kansas sparrow is plain brownish gray with black spots on the back, an unmarked breast, and a relatively long tail. In flight, it shows white on the tips of the tail feathers. Size: Length 6 inches; wingspan 8 inches.

Habitat and Distribution: This sparrow is found in the semiarid grasslands of southwestern Kansas, especially hilly areas where sage and yucca are present. It also likes sandy prairies with thickets of aromatic sumac and other shrubs. It occurs mostly southwest of a line between Cheyenne and Barber Counties. Populations fluctuate from year to year in response to rainfall. During drought years, they are found farther east and north in the state. There are records from 48 counties in south-central and western Kansas.

Seasonal Occurrence: Summer resident. Arrives in April and departs in September. Most eBird records are from May and June, when males are singing. There are confirmed nesting records from 15 counties in the southwestern quarter of the state.

Field Notes: In late spring and early summer, the display flight of the male can be observed in areas with Cassin's preferred habitat. He flies 20 to 30 feet in the air and then slowly descends in a stiff-winged fluttering flight while singing a beautiful trilling song. After the nesting season, Cassin's fall silent and are easy to overlook.

American Tree Sparrow

Spizelloides arborea

Field Identification: This common winter sparrow has an unmarked breast with a single dark central spot, a grayish face with a red crown, a brown back, a long tail, and usually some brownish color on the flanks. Size: Length 6 inches; wingspan 9 inches.

Habitat and Distribution: Found statewide in prairies with interspersed thickets, weedy fields, and brushy areas. There are records from all 105 counties.

Seasonal Occurrence: Winter resident. Arrives in late October and departs in mid-March.

Field Notes: Winter flocks often number in the hundreds of birds and frequently include other wintering sparrows, especially Harris's Sparrows and Dark-eyed Juncos. Their tinkling call notes fill the air on chilly winter days. Despite being common in winter, they are not typically seen at bird feeders unless major ice or snowstorms force them to seek secondary food sources.

Spizella passerina

Chipping Sparrow

Field Identification: This sparrow has a cap that is bright red in spring and summer and dull brown and streaked in fall and winter. Look for the white line above the eye and a dark line through the eye. The bright white line above the eye becomes dull in fall. It is also streaked brown above, with thin white wing-bars, a gray rump, and unstreaked grayish-white underparts. Size: Length 5.5 inches; wingspan 8.5 inches.

Habitat and Distribution: Found statewide in towns, parks, woodlands, agricultural areas, and a variety of other habitats. Nesting birds typically select residential areas with scattered trees, especially pines. In the fall, they prefer weedy fields and other rural habitats. There are records from all 105 counties.

Seasonal Occurrence: Migrant and summer resident. Spring migrant from mid-April through May; fall migrant from late September through early November. This is a common nesting species in northeastern Kansas and sporadically elsewhere in the eastern half of the state. Confirmed nesting has been recorded in 32 counties, all in the eastern half.

Field Notes: In the spring, this is one of the most common and conspicuous migratory songbirds in Kansas. Their mechanical, trilling songs fill the morning air as flocks of migrating birds feed on the seeds of dandelions and other plants.

Clay-colored Sparrow

Spizella pallida

Field Identification: This species resembles the Chipping Sparrow, to which it is closely related. It differs in that the rump is brown, the crown is brownish with a median stripe, there is no black line through the eye, and there is a broad ear-patch that is partially bordered with black or brown. In the fall, they are brownish on the head and breast, unlike the grayer Chipping Sparrows. Size: Length 5.5 inches; wingspan 7.5 inches.

Habitat and Distribution: Statewide in towns, parks, open woodlands, and brushy areas. There are records from all 105 counties.

Seasonal Occurrence: Spring migrants are seen from mid-April through mid-May, fall migrants from early September through late October.

Field Notes: Migrating Clay-colored Sparrows are often seen in mixed flocks with Chipping Sparrows in both spring and fall. During the spring migration, listen for their song, which consists of two or three slow, monotone buzzing notes. In spring, they usually arrive about a week later than Chipping Sparrows. They are often seen feeding on seeds on the ground.

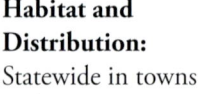

Spizella breweri

Brewer's Sparrow

Field Identification:
This is a close relative of the Chipping and Clay-colored Sparrows, which it resembles in size and shape. It has a pale face without any strong pattern. The eye-ring is a useful field mark. The crown is brown with black streaks and does not have the white median stripe seen on Clay-colored. Size: Length 5.5 inches; wingspan 7.5 inches.

Habitat and Distribution:
Found only in the western quarter of Kansas. During migration, it can be seen in sage habitats as well as brushy and weedy areas. There are records from 25 western counties.

Seasonal Occurrence: Migrant and rare summer resident. Spring migrants are seen in April and May, fall migrants primarily in September and early October. In summer, a few remain to nest in high-quality sand sage habitat in Morton County. Has also been confirmed as a nesting species in Finney County at least once.

Field Notes: This is one of the least-likely-to-be-seen species in Kansas that was included in this book. Because it looks a lot like Chipping and Clay-colored Sparrows, it is easy to overlook and is probably more common in migration than the limited number of records suggests. In western Kansas, check flocks of migratory sparrows carefully for this species. The small nesting population in Morton County represents the southeastern extreme of the nesting range for this species in North America.

Field Sparrow

Spizella pusilla

Field Identification: A sparrow of brushy habitats, Field Sparrow has a prominent eye-ring, pink bill, and red crown. They are brown above, with unmarked grayish-white underparts and a long tail. Size: Length 6 inches; wingspan 8 inches.

Habitat and Distribution: Found in grasslands, prairies, and pastures that are interspersed with thickets. It is common in the eastern two-thirds of Kansas and rare and local in the west. There are records from all 105 counties.

Seasonal Occurrence: Migrant, common summer resident, and rare winter resident. Arrives in late March and departs in late October. It is rare but regular in winter. There are confirmed nesting records from 57 counties. Only a few counties in the western third of the state have nesting records.

Field Notes: After arriving at their nesting territories in early spring, Field Sparrows immediately begin to give their ascending whistled song from the tops of dogwood and plum thickets. They continue to sing until early August. During migration and winter, they often join other species of sparrows in mixed-species flocks. In winter, beware of confusion with the more abundant American Tree Sparrow. If it has an eye-ring and a pink bill, it is a Field Sparrow.

Pooecetes gramineus

Vesper Sparrow

Field Identification: This streaked sparrow of open country has a brown back with black streaks and white underparts with light streaking on the breast and flanks. The best field marks are the white eye-ring and white outer tail feathers, which are obvious on flying birds. At close range, a reddish shoulder is visible. Size: Length 6 inches; wingspan 10 inches.

Habitat and Distribution: Statewide in grasslands and agricultural areas. There are records from all 105 counties.

Seasonal Occurrence: Spring migrants are found from March through early May with a peak in early April, fall migrants from September through mid-November with a peak in the first two weeks of October. A few remain as a nesting species in Brown and Doniphan Counties in the extreme northeast and Morton County in the extreme southwest. There are nesting records from five counties.

Field Notes: Vesper Sparrows are one of the earlier-arriving migratory sparrows in spring. They are often seen in large numbers during the fall migration in western Kansas. In mid-September, hundreds can sometimes be seen during a day on the Cimarron National Grassland. They feed along the roads and fly a short distance when disturbed by vehicles, sometimes perching briefly on low fence wires, where they can be easily studied. During migration, they are often seen in flocks that include other species of sparrows, especially Lark and Savannah Sparrows.

Lark Sparrow

Chondestes grammacus

Field Identification: The intricate head pattern of black, white, and red along with the single spot on the unstreaked breast make this species easy to identify. The white-tipped tail feathers are a useful field mark on flying birds. Size: Length 6.5 inches; wingspan 11 inches.

Habitat and Distribution: Migrating birds occur in a variety of habitats, including parks and other urban situations. Lark Sparrows nest throughout Kansas in shortgrass habitats. In summer, they are most numerous in the western part of the state. They do not inhabit dense grasslands but prefer more sparsely vegetated areas, often associated with some combination of habitat disturbance, grazing, mowing, rocky areas, or poor soils. There are records from all 105 counties.

Seasonal Occurrence: Migrant and summer resident. Spring migrants arrive in April, and in fall, they depart in September and October. This is a widespread nesting species in Kansas. Has been confirmed as a breeding species in 95 counties.

Field Notes: In the Red Hills region, where appropriate habitat is widespread, this is one of the most abundant species in summer. The song is complex and variable and can be heard at a considerable distance. In spring and early summer, look for males noisily performing their courtship displays on the ground, often on dusty rural roads.

Calamospiza melanocorys

Lark Bunting

Field Identification: Summer males are all black, and winter males are brownish with prominent black streaking. Males have large white wing-patches at all seasons. Females look like winter males but have lighter streaking and less white in the wing. Size: Length 7 inches; wingspan 10.5 inches.

Habitat and Distribution: Found in grasslands and fallow fields of western Kansas and east to Barber and Ellsworth Counties. Migrants are occasionally seen farther east. There are records from 84 counties.

Seasonal Occurrence: Migrant and summer resident. Arrives in April and departs in September. Drought years often push nesting populations eastward to the edge of their range. During some years, small numbers remain for the winter in southwestern Kansas. There are confirmed nesting records from 34 counties, almost all of which are in the western half of the state.

Field Notes: Lark Buntings are birds of the western Kansas High Plains, where they nest in loose colonies. Like Cassin's Sparrows, they perform unique display flights, singing as they climb into the air and slowly descend. During migration, they are sometimes seen in flocks of impressive size, but more often in smaller flocks of 10 to 20 birds.

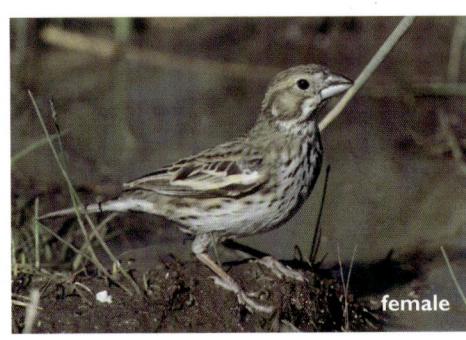

female

Savannah Sparrow

Passerculus sandwichensis

Field Identification: This streaked sparrow of grasslands has a short, notched tail; fine streaking on flanks and breast; and a white belly. There are several subspecies that vary in appearance. The back and wing colors vary from gray to warm brown. The yellow line above the eye can appear pale and washed out or bright. Look for the short tail and the yellowish line above the eye to separate this species from the Vesper Sparrow. Size: Length 5.5 inches; wingspan 6.75 inches.

Habitat and Distribution: Found statewide in grasslands and agricultural areas. Has been recorded in all 105 counties.

Seasonal Occurrence: Spring migrants are seen from mid-March through early May with a peak in late April, fall migrants from mid-September through early November with a peak in mid-October. Small numbers remain for the winter in the southernmost counties.

Field Notes: Savannah Sparrows are common migrants found in habitats similar to those preferred by Vesper Sparrows. The two species often occur together as they migrate through Kansas. Savannahs are often tame and easily viewed, but when they are in taller grasslands, they fly for short distances and then plop back into the grass, making it difficult to see them well.

Ammodramus savannarum

Grasshopper Sparrow

Field Identification: This grassland sparrow has a flat head, large bill, short tail, and buff face. A yellow-orange spot is often visible in front of the eye. Another yellow spot is often visible at the bend of the folded wing on perched birds. The black crown has a narrow white line down the center. The breast is buff and unstreaked, and the belly is white. Size: Length 5 inches; wingspan 7.75 inches.

Habitat and Distribution: Found statewide, exclusively in grassland habitats. Has been recorded in all 105 counties.

Seasonal Occurrence: Summer resident. Arrives in April and departs in October. There are confirmed nesting records from 91 counties.

Field Notes: Look for these prairie birds perched atop clumps of grasses or on fence wires. The high, thin, trilling song sounds more like an insect than a bird and escapes the hearing of some people. Grasshopper Sparrows are one of the most widespread nesting sparrows in Kansas. Despite concerns about habitat loss and declining populations, Kansas remains a stronghold for this and other grassland species in need of conservation.

Henslow's Sparrow

Centronyx henslowii

Field Identification: This short-tailed sparrow has a flat head, large bill, and narrow necklace of streaks on a buffy chest. The head is mostly olive in color, and the wings have reddish tones. The song is a faint two-note *tsi-lick*, which carries for a surprising distance. Size: Length 5 inches; wingspan 6.5 inches.

Habitat and Distribution: Found in tall, dense, unburned grasslands of eastern Kansas and west to Harvey and Dickinson Counties. There are records from 54 counties, all in the eastern half of the state and mostly from the Flint Hills eastward.

Seasonal Occurrence: Summer resident. Arrives in April and departs in October. The greatest number are reported in late May and early June, when males are singing. Confirmed nesting has been reported in 14 counties and probable nesting in 10 additional counties.

Field Notes: Henslow's Sparrows require specific grassland conditions for nesting. They prefer prairies that have been unburned or ungrazed for at least two years. This habitat contains the accumulated dead vegetation that they require for nesting. Since this habitat is found in different locations from year to year, Henslow's Sparrows are as well. Konza Prairie, near Manhattan, is a reliable location for them, as is Shawnee SFL. Of all grassland birds, Henslow's Sparrows may have suffered the most dramatic national population decline. Kansas populations have done better in comparison.

Ammospiza leconteii

LeConte's Sparrow

Field Identification: The LeConte's Sparrow has a flat head, small bill, short tail, necklace of fine streaks on the buffy breast, white belly, streaked sides, and a purplish nape with fine streaks. The face is yellow orange with gray cheeks, and the black crown is divided by a white stripe. Size: Length 5 inches; wingspan 6.5 inches.

Habitat and Distribution: Found in central and eastern Kansas. Prefers wet meadows and grassy wetlands and sometimes drier tallgrass prairies. There are records from 78 counties.

Seasonal Occurrence: Spring migrant in April and May; fall migrant in October and November. Small numbers winter in wetlands of southern and eastern Kansas.

Field Notes: LeConte's Sparrows are seen far more often in fall than in spring. At grassy wetlands, such as the Baker Wetlands, Gurley Salt Marsh WA, Marais des Cygnes WA, Quivira NWR, and Slate Creek WA, they can be numerous in October and early November. They share habitat with Sedge and Marsh Wrens. In the fall, they are responsive to "spishing" sounds and may closely approach the observer.

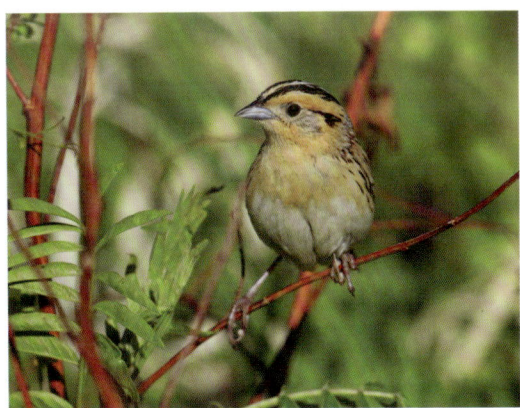

295

Nelson's Sparrow

Ammospiza nelsoni

Field Identification: This wetland species resembles LeConte's Sparrow but has an overall darker appearance. The face and breast are deeper orange, and the bill is longer. The crown is solid gray, lacking the white stripe of LeConte's. The nape is also solid gray, not purplish with streaks, as on LeConte's. Size: Length 5 inches; wingspan 7 inches.

Habitat and Distribution: Found in central and eastern Kansas. Inhabits wetlands with dense grass, especially prairie cordgrass. There are records from 38 counties, almost all from the eastern half of the state.

Seasonal Occurrence: Rarely seen in spring, with only a few records from late May. Fall migrants are more commonly observed, mostly in October and early November.

Field Notes: Nelson's Sparrows are uncommon migrants that are regularly seen during the fall at a few locations, including the Baker Wetlands, Cherokee Lowlands, Quivira NWR, and Slate Creek WA. They need to be carefully distinguished from LeConte's Sparrow. Pay special attention to the crown and nape markings. This species is seldom observed in the spring.

Passerella iliaca

Fox Sparrow

Field Identification: This is a large, colorful, long-tailed sparrow with a gray head. The crown, facial markings, wings, back, and tail are all reddish. The breast is white with bold reddish streaks and a central spot. Size: Length 7 inches; wingspan 10.5 inches.

Habitat and Distribution: Found in central and eastern Kansas. Prefers thickets and brushy areas, usually in or near wooded areas, especially where coralberry is abundant. It is rare in the west. There are records from 94 counties.

Seasonal Occurrence: Migrant and winter resident. Arrives in mid-October and departs in April. It is most numerous in March and November as migrants increase the numbers of wintering birds.

Field Notes: Fox Sparrows are always a treat to find on a winter day in a brushy woodland thicket. They are usually seen singly or in flocks of four or five birds. They frequently join other sparrow species in loosely cohesive foraging flocks.

Song Sparrow

Melospiza melodia

Field Identification: This long-tailed and rather dark sparrow has streaked flanks and breast, with a central breast spot. It also has black markings on the face and sides of the throat, a streaked back, and rusty wings. The coarse call note resembles that of the House Sparrow. Size: Length 6 inches; wingspan 8 inches.

Habitat and Distribution: Found statewide in cattail marshes as well as brushy and weedy areas. There are records from all 105 counties. Widespread winter resident, especially in the eastern half of the state. In summer, some remain to nest in the Glaciated Region in the northeastern corner of the state.

Seasonal Occurrence: Migrant, winter resident, and local summer resident. Arrives in late September and departs in April. Most numerous in late March and early November, when migrants supplement the wintering population. Winter populations fluctuate from year to year. There are confirmed nesting records from 10 counties in northeastern Kansas and along the Nebraska border west to Jewell County.

Field Notes: Song Sparrows are easily attracted by squeaking and "spishing" sounds. They actively twist and turn when flying through the cattails and weeds, eventually flying to another stand of vegetation with the tail held upward. As the days grow longer in early spring, their melodious song is often heard.

Melospiza lincolnii

Lincoln's Sparrow

Field Identification: The Lincoln's Sparrow is similar to the Song Sparrow, but the face is grayer, the breast is buffier, and the streaking on the breast and flanks is finer. Size: Length 6 inches; wingspan 7.5 inches.

Habitat and Distribution: Found statewide in weedy areas and densely brushy habitats, including prairies, urban parks, and woodland edges. There are records from all 105 counties.

Seasonal Occurrence: Spring migrants are found from March through May with a peak in late April and early May. Fall migrants are found from September through November with a peak in October. A few remain for the winter in eastern Kansas.

Field Notes: Lincoln's Sparrows are common migrants but are shy and retiring and so are not as conspicuous as other sparrows. Their wet-sounding call note is unlike the hoarse note of the Song Sparrow.

Swamp Sparrow

Melospiza georgiana

Field Identification: This wetland sparrow is dark overall, with dark-reddish wings and reddish-orange flanks. The face and underparts are gray, contrasting with the white throat. The metallic *chip* notes are often heard before the birds are seen. Size: Length 6 inches; wingspan 7 inches.

Habitat and Distribution: Found in central and eastern Kansas, almost always in wetlands or dense vegetation near water, especially cattails and sedges. There are records from 95 counties. Scarce in the western counties, where suitable habitat is limited.

Seasonal Occurrence: Migrant and winter resident. Arrives in October and departs in April. Most often seen in migration, especially in the fall. Some remain through the winter at wetland locations, mostly in the eastern half of Kansas.

Field Notes: Swamp Sparrows and Song Sparrows are often found in the same wetland locations. Unlike the less selective Song Sparrow, Swamp Sparrows are restricted to this habitat and are never found far from water. They are difficult to see well, but with patience, you can get a good look at this attractive sparrow.

Zonotrichia albicollis

White-throated Sparrow

Field Identification: White-throated Sparrows always have a neatly outlined white throat and a yellow spot above the base of the bill. The head is striped black and white, sometimes black and tan. They are grayish below, with brown wings and a striped back. Size: Length 7 inches; wingspan 9 inches.

Habitat and Distribution: Statewide in brushy areas associated with woodlands. Often seen in parks and yards, especially during migration. There are records from 103 counties. Much less numerous in western Kansas at all seasons.

Seasonal Occurrence: Migrant and winter resident. Arrives in October and departs in May. Most numerous during spring and fall migration in eastern and central Kansas. Some remain for the winter, especially in the eastern half.

Field Notes: These attractive sparrows are usually found in small flocks feeding on the ground under bushes or bird feeders. Their song is a series of clear notes often represented as *Old Sam Pea-bo-dy Pea-bo-dy Pea-bo-dy* and is commonly heard during the spring migration and occasionally even on cold winter days.

White-crowned Sparrow

Zonotrichia leucophrys

Field Identification: This distinctive sparrow has a head striped black and white, with a pink bill. It has brown wings and upper back, with gray underparts, rump, and tail. Juveniles are often seen in fall and winter. They have the same head pattern as adults, but the black on the head of adults is replaced with reddish brown on the juveniles. Size: Length 7 inches; wingspan 9.5 inches.

Habitat and Distribution: Found statewide in open country, especially in weedy or brushy fields. Also found in woodland edges, yards, and parks. There are records from all 105 counties.

Seasonal Occurrence: Migrant and winter resident. Fall migrants are most abundant in October and November. In late April and early May, northbound migrants are numerous. White-crowned Sparrows winter in large flocks in western Kansas and in much smaller numbers in the east.

Field Notes: White-crowned Sparrows are closely related to Harris's Sparrows, and during the winter, the two species are often observed together in mixed flocks. In western Kansas, winter flocks sometimes number in the hundreds.

immature

Zonotrichia querula

Harris's Sparrow

Field Identification: This is the largest sparrow in Kansas. Winter adults have a brown face with a mottled black crown, throat, and upper breast; a gray rump; brown wings; a white belly; and a pink bill. Immature birds are often seen in winter and are browner on the head, with more limited black markings on the breast and crown. Adults molt into breeding plumage

in spring before migrating north and have a solid black crown, a silvery-gray face, and an overall "cleaner" appearance. The black facial plumage gradually increases as the birds age. Size: Length 7.5 inches; wingspan 10.5 inches.

Habitat and Distribution: Found statewide in hedgerows, thickets, weedy areas, and woodland edge habitats. Occurs statewide during the winter but is most numerous in the central and eastern counties. There are records from all 105 counties.

Seasonal Occurrence: Winter resident. Arrives in October and departs in early May.

Field Notes: The winter range of Harris's Sparrows lies entirely within the Great Plains states. In most years, one of the south-central Kansas Christmas Bird

Counts records the highest number of this species in the nation. Their mournful, whistled songs interspersed with sharp *chink* notes enliven the winter landscape. Harris's Sparrows are exceptionally responsive to "spishing" sounds, flying to the tops of thickets to investigate. In early spring, they sing in earnest. In April and early May, Kansas populations swell as they are joined by migrants from the south.

immature

303

Dark-eyed Junco

Junco hyemalis

Slate-colored

Field Identification: Several subspecies of this sparrow with varying plumages occur in Kansas. All have dark tails with prominent white outer feathers. The Slate-colored is by far the most abundant subspecies and is uniformly slate gray with a white belly. The Oregon subspecies has a cleanly defined black hood, reddish back, reddish sides, and a gray rump. The Pink-sided subspecies has a slate-gray head, with pinkish-red back and flanks. The pink on the flanks wraps around the base of the wings. The Gray-headed subspecies is lighter gray than the Slate-colored, with black at the base of the bill and a bright red back. Size: Length 6.25 inches; wingspan 9.25 inches.

Habitat and Distribution: All juncos are found in a variety of habitats, usually with some brush or other nearby cover. The abundant Slate-colored subspecies occurs statewide. The Oregon subspecies is most likely to be found in western Kansas but can be found statewide in much lower numbers. The Gray-headed and Pink-sided subspecies are rare but regular in the far west. There are records from all 105 counties.

Slate-colored

Junco hyemalis

Dark-eyed Junco

female

Slate-colored

Oregon

Seasonal Occurrence:
Winter resident. Most arrive in mid-October and depart in mid-April. A few arrive as early as September and remain as late as mid-May.

Field Notes: Juncos, sometimes called snowbirds, are one of the most abundant birds in Kansas during the winter and are frequent visitors to bird feeders. Sorting out the various subspecies can be challenging, especially since they often hybridize and produce a variety of intermediate plumages. Their call notes are somewhat like those of the Northern Cardinal, but higher-pitched.

Gray-headed

Pink-sided

© David Butel

305

House Sparrow

Passer domesticus

Field Identification: The male in spring and summer has a black throat and bib, gray crown, bright chestnut nape and wings, and black-and-tan-striped back. All colors are muted in the winter. The female has a reddish-brown cap, light eye-line, striped back, and grayish-white throat and underparts. Size: Length 6 inches; wingspan 9.5 inches.

Habitat and Distribution: Found statewide, almost exclusively in association with humans in cities and towns, on farms, or near bridges and railroad tracks. There are records from all 105 counties.

Seasonal Occurrence: Permanent resident. Breeding has been confirmed in all 105 counties.

Field Notes: Along with the Rock Pigeon and the European Starling, this species was deliberately introduced to North America in the nineteenth century and swiftly became one of the most abundant bird species of the urban landscape. House Sparrows do no great damage to native bird populations and probably fill an ecological niche that would otherwise be unoccupied.

female

Cyanocitta cristata

Blue Jay

Field Identification: Unmistakable for any other species in Kansas, the Blue Jay is sky blue above with white wing markings, white below with black markings across the breast and a tall blue crest. Size: Length 11 inches; wingspan 16 inches.

Habitat and Distribution: Found statewide in woodlands and towns. There are records from all 105 counties. Most common in the east, gradually becoming less common westward.

Seasonal Occurrence: Summer resident across most of the state, with confirmed breeding records from 93 counties. Loose flocks of migrants move northward in late April and early May, southward in late September and early October. During the winter, most Blue Jays are seen in central and eastern Kansas, especially in oak forest habitats and in cities and towns.

Field Notes: While five species of jays have been recorded in Kansas at least once, most are rare winter visitors to the far western counties. This is the only jay likely to be seen in most parts of Kansas. Blue Jays get a lot of bad press because of their raucous calls and their habit of occasionally eating the eggs and nestlings of other songbirds, but most people enjoy their bright coloring and inquisitive habits. They are frequent visitors to bird feeders. Blue Jays often act as "sentries," announcing the presence of threatening predator species such as hawks, owls, and house cats.

Woodhouse's Scrub-Jay

Aphelocoma woodhouseii

Field Identification: This slender jay does not have a crest. The face, crown, wings, and tail are a subdued shade of blue. The back is gray, and the underparts are light gray. The throat and upper breast are white with dark gray streaks. Size: Length 11 inches; wingspan 15.5 inches.

Habitat and Distribution: There are records from 13 counties in the southwestern corner of Kansas and 4 additional counties from farther east. Most records are from brushy wooded habitats along the Arkansas and Cimarron Rivers or their tributaries. Also seen occasionally in cities and towns.

Seasonal Occurrence: This wanderer from the west is not seen every year in Kansas. It inhabits pinyon-juniper and scrub-oak habitats in Colorado, New Mexico, and Texas. Food-crop failures may trigger "invasion years," when some birds reach the southwestern counties of Kansas. Sightings occur between early September and early May but are most likely in late September and late October.

Field Notes: This is one of several irruptive western bird species that move into Kansas in some years in fall and winter. It is most likely to be seen on the Cimarron National Grassland in Morton County; along the Arkansas River in Hamilton County; or in towns such as Hugoton, Garden City, and Satanta.

Pica hudsonia

Black-billed Magpie

Field Identification:
Magpies are large and conspicuous, with a long, streaming tail; black head; and white underparts. The wings and tail are black with an iridescent blue-green sheen. In flight, the wings show flashy white primary feathers. The nasal, chattering calls are loud and obvious. Size: Length 19 inches; wingspan 25 inches.

Habitat and Distribution:
Found in open country and farmlands in the western half of Kansas. Occasionally wanders to the south and east during fall and winter. Requires isolated stands of trees for nesting. There are records from 83 counties. Does not occur in the southeastern counties.

Seasonal Occurrence: Uncommon permanent resident. There are confirmed breeding records from 54 counties, mostly in the western half of the state.

Field Notes: This colorful species brightens the landscape of western Kansas. Their huge, domed stick nests are often grouped together in loose nesting colonies. Magpies tend to travel in social groups that are both conspicuous and vocal. During the early 2000s, West Nile virus devastated the populations of several species of birds, including magpies, crows, and jays. Magpie numbers are slowly recovering, but they are still far less common than they were formerly.

American Crow

Corvus brachyrhynchos

Field Identification: These larger relatives of the jays are entirely coal black, including eyes, bill, and legs. Their bill is long and stout. Size: Length 17 inches; wingspan 39 inches.

Habitat and Distribution: Found statewide in a variety of rural and urban habitats. There are records from all 105 counties.

Seasonal Occurrence: Migrant and permanent resident. Some are present all year in central and eastern Kansas. In spring and fall, it is an abundant migrant, and flocks of thousands are seen. Spring migrants are most abundant in March and April and fall migrants in October. Many remain for the summer to nest. Confirmed nesting has been reported from 90 counties. Nearly absent from most of western Kansas during the winter.

Field Notes: American Crows are adaptable, intelligent, and familiar to most Kansans. They have benefited from changes brought by the "settlement" of the Great Plains. In winter, they gather in large roosts in some cities of central Kansas. They fly to surrounding rural areas during the day to feed on waste grain, returning to the warmth and safety of the city for the night. The roost at Wichita, not present every year, has sometimes been estimated in the tens of thousands of birds. Like Blue Jays, crows act as watchdogs and will noisily mob Great Horned Owls and other avian predators.

Corvus ossifragus

Fish Crow

Field Identification: The Fish Crow is nearly identical to the American Crow but has a proportionately longer tail and longer wings. The bill averages a bit smaller. It flies with quicker wingbeats than the American Crow. The voice is the most reliable difference and is a quieter *cah-cah* that is lower on the second note, with a much more nasal quality than the throaty *caw-caw* of the American Crow. Size: Length 19 inches; wingspan 36 inches.

Habitat and Distribution: Usually seen along wooded rivers, streams, and lakes in southeastern and eastern Kansas. Steadily expanding its range to the west and north. There are records from 44 counties, currently west to Reno County and north to Marshall County.

Seasonal Occurrence: Summer resident. Arrives in spring in mid-March and departs during August and September. A few sometimes remain later into the fall. Confirmed breeding has been reported from only six counties, but it probably breeds in other eastern and southeastern counties.

Field Notes: Fish Crows first started to appear in Kansas in the late 1980s and are expanding their range each year, generally following riparian corridors. They are expected to continue their range expansion in Kansas in future years. Young American Crows can sound somewhat like Fish Crows, but their calls do not have the same nasal tone.

Bobolink

Dolichonyx oryzivorus

Field Identification: Breeding males are black with a creamy yellow nape, a white rump, and large white wing-patches. Females are streaked black and brown above, are buffy white below, and have a black crown with a white median stripe. Fall birds are not often seen in Kansas. They look like spring females, but the underparts are a warmer yellow-buff color. Size: Length 7 inches; wingspan 11 inches.

Habitat and Distribution: Migratory species most often seen in eastern and central Kansas. They are rare in the west. In spring, they are almost always observed in alfalfa fields. In some years, they nest in wet meadows at the central wetlands and in tallgrass prairies of extreme northeastern Kansas. There are records from 87 counties.

Seasonal Occurrence: Migrant and sporadic summer resident. Spring migrants are seen almost exclusively in May. There are nesting records from Cheyenne Bottoms, Quivira NWR, and seven northeastern counties. The nesting population departs in July. Migrants are seen much less frequently in fall but can be seen through mid-October.

Field Notes: In early May, flocks of several hundred Bobolinks are sometimes encountered feeding on insects in alfalfa fields. The large, grassy field at the northwestern corner of Quivira is known as the "Bobolink Field" by birders. It is the southernmost breeding site for the species in the US. They do not nest there every year. During May and June, the males fly up from the grass and perform a showy display while singing their beautiful song.

female

© David Seibel

Agelaius phoeniceus

Red-winged Blackbird

Field Identification: The male is all black except for the red-and-yellow blaze on the wing. The female does not look like a blackbird except for the sharply pointed bill and is streaked above and below, with a white eye-line. Single females are not always immediately recognized as Red-winged Blackbirds and are sometimes a source of confusion. The sharp-pointed bill separates them from other possible species. Size: Length 9 inches; wingspan 13 inches.

Habitat and Distribution: Found statewide in wetlands, agricultural areas, feedlots, roadsides, and pastures. There are records from all 105 counties.

Seasonal Occurrence: Permanent resident. During winter, the population is at its largest, with flocks from northern states and Canada joining residents. Common statewide in summer. Nesting has been confirmed in all 105 counties.

Field Notes: This is one of the most abundant birds in Kansas at all seasons. Wintering birds gather in immense flocks, usually at large wetlands with abundant cattails. These flocks can number in the millions and are an impressive sight as they disperse in the morning to feed in agricultural areas and then return to the roost in the evening.

female

Eastern Meadowlark

Sturnella magna

Field Identification: This plump, short-tailed bird of open country has a yellow breast with a prominent black "V" marking. The outer three tail feathers are white, and these broad white tail-sides are visible on flying birds. Vocalizations are useful for separating the two meadowlark species. The song is usually a simple four-note phrase, *see me. . . . SEE you*, highest and most strident on the third note. The call note is a guttural *chup*. Size: Length 9.5 inches; wingspan 14 inches.

Habitat and Distribution: Both meadowlark species are grassland birds. The Eastern prefers taller, moister prairies than those preferred by the Western Meadowlark. There are records from 99 counties. Scarce in the northwestern counties.

Seasonal Occurrence: Permanent resident. During winter, additional individuals move into Kansas from northern states. There are confirmed nesting records from 63 counties in central and eastern Kansas.

Field Notes: Eastern Meadowlarks are among the most visible birds of Kansas grasslands and cultivated areas. They spend most of their time on the ground but are often seen perched on fences and utility lines.

Sturnella neglecta

Western Meadowlark

Field Identification: Closely resembles the Eastern Meadowlark but differs subtly in plumage. The white edges of the tail are narrower on the Western Meadowlark because only the two outer feathers are white. On summer adults, the yellow throat plumage is more extensive. The vocalizations of the two meadowlark species differ. The Western's song consists of a few clear descending notes, followed by an odd gurgling that ends abruptly. The call notes differ as well. The Western's call note is a buzzy *zert* sound. Size: Length 9 inches; wingspan 14 inches.

Habitat and Distribution: Prefers drier, shorter grassland habitats than those preferred by the Eastern Meadowlark. There are records from all 105 counties.

Seasonal Occurrence: Permanent resident. Like the Eastern Meadowlark, many move into Kansas from the north during the winter. There are confirmed nesting records from 69 counties. Does not nest in the southeastern counties.

Field Notes: In 1925, schoolchildren elected the Western Meadowlark as the state bird of Kansas. In 1937, the Kansas Legislature made their choice official. The two meadowlarks occur together across most of the state, but in summer, the Eastern is absent from the arid shortgrass prairie of northwestern Kansas, and the Western similarly shuns the comparatively moist southeastern counties in summer.

Yellow-headed Blackbird

Xanthocephalus xanthocephalus

Field Identification: The males are unmistakable with the yellow head and breast, black body and wings, and bold white wing-patch. The females and young males have a dull yellow face, breast, and eye-line, with the remainder of the plumage dark brown. The unusual song of the males has been compared to a loud, squeaky barn-door hinge. Size: Length 10 inches; wingspan 15 inches.

Habitat and Distribution: Migrants are found statewide in wetlands and rural areas, especially cattle feeding areas. Nests colonially in cattail marshes in summer. There are records from all 105 counties. Most are found in central and western Kansas.

Seasonal Occurrence: Migrant and summer resident. Arrives in mid-April and departs in September. Migrants are most numerous and conspicuous in late April and early May. There are confirmed nesting records from 16 counties spread widely across the state. Occasionally a few linger into November and December.

Field Notes: This is the most colorful blackbird in Kansas. In the spring, look for large, noisy flocks feeding in cattle feedlots, often with other blackbirds. In the summer, they nest in large colonies at Cheyenne Bottoms, Quivira NWR, and a few smaller wetlands.

female

© Judd Patterson

Euphagus carolinus

Rusty Blackbird

nonbreeding male

Field Identification: The plumage of the breeding male Rusty Blackbird is all black with a yellow eye. Most that are seen in Kansas are in the nonbreeding winter plumage and have pale golden-brown edging on the feathers. The plumage of the breeding female is grayish with a yellow eye. Nonbreeding females are golden brown with a small black mask. Vocalizations are complex and include high-pitched squeaking notes. Size: Length 9 inches; wingspan 14 inches.

Habitat and Distribution: Found in moist woodlands, mostly in the eastern half of Kansas. There are records from 97 counties. Rare in western Kansas, where its habitat is limited or absent.

Seasonal Occurrence: Winter resident. Arrives in late October and departs in April.

Field Notes: Rusty Blackbirds are usually found in small, vocal flocks. Because their habitat preference is specialized, they are only occasionally observed in mixed-species blackbird flocks. Rusty Blackbirds nest in the northern boreal forests, and like several other boreal nesting species, their numbers have declined significantly in the past 25 years.

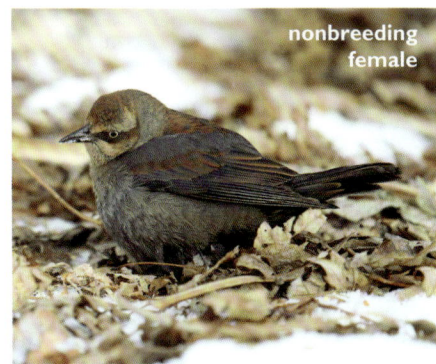

nonbreeding female

317

Brewer's Blackbird

Euphagus cyanocephalus

Field Identification: The male is black with an iridescent sheen from late winter through the summer and duller in color in other seasons. The female is dark gray with a lighter gray head and dark eye. Brewer's and Rusty Blackbirds are similar in appearance and only occasionally occur together. The Rusty Blackbird is usually found in small flocks along wooded streams; the Brewer's is fond of open farm country. Size: Length 9 inches; wingspan 15 inches.

Habitat and Distribution: Found statewide in rural areas. There are records from all 105 counties. Most numerous in central and western Kansas. Look for them around farms, feedlots, and anywhere cattle are present.

Seasonal Occurrence: Migrant and winter resident. Fall migrants occur from October through November. Many remain through the winter months. In spring, northbound migrants are numerous in March and April.

Field Notes: Brewer's Blackbirds can be found in large numbers in the winter, often in flocks with other blackbird species. It is usually thought of as a western species, but large flocks are seen regularly in the east, especially in the Flint Hills.

female

Quiscalus quiscula

Common Grackle

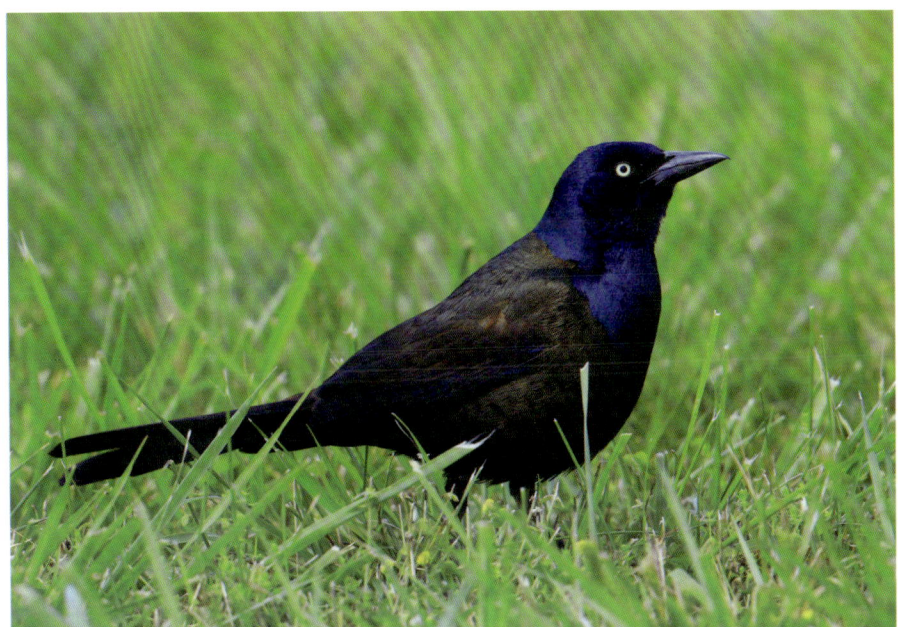

Field Identification: Males are iridescent bronze on most of the body, with a deep blue or purplish sheen on the head, iridescent purple wings, and a long tail that flares at the end. Females are similar, but the iridescence is more subtle and muted. Size: Length 12 inches; wingspan 17 inches.

Habitat and Distribution: Found statewide in a variety of habitats; often abundant in towns and on farms. There are records from all 105 counties.

Seasonal Occurrence: Permanent resident. Abundant in the summer, decidedly rare in the winter months. Migrants are abundant in March and April and again in October and November. Widespread in summer, with confirmed nesting in all 105 counties.

Field Notes: The arrival of large numbers of grackles in March is one of the first signs of spring in urban areas. Males perform an entertaining courtship display on urban lawns, throwing their heads back, puffing out their feathers, and making high-pitched squealing calls as they attempt to impress the females.

female

Great-tailed Grackle

Quiscalus mexicanus

Field Identification: This is the largest blackbird in Kansas. The male is iridescent purple, with a long bill and a ridiculously long wedge-shaped tail. Females are grayish brown with a pale throat and eye-line. Their loud, chattering, whistling calls are given year-round. Size: Length 18 inches; wingspan 23 inches.

Habitat and Distribution: Found statewide in cattail marshes, cattle feedlots, and urban areas. There are records from all 105 counties.

Seasonal Occurrence: Permanent resident. Widespread in summer. There are confirmed nesting records from 60 counties. In winter, most depart from northern and western Kansas.

Field Notes: Great-tailed Grackles are a relatively recent addition to the bird life of Kansas. In the 1960s, they began to colonize the state from the southwest,

and they are now established in many areas. They are also a familiar sight in urban areas, where they often select groves of arborvitae or Bartlett Pear trees, even those located in busy retail shopping center parking lots, as nesting sites. They often forage in parking lots for waste food and dead insects that they pick from the grilles of vehicles. In winter months, they gather into large flocks of hundreds or even thousands of birds that typically are seen in south-central cities and towns such as Dodge City or Wichita.

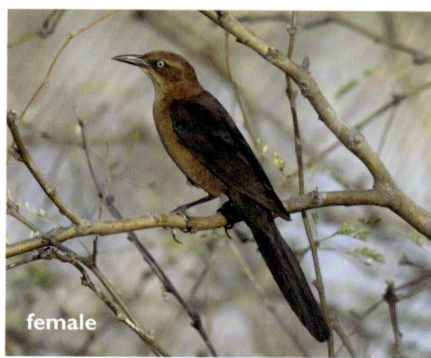

female

Sturnus vulgaris

European Starling

Field Identification: In summer breeding plumage, European Starlings are dark-bodied with white spots on the back, a purplish sheen on the head, and a yellow bill. In winter plumage, the entire body is spotted with white, the head is mostly gray, and the bill is black. Size: Length 8.5 inches; wingspan 16 inches.

Habitat and Distribution: Found statewide, usually near towns, farms, feedlots, or other man-made habitats. It is common in all 105 counties.

Seasonal Occurrence: Permanent resident. There are confirmed nesting records from all 105 counties.

Field Notes: Starlings were introduced to North America in the late nineteenth century and within a few decades had expanded their range across the entire continent. They have become one of the foremost nuisance species in North America. Huge flocks gather in the fall and winter at feedlots, on grainfields, and in urban areas. Starlings nest in and on buildings. They compete with native species for nest sites in tree cavities.

nonbreeding

Brown-headed Cowbird

Molothrus ater

Field Identification: The male is glossy black with a brown head. The female is light brown with a pale throat. Their bills are shorter and more conical than those of other blackbirds. Size: Length 7 inches; wingspan 12 inches.

Habitat and Distribution: Found statewide in grasslands, brushy areas, and woodland edges. True to their name, they often feed on the ground near herds of cattle. There are records from all 105 counties.

Seasonal Occurrence: Abundant migrant and summer resident, uncommon in winter. Spring migrants arrive in early March, and most have departed by late October. Common in summer. There are confirmed breeding records from 95 counties. Most depart southward for the winter, but some remain in the southern half of the state during the winter months.

Field Notes: Brown-headed Cowbirds do not build their own nests but lay their eggs in the nests of other birds, leaving the host species to raise the young cowbirds as their own. They were originally restricted to the Great Plains and were a nomadic species, following the herds of bison. Changes brought by European settlement created new habitats, allowing their population to expand. During migration, especially in the fall, flocks of thousands can be observed.

female

Icterus spurius

Orchard Oriole

Field Identification: The adult male has a black head, back, wings, and tail and a brick-red chest, belly, and wing marking. The female is lemon yellow below, with a grayish back and white wing-bars. One-year-old males resemble females but have a black mask and throat. Size: Length 7 inches; wingspan 10 inches.

Habitat and Distribution: Found statewide in brushy woodland edges, hedgerows, and groves of small or medium-sized trees surrounded by open areas. There are records from all 105 counties. Most common in western Kansas, especially in summer.

Seasonal Occurrence: Migrant and summer resident. Spring migrants are almost all seen in May. Widespread in summer. There are confirmed nesting records from 91 counties. Many breeding individuals have departed by early August. Southbound migrants from northern populations continue to be seen through mid-September.

Field Notes: Orchard Orioles are often heard singing before they are seen. They often sing while hidden in dense foliage. They are easily found in western Kansas where nesting trees and adjacent open areas are available. They feed on insects in weedy fields and crops, especially alfalfa. One-year-old males often associate with paired adults and assist in defending the nesting territory.

female

immature male

Baltimore Oriole

Icterus galbula

Field Identification:
The male has a black head, back, and wings and orange underparts, rump, and tail-sides. Males have one orange and one white wing-bar. The female is light orange below and light brown on the head and back, with white wing-bars.
Size: Length 9 inches; wingspan 11 inches.

Habitat and Distribution: Found statewide in urban yards, parks, and open woodlands, especially those with cottonwoods. Common in much of the state but less so in western Kansas. There are records from all 105 counties.

Seasonal Occurrence: Migrant and summer resident. Spring migrants arrive in late April and have mostly passed through by late May. Common in the summer. Fall migrants peak in late August and early September, but a few linger until the end of September. Nests in most of the state, but in the westernmost counties, it is seen only during migration. Confirmed nesting has been recorded in 91 counties.

Field Notes: This is one of the most colorful birds found in Kansas and is common in cities and towns. When wild fruits such as mulberries or hackberries ripen, small flocks of orioles congregate to devour the fruit. Orioles can be attracted to special nectar feeders and to feeders that offer orange halves and grape jelly. Their abandoned socklike nests are obvious after the leaves have fallen in autumn.

female

immature

Icterus bullockii

Bullock's Oriole

Field Identification: The male has a black back, crown, and throat with an orange face and underparts. There is a black line through the eye with an orange line above it. The large white wing-patch is prominent. The female has a yellow-orange head and breast, gray wings, gray back, and white belly. Size: Length 9 inches; wingspan 12 inches.

Habitat and Distribution: Nests in riparian woodlands and towns in the western counties. Migrants are occasionally seen farther east in both spring and fall migration. There are records from 57 counties in central and western Kansas.

Seasonal Occurrence: Migrant and summer resident. Arrives in May and departs by mid-September. There are confirmed nesting records from 22 western counties.

Field Notes: The Bullock's Oriole replaces the closely related Baltimore Oriole as a nesting species in westernmost Kansas. The ranges of these two orioles meet in Kansas, and hybrids frequently occur that have mixed markings of both species. Bullock's Orioles are common along the Arkansas and Cimarron Rivers and their tributaries, where their bright colors and pleasing song make this bird a refreshing sight.

female

immature male

Scarlet Tanager

Piranga olivacea

Field Identification: The male in breeding season is a deeper shade of red than the Summer Tanager, with jet-black wings and tail. Females and fall males are yellowish green with dark wings. Size: Length 7 inches; wingspan 11.5 inches.

Habitat and Distribution: In summer, it nests locally in mature deciduous hardwood forests of eastern Kansas. In migration, it can be observed in a variety of wooded habitats, including cemeteries and parks. Migrants occur mostly in central and eastern Kansas, rarely farther west. There are records from 71 counties.

Seasonal Occurrence: Migrant and summer resident. Spring migrants arrive in late April and are most common in the first three weeks of May. Fall migrants are seen mostly in September. Confirmed nesting has been recorded in 11 counties, mostly in the northeastern corner but also west to Cloud and Cowley Counties.

Field Notes: Male Scarlet Tanagers are spectacular in appearance and never fail to please birders. During the breeding season, they are considerably less widespread and numerous than Summer Tanagers but can be found at a few locations in the Glaciated Region, such as the woodlands near Wathena or Wyandotte County Lake Park.

female

Piranga ludoviciana

Western Tanager

Field Identification: The breeding male Western Tanager has a bold black-and-yellow pattern and red head. The black wings have one bright yellow wing-bar and one white or pale yellow wing-bar. Females and fall males are yellow green, with black back and wings and yellow and white wing-bars. Size: Length 7 inches; wingspan 11.5 inches.

Habitat and Distribution: A migrant seen in woodlands and towns in the western counties. A rare stray to the eastern half of the state during migration. There are records from 43 counties.

Seasonal Occurrence: Spring migrant in May; fall migrant in September. More likely to be seen in the fall than in the spring.

Field Notes: This tanager is one of the sought-after western species that draws birders to western Kansas during spring and fall migration. They can be expected anywhere in the counties near the Colorado border, most often in towns. The most popular location for finding them is Morton County during September, both in the town of Elkhart and in the woodlands along the Cimarron River.

female

© Judd Patterson

Summer Tanager

Piranga rubra

Field Identification: The male in breeding season is entirely red with a long, thick bill. The female is typically greenish above and dull yellow below. Some females have a slight orange wash. Some spring males look "in-between," mostly yellow green with blotchy red areas or red with greenish wings. A sharp *spi-tuck* call is often heard. Size: Length 8 inches; wingspan 12 inches.

Habitat and Distribution: Nesting birds are found from the Flint Hills eastward in wooded areas. They are strongly associated with oak forest but can also be seen in other woodland communities. In the Flint Hills, they are often found in isolated stands of oaks surrounded by prairies. Migrants are scarce but regular throughout the west during migration. There are records from 89 counties.

Seasonal Occurrence: Migrant and summer resident. Spring migrants arrive in late April and May. Some continue north, and others remain to nest. There are confirmed nesting records from 27 eastern counties. Fall migrants are observed mostly in September.

Field Notes: Summer Tanagers are fairly common in areas of eastern Kansas with appropriate habitat. Their rich, warbling song and guttural call notes announce their presence in the dense canopy.

female

Cardinalis cardinalis

Northern Cardinal

Field Identification: Both sexes are crested, with a black mask and red bill. The male is crimson, and the female is brown with red shafts in the wings. Juveniles look like females but have a black bill. Size: Length 9 inches; wingspan 12 inches.

Habitat and Distribution: Found statewide in woods, brushy areas, parks, and yards. Considerably less common in western Kansas, but the population there is slowly increasing. There are records from all 105 counties.

Seasonal Occurrence: Permanent resident. There are confirmed nesting records from 78 counties. All the counties that lack nesting records are in western Kansas.

Field Notes: Northern Cardinals are a familiar bird to most Kansans. They are frequent visitors to suburban yards and bird feeders. They readily nest in yards and parks. As trees and shrubs mature on the High Plains of western Kansas, cardinals have begun to colonize that part of the state, especially in cities and towns.

female

Rose-breasted Grosbeak

Pheucticus ludovicianus

Field Identification: Males have a large conical beak, a black head and upper-parts, a red breast, and white underparts. White wing-patches are prominent on flying males. Females have a striped face, brown back, and white breast with heavy streaking. Fall males are like the female Black-headed Grosbeak but differ in that the bill is all pale and they have at least some streaking on the center of the breast. Size: Length 8 inches; wingspan 13 inches.

Habitat and Distribution: This species nests in central and eastern Kansas in riparian woodland and upland hardwood forest. Migrants are seen in a variety of habitats during migration. It is a wide-spread migrant in eastern Kansas and a rare but regular migrant in the west. There are records from 103 counties.

Seasonal Occurrence: Migrant and summer resident. Spring migrants are seen in late April through mid-May, fall migrants in September through early October. Confirmed nesting has been reported from 32 counties, mostly in northeastern and north-central Kansas west to Smith County but also in a handful of south-central counties.

Field Notes: Rose-breasted Grosbeaks can be frustrating to see well. They often sing their rich, warbling song from the densest canopy in the forest. They land in what appear to be small clumps of leaves and seemingly disappear. Learn their metallic *chink* note, as it often is the first sign of their presence.

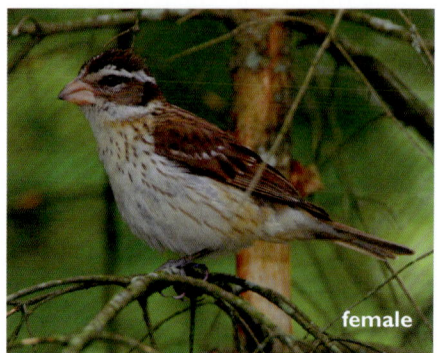

female

Pheucticus melanocephalus

Black-headed Grosbeak

Field Identification: The male is orange with a black head, back, and wings and white wing-patches. The female resembles the Rose-breasted female, but the underparts are buffier and the streaking is much finer and restricted to the flanks. In both sexes, the upper mandible of the bill is dark, unlike the entirely pale bill of the Rose-breasted Grosbeak. Size: Length 8 inches; wingspan 12 inches.

Habitat and Distribution: This is a nesting species in the riparian woodlands of central and western Kansas. It is a widespread migrant in the west and a rare migrant in the eastern counties. There are records from 83 counties.

Seasonal Occurrence: Spring migrants are seen from late April through mid-May. Fall migrants are seen in late August and early September. There are confirmed nesting records from 16 north-central and northwestern counties

as far east as Cloud and Stafford Counties.

Field Notes: This is one of the few species that is more common in northwestern Kansas than in any other area of the state. Black-headed and Rose-breasted Grosbeaks are closely related, and hybrids between the two are sometimes seen, especially in the central part of the state, where the nesting ranges of the two species converge in counties such as Smith and Rush Counties.

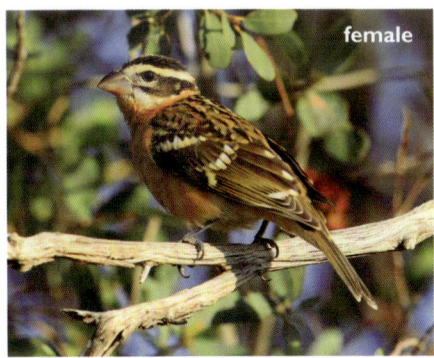

female

Blue Grosbeak

Passerina caerulea

Field Identification: Males are all blue with reddish-brown wing-bars and a large conical bill. Females are entirely grayish brown with light buffy wing-bars. The burry and warbling song is often the first clue to their presence. A hard, metallic call note is often heard. Size: Length 7 inches; wingspan 11 inches.

Habitat and Distribution: Found in brushy areas, edge habitats, fencerows, and scrubby vegetation. Occurs statewide, but more numerous in the southern half of the state. There are records from all 105 counties.

Seasonal Occurrence: Summer resident. Arrives in mid-May, later than most other summer species, and departs in mid-September. Widespread as a nesting species, with confirmed nesting records from 56 counties and probable nesting records from an additional 30 counties.

Field Notes: Look for these colorful birds in thickets and dense edge vegetation. In areas of appropriate habitat, such as the Cross Timbers, Flint Hills, and Red Hills, they are common, and many can be seen in a single day's drive on back roads. They are often seen feeding in tall weeds near roadsides.

female

Passerina cyanea

Indigo Bunting

Field Identification: The summer male is entirely deep blue, giving way to a patchy blue-and-gray mix in the fall. The female is grayish brown with blurry streaking below (see comments for Lazuli Bunting). It is possible to confuse both males and females of this species with the Blue Grosbeak, but size differences in the bill will resolve any identification confusion. Also notice the reddish-brown wing-bars on the male Blue Grosbeak, which the Indigo Bunting lacks. Size: Length 5.5 inches; wingspan 8 inches.

Habitat and Distribution: In summer, found in riparian woodlands, hedgerows, edge habitats, and towns. Migrants are found statewide and can be seen in a wide variety of habitats. Abundant in summer in central and eastern Kansas, less common in the west. There are records from all 105 counties.

Seasonal Occurrence: Migrant and summer resident. Spring migrants are seen from late April through late May. There are confirmed nesting records from 51 counties, mostly in the eastern half of the state. Summer residents continue to sing through August. Fall migrants arrive in September and are seen into mid-October.

Field Notes: Indigo Buntings are among the most common summer species of eastern Kansas. They sing from high, exposed tree perches throughout the day, even in the hottest weather.

female

333

Lazuli Bunting

Passerina amoena

Field Identification: The male has a blue head, upperparts, and wings; orange bib; white belly; and prominent wing-bars. The female is uniformly grayish brown, with narrow wing-bars. Females can be difficult to distinguish from female Indigo Buntings. Indigo females usually have a paler throat and vague streaking on the breast. Lazuli females have a grayer throat and lack any markings on the underparts. Size: Length 5.5 inches; wingspan 9 inches.

Habitat and Distribution: Migrants can be found in towns and woodlands statewide. It is a common migrant in the west and a rare but regular migrant in the east. There are records from 96 counties.

Seasonal Occurrence: Spring migrants are seen in late April and May with a peak in mid-May. Fall migrants are seen in August and September with a peak in early September. Sightings are more frequent in the spring than in the fall. There is one confirmed breeding record from Morton County.

Field Notes: This is a species of the western US. Lazuli Buntings migrate through the state but for the most part do not remain in Kansas to nest. Spring migrants have been reported more often in eastern Kansas in recent decades than in the past.

female

© Judd Patterson

334

Passerina ciris

Painted Bunting

Field Identification: Arguably the most colorful bird in Kansas, the male is crimson below, with a green back, a blue head, and a red eye-ring. The female is uniformly bright green above and yellow green below. The warbling song is reminiscent of that of the Blue Grosbeak but is higher-pitched and more abbreviated. Size: Length 5.5 inches; wingspan 9 inches.

Habitat and Distribution: Found in brushy or open wooded hillsides in southern and eastern Kansas. During migration, it sometimes occurs west and north of the breeding range. There are records from 73 counties. Summer resident in most counties in the southeastern third of the state, south and east of a line between Clark and Riley Counties.

Seasonal Occurrence: Summer resident. Arrives in early May and departs in late August. It nests west to Clark County and north to Riley County. Confirmed nesting has been reported in 18 counties.

Field Notes: Many Kansans are surprised to learn that this colorful bird can be seen in their state. Even people who live in areas where they are relatively common are often unfamiliar

with them. The males usually sing from a partially concealed perch near the top of a leafy tree, rarely perching in the open. As with many birds, knowing their song is useful in locating them. They are most abundant in the Red Hills and Cross Timbers regions.

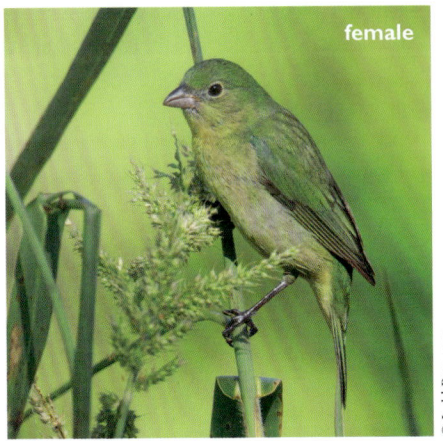

female

© Judd Patterson

335

Dickcissel

Spiza americana

Field Identification: The male has a yellow throat and breast with black bib, yellow eye-line, russet wings, and gray breast and nape. The female is similar but does not have the black bib. Some females have a yellow wash on the chest; others show no yellow. Size: Length 6 inches; wingspan 10 inches.

Habitat and Distribution: Found statewide in tall and mixed-grass prairies, agricultural fields, and weedy areas. Less numerous in the west than in the rest of the state. There are records from all 105 counties.

Seasonal Occurrence: Migrant and summer resident. Arrives in early May and departs in October. There are confirmed nesting records from 96 counties. In August and September, they gradually gather into foraging flocks before departing southward. These flocks sometimes number in the hundreds.

Field Notes: Dickcissels are one of the most abundant summer birds in Kansas despite their habit of nesting in wheat and alfalfa, which are harvested during their nesting season. The males sing from atop tall plants, fence posts, or thickets. Scanning across a wheat field will often reveal singing males perched at frequent intervals. During migration, listen for their distinctive single *buzz* notes, which they often give while in flight.

female

© David Seibel

Haemorhous purpureus

Purple Finch

Field Identification: The male is raspberry purple on the head, breast, back, and wing edges. Unlike the more abundant House Finch, males lack obvious streaking on the flanks. The female has a brown back and head with white stripes on the face and white underparts with heavy dark streaks. Female House Finches lack strong facial markings and are dingier below. Size: Length 6 inches; wingspan 10 inches.

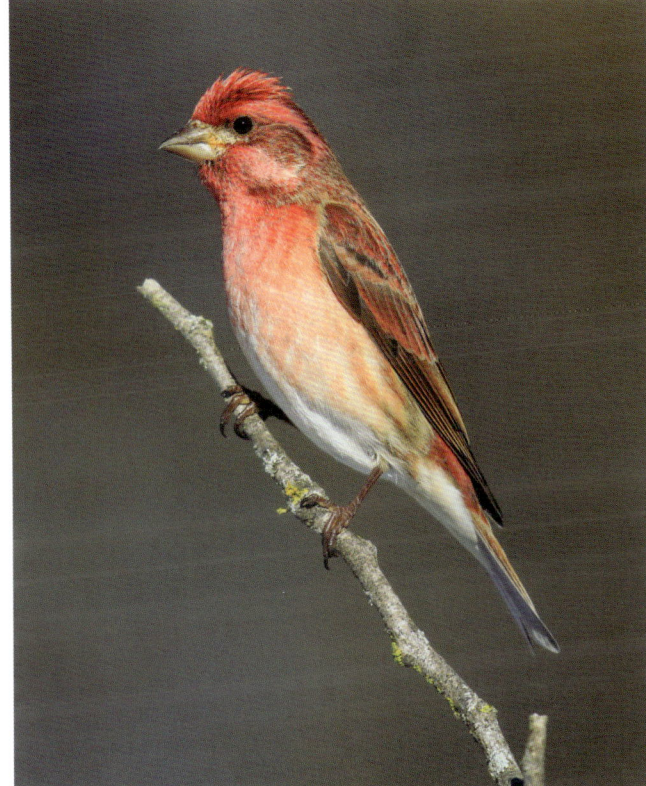

Habitat and Distribution: Can be seen statewide, but most likely to be seen in the eastern half of Kansas. Prefers forests, brushy habitats, and towns. There are records from 83 counties. In western Kansas, there are about 20 counties where this species has never been recorded.

Seasonal Occurrence: Winter visitor. Abundance varies from year to year. Occurs in Kansas between November and March.

Field Notes: This is one of several "northern finches" that birders hope to find in the winter. They can often be seen feeding on the seeds of Green Ash trees. They readily visit bird feeders, sometimes in small flocks. Purple Finches are scarce in most years but occasionally invade Kansas in large numbers, especially east of the Flint Hills.

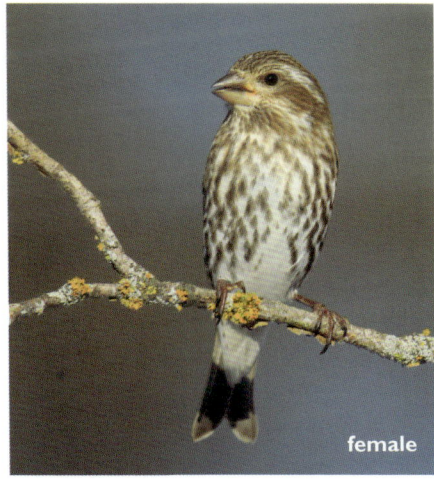

female

337

House Finch

Haemorhous mexicanus

Field Identification: The male is red on the crown, throat, breast, and rump and otherwise brown with weak wing-bars. The red plumage is more crimson and more restricted than that of the Purple Finch. The female is gray brown above and has dingy underparts with obscure streaking, lacking the strong facial pattern and well-defined breast streaks of the female Purple Finch. Size: Length 6 inches; wingspan 9.5 inches.

Habitat and Distribution: Found statewide in weedy fields, in towns, and around farms. Most likely to be seen near human habitations. There are records from all 105 counties.

Seasonal Occurrence: Permanent resident. Confirmed nesting has been recorded in 78 counties, and it probably nests in all counties.

Field Notes: House Finches expanded their range in North America significantly in the twentieth century. Until the 1970s, they were found only in southwestern Kansas, but they are now abundant throughout the state. They frequently nest in hanging plant baskets and are often the most numerous species at bird feeders. During the winter, they often gather in flocks of several hundred birds and wander widely in search of food.

female

Spinus pinus

Pine Siskin

Field Identification: This small brown finch has a sharp, pointed bill; notched tail; and heavy streaking on most of the body. Adults have a long yellow wing-stripe, which appears as a wing-bar on perched birds, and wing and tail feathers with yellow edging. The distinctive call is an ascending high-pitched *buzz*. Size: Length 5 inches; wingspan 9 inches.

Habitat and Distribution: Found statewide in pines, other conifers, weedy fields, and sunflower patches as well as at bird feeders. There are records from all 105 counties.

Seasonal Occurrence: Migrant and winter resident. Summer resident in some years. Fall arrival begins in October, and in spring, most have departed by late May. A few pairs remain to nest in some years, and there are confirmed nesting records from 26 counties spread across the state.

Field Notes: These small, energetic finches often associate with American Goldfinches, to which they are closely related. Like several other finch species, they can be abundant one year and virtually absent the next. During "invasion years" when siskins move into Kansas in large numbers, they noisily swarm sunflower and thistle feeders.

American Goldfinch

Spinus tristis

breeding male

Field Identification: The summer male is bright yellow with black wings, tail, and crown. Females and all winter birds have black wings and yellow wing-bars and are grayish with a variable amount of yellow on the face and breast. They have an undulating flight and, while flying, often give a three- or four-note call every few seconds. Size: Length 5 inches; wingspan 9 inches.

Habitat and Distribution: Found statewide in prairies, weedy fields, brushy areas, open woodlands, cities, and towns. There are records from all 105 counties.

Seasonal Occurrence: Permanent resident, but populations vary seasonally and from year to year. Confirmed nesting has been recorded in 49 counties and probable nesting in over 30 additional counties. Most counties lacking nesting records are in the west.

Field Notes: This colorful finch is found in open country throughout Kansas. American Goldfinches nest later in the summer than almost any other songbird, often laying eggs in early August. Like Pine Siskins, they sometimes visit feeders in large flocks and consume significant amounts of sunflower and thistle seed.

nonbreeding

Loxia curvirostra

Red Crossbill

Field Identification: Both sexes have dark wings and large bills with the mandibles crossed on the tips. Males are brick red; females are yellowish. They are usually seen in flocks and communicate frequently with hard call notes. Size: Length 6 inches; wingspan 11 inches.

Habitat and Distribution: Found statewide in pines and other coniferous trees and sometimes in weedy fields. There are records from 87 counties.

Seasonal Occurrence: The occurrence of this species is exceptionally erratic and unpredictable. It can appear at any time and has been observed in Kansas in all months of the year. It is most likely to be seen in the fall and winter months. On rare occasions, it has remained in Kansas to nest, and there are confirmed nesting records from three counties.

Field Notes: The erratic movements of Red Crossbills are triggered by the success or failure of pine-cone crops within the boreal forests of Canada and the Rocky Mountains. When these crops fail, they wander onto the plains in search of food. Look for them in areas with abundant pines. They use their specialized bills to extract seeds from the cones. Arborvitae trees sometimes attract feeding flocks. Sites that have attracted crossbills during invasion years include the pines below the dam at Perry Reservoir, Sim Park in Wichita, and the Fort Hays State University Experimental Farm.

female

Kansas Birding Hotspots

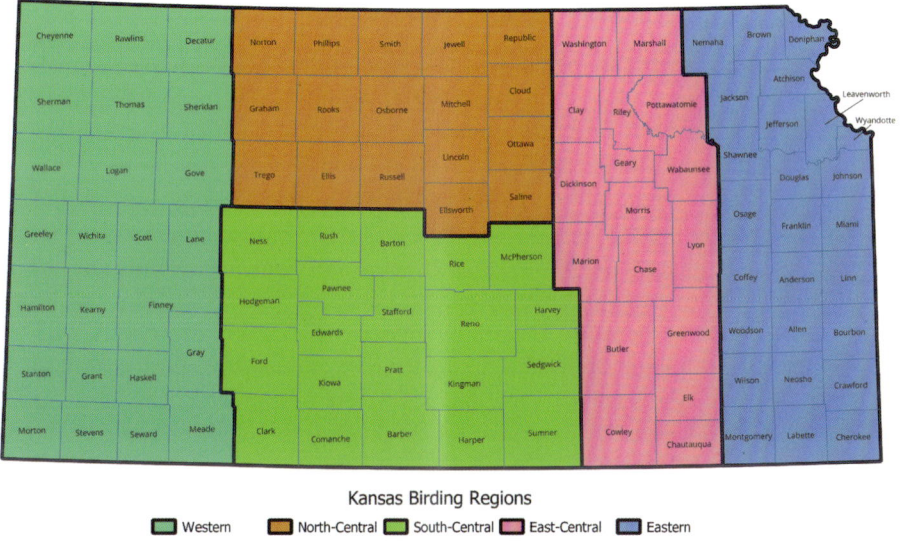

Kansas Birding Regions

Western   North-Central   South-Central   East-Central   Eastern

The species accounts in this book show the diversity of Kansas birdlife. Many people with a growing interest in Kansas birds begin by learning the birds they see in their yards and then become interested in venturing a bit farther from home to nearby birding destinations. Some eventually travel farther afield in search of new and interesting bird species. To assist with these explorations, we have highlighted 100 of the best birding locations in Kansas. The above map of the entire state shows all 100 of these birding hotspots.

To make this book useful throughout Kansas, we have divided the state into five regions and selected roughly 20 of the best birding hotspots within each region. Each of the five regions is covered in a separate chapter, which includes a more detailed regional hotspot map showing important highways, rivers, and reservoirs. Some of these are true destinations that require a full day or even several days to explore fully. Others are excellent birding locations that can be covered adequately in a visit of several hours. Some of them require only an hour or two to explore and are best thought of as places to visit if your travels take you nearby. All are on publicly owned lands or can be viewed from public roads without trespassing.

It is impossible to discuss birding hotspots without some mention of the eBird birding website and phone app. Over the past 20 years, eBird has risen to prominence on a global scale and is now integral to our understanding of bird populations and distribution. Created and maintained by Cornell University, eBird is a powerful tool with multiple uses. First and foremost, it is an ever-growing repository of ornithological data. The companion phone app allows birders to record their sightings and upload them in real time, where they are then instantly available on the internet. eBird also allows you to store all your personal bird sightings and then retrieve them in a variety of formats, including seasonal and geographical occurrence.

Used with permission, here is the introduction to eBird that you will find on the eBird website https://ebird.org/about:

eBird began with a simple idea—that every birdwatcher has unique knowledge and experience. Our goal is to gather this information in the form of checklists of birds, archive it, and freely share it to power new data-driven approaches to science, conservation, and education. At the same time, we develop tools that make birding more rewarding. From being able to manage lists, photos, and audio recordings, to seeing real-time maps of species distribution, to alerts that let you know when species have been seen, we strive to provide the most current and useful information to the birding community.

eBird is among the world's largest biodiversity-related science projects, with more than 100 million bird sightings contributed annually by eBirders around the world and an average participation growth rate of approximately 20% year over year. A collaborative enterprise with hundreds of partner organizations, thousands of regional experts, and hundreds of thousands of users, eBird is managed by the Cornell Lab of Ornithology.

eBird data document bird distribution, abundance, habitat use, and trends through checklist data collected within a simple, scientific framework. Birders enter when, where, and how they went birding, and then fill out a checklist of all the birds seen and heard during the outing. eBird's free mobile app allows offline data collection anywhere in the world, and the website provides many ways to explore and summarize your data and other observations from the global eBird community.

eBird is available worldwide, comprising a network of local, national, and international partners. eBird directly collaborates with hundreds of partner groups for regional data entry portals, outreach, engagement, and local impact.

Data quality is of critical importance. When entering sightings, observers are presented with a list of likely birds for that date and region. These checklist filters are developed by some of the most knowledgeable bird distribution experts in the world. When unusual birds are seen, or high counts are reported, the regional experts review these records.

eBird data are stored across secure facilities, archived daily, and are freely accessible to anyone. eBird data have been used in hundreds of conservation decisions and peer-reviewed papers, thousands of student projects, and help inform bird research worldwide.

It is beyond the scope of this book to explain everything that you can do with eBird. Even if you do not elect to use the software to record your own sightings, you can use it as a powerful tool to assist you in your birding explorations. If you are planning a birding trip, a variety of information can be sourced using the "Explore" tool available at https://ebird.org/explore. From this page, in the "Explore Regions" field, enter the county that you want to learn more about. This will then present you with a wealth of information, including the total number of species ever reported in that county to eBird, a list of recent visits by birders, and a list of the "Top Hotspots" in that county. Try clicking on one of the "Top Hotspot" links. This brings you to a submenu showing the total number of species ever reported at that specific hotspot, the total number of checklists ever submitted for that hotspot, and recent checklists submitted for that hotspot. Use the "map" and "direction" buttons, which allow you to pinpoint and navigate to the hotspot using a smart phone.

The Kansas birding hotspot accounts in this book are intended to highlight some of the birds of interest, impart a sense of what birds to expect seasonally, and provide logistical information for visiting birders. The eBird Hotspot names shown in this book are exactly as shown on the website.

Western Region

Western Kansas provides unique birding opportunities. Many bird species that are scarce or absent in the rest of the state occur regularly in the west. These include species such as Scaled Quail, Lesser Prairie-Chicken, Ferruginous Hawk, Golden Eagle, Western Wood-pewee, Woodhouse's Scrub-Jay, Curve-billed Thrasher, Chestnut-collared Longspur, Thick-billed Longspur, Lark Bunting, Cassin's Sparrow, Brewer's Sparrow, Bullock's Oriole, Townsend's Warbler, and Western Tanager.

Cities and towns in western Kansas act as oases of trees and water amid vast expanses of semiarid grasslands and croplands. Because of this, bird diversity in the west is often highest in or near these urban areas. Urban parks, cemeteries, and water treatment plants often attract interesting birds. Simply walking down alleys in western Kansas towns can be a good strategy for finding interesting birds. We have included some of the urban hotspots that are most popular with birders, but many other towns offer good birding. In addition to the hotspots in this book, and especially during migration, check city parks and cemeteries in towns such as Cimarron, Dighton, Goodland, Johnson City, Lakin, Leoti, Satanta, Sharon Springs, Sublette, and Tribune for interesting migrants. Not all water treatment plants are accessible for viewing, but any that can be easily viewed without trespassing should be checked, as they act as magnets for many species of birds.

Prairie dog towns have been steadily eliminated from the western Kansas landscape over many years, which is saddening to many wildlife conservationists. Those that remain attract numbers of Burrowing Owls, Ferruginous Hawks, and Golden Eagles.

A special note is in order regarding Lesser Prairie-Chickens. Western Kansas has the largest and healthiest populations of this species in the world. Almost all of these populations are on privately owned ranches from Gove and Logan Counties southward to Oklahoma. Because of this private ownership, you will not find them mentioned much in the hotspots accounts in this book. For decades, birders have come to Kansas from around the nation and the world to view these unique prairie grouse displaying on their leks in the spring. For many years, the Cimarron National Grassland was the place to go see them, but years of drought have made that population a shadow of its former size. As of this writing, a consortium of ranchers and other interested parties have a thriving commercial operation offering the use of

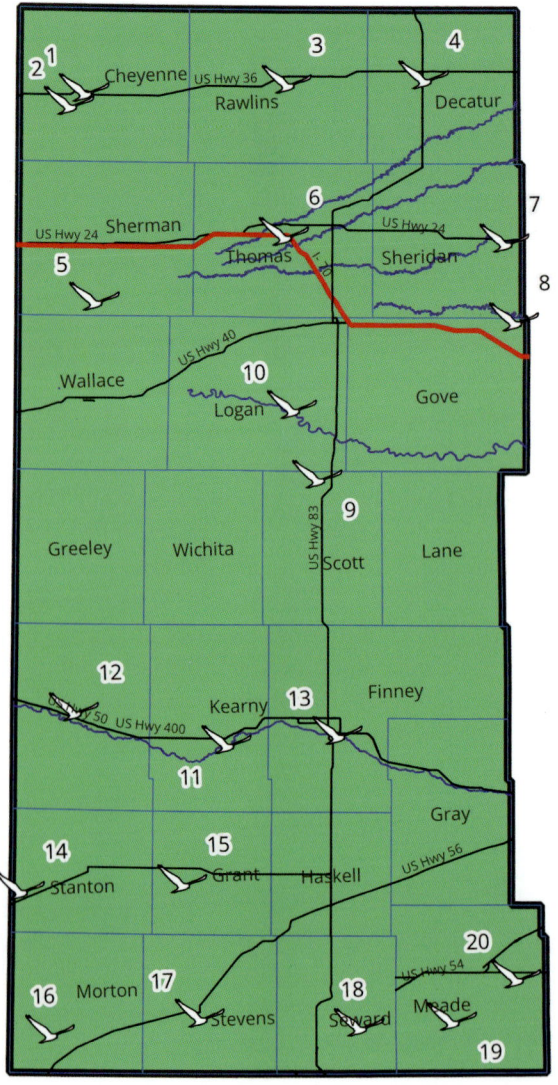

viewing blinds situated near booming grounds (leks) in Gove and Logan Counties. During March and April, many of the major national birding tour companies bring groups of birders to these blinds as part of their "Colorado Grouse" tour packages. Currently, Audubon of Kansas is hosting a "Lek Trek" birding festival in early spring, focused on observing both Lesser and Greater Prairie-Chickens. Future changes in these tourism efforts, changes in landownership, and the potential listing of the Lesser Prairie-Chicken as an endangered species could all impact the availability of these viewing opportunities in the future. For these reasons, we have not included detailed information on these currently existing offerings. As long as they continue, information will be readily available on the internet and on social media. It should also be noted that eBird has elected to conceal the locations where sensitive and endangered species have been found, including Lesser Prairie-Chicken leks.

Western Region Hotspots

1. St. Francis Area
2. Rural Cheyenne County
3. Atwood Area
4. Oberlin Area
5. Soldiers Memorial Park
6. Colby Area
7. Sheridan State Fishing Lake
8. Sheridan Wildlife Area
9. Historic Lake Scott State Park
10. Little Jerusalem Badlands State Park and Smoky Valley Ranch
11. Arkansas River Scenic Drive
12. Hamilton County Wildlife Area
13. Garden City Area
14. Manter Dam Recreation Area
15. Ulysses Area
16. Cimarron National Grassland
17. Hugoton Area
18. Arkalon Park
19. Meade State Park
20. Lakeview Playa

1. St. Francis Area

County: Cheyenne

eBird Hotspot names: St. Francis River Walk, Keller Lake

Hotspot type: Worth a stop

If your mental image of western Kansas is shaped largely by hours of windshield time through the endless wheat desert on I-70, a visit to St. Francis and other Cheyenne County hotspots will change your perspective. The South Fork of the Republican River runs out of Colorado with a surprisingly strong flow and crosses Cheyenne County diagonally. The river is flanked by verdant woodlands and grasslands. The rolling hills and riparian habitat in the river valley combine to make this a visually appealing landscape.

Located in the town of St. Francis is the **St. Francis River Walk** and the adjacent **Keller Lake**. These are open to the public. The city has developed a 1-mile hiking trail along the river that begins at the northwestern corner of St. Francis at the bridge over the Republican River on River Street. There is a parking area just east of the bridge for the River Walk, or you can cross the bridge and use the parking lot for Keller Pond. The 1-mile River Walk Trail connects the two parking lots, so you can walk the trail as a loop that begins and ends from where you parked. Allow two hours to bird these locations adequately.

The River Walk trail follows the east bank of the river through a mature riparian corridor. Over 125 species of birds have been recorded at the River Walk. There are stands of cattail marsh along the banks of the river that attract wetland species such as Wood Duck, Black-crowned Night Heron, Marsh Wren, and Common Yellow-throat. The woodlands along the river have Eastern Screech-Owl and Great Horned

Owl; Red-bellied, Hairy, and Downy Woodpeckers; Black-capped Chickadee; White-breasted Nuthatch; and Northern Cardinal year-round. In summer, expect Red-headed Woodpecker, Yellow-billed Cuckoo, Great Crested Flycatcher, Eastern and Western Kingbirds, Eastern Bluebird, Warbling Vireo, Orchard Oriole, and Baltimore Oriole. Winter species are typical for western Kansas and include Golden-crowned Kinglet, Brown Creeper, and Townsend's Solitaire. A variety of interesting migrants have been seen in spring and fall, including at least 10 species of warblers. Western Wood-Pewee is one of the more abundant migrant songbirds and possibly nests at this location.

The River Walk trail eventually reaches a footbridge crossing the river. On the west bank, the trail follows the river south and then veers west to the parking area for Keller Lake. The land is owned by the Keller family but is perpetually leased to the city of St. Francis. Starting in 2015, the city made some extensive improvements to the property that eliminated some of the wild vegetation and cattails that tended to attract birds. However, the lake still attracts a good variety of waterfowl, wading birds, and shorebirds. Twenty-one species of waterfowl have been recorded at the lake, mostly in early spring and late fall. Other water-dependent species such as Common Loon, White Pelican, Forster's Tern, White-faced Ibis, and Osprey have been reported at the lake. From the lake, you can return to your vehicle by walking across the bridge on the River Road to the River Walk trailhead.

2. Rural Cheyenne County

County: Cheyenne

eBird Hotspot Names: Arikaree Breaks—Rd. 2, Cowpe Creek—Road 3, S Fork Republican River—Road G & CO State Line, St Francis WA

Hotspot type: Half-day tour

If you have taken the time to visit St. Francis for birding, several other birding hotspots in the valley of the Republican River southwest of town may be of interest to you. The 480-acre **St. Francis WA** is administered by the Kansas Department of Wildlife and Parks (KDWP). At the west edge of St. Francis, turn left (south) onto Road 13, which curves and becomes RS 877 (known informally as the River Road). Proceed southwest for 3 miles to the intersection with Cheyenne County Road 11. From the corner, go north on Road 11 for a half mile to the parking area. The property is bisected by the South Fork of the Republican River. There is a long fishing pit adjacent to the parking area. You are free to hike up to a half mile east or west, but most birder visits to this location consist of viewing birds in the trees and grasslands along or near the entrance road and parking area as well as scanning the fishing pit for waterbirds. This is a much more open area than the River Walk. During migration, you may encounter flocks of Clay-colored, Chipping, Savannah, and other sparrows, which sometimes include Brewer's Sparrow. Other migrants you might find in spring and fall include Olive-sided Flycatcher, Spotted Towhee, Lark Bunting, Lazuli Bunting, and Blue Grosbeak.

To continue your tour, return to the River Road and proceed southwest for 9 miles to the intersection with County Road 3. This drive passes though the scenic river valley. Black-billed Magpie is fairly reliable on this stretch of the River Road. There are also several ponds of decent size along the road where waterfowl and shorebirds are possible, depending on the season. Go south on Road 3 for a half mile to reach a good stand of mature riparian timber along the South Fork of the Republican. This is the eBird hotspot known as **Cowpe Creek—Road 3**. Eighty-eight species have been reported at this location to eBird. Park and walk along the road through this appealing open woodland. You will find several wetlands along the road where Virginia Rail has been observed multiple times.

From here, continue south for 0.8 miles to the fork in the road. Take the right fork onto Road H, then County Road G and continue for another mile to the state line. For the final half mile, you will pass through an extensive stand of trees before reaching the large Colorado South Republican State Wildlife Area, which begins at the state line and is often visited by Colorado birders seeking eastern species. There is typically good birding on this stretch. This location is known as the **S Fork Republican River—Road G & CO State Line** area in eBird.

If you still have time to spare, a drive to the northwestern corner of the county will bring you to the rugged **Arikaree Breaks**, which is an area of deeply dissected plateaus associated with the Arikaree River. This is more of a scenic stop than a birding stop. The rocky slopes along County Roads 2 and 3 are covered with yucca, sagebrush, and thickets of aromatic sumac. Bird diversity is not huge, but numbers of Rock Wren inhabit the rock formations. There is a low-water crossing of the river near the Colorado-Kansas-Nebraska three-state marker that sometimes has a few birds of interest in the sparse woodlands. There is a substantial prairie dog town at the three-state marker where Burrowing Owl is possible.

3. Atwood Area

County: Rawlins

eBird Hotspot names: Atwood Lake, Atwood WTP

Hotspot type: Worth a stop

All the western counties located along the Nebraska border feature wooded streams that drain north to the Republican River. In Rawlins County, the primary stream system is the ancient Beaver Creek watershed, which bisects the county diagonally. The county seat, Atwood, is located along the creek. If your travels take you down US 36, a couple of stops in Atwood are worth the short amount of time required to check for birds. On the west edge of the town is **Atwood Lake**, formed by a dam on Beaver Creek. The lake and surrounding Atwood Lion's Park cover 43 acres. The lake itself has 28 surface acres of water. From the intersection of US 36 and K-25, go north one block and turn right (east) onto Lake Road, which encircles the lake. There are pullouts and parking areas from which you can scan the lake. Twenty-two species of waterfowl have been seen at the lake, along with Pied-billed, Eared, and Western Grebes; American Coot; American White Pelican; Black and Forster's Terns; and other waterbirds. During migration, all six species of swallows can be seen foraging over the surface of the lake. At the northwestern corner of the lake is a footpath through a wooded area along the edge of the adjacent golf course where a diverse list of passerine migrants has been found, including nine species of warbler, Western Tanager, Black-headed Grosbeak, and Lazuli Bunting. In years with above-average rainfall, the low area on the west side of K-25 becomes a mudflat area that attracts shorebirds and wading species. On this lake property, 142 species of birds have been reported. A half mile east of Atwood, you can turn north on Road 21.5 and go north for another half mile to reach the **Atwood WTP** (water treatment ponds). You can view most of the ponds from the road. The same waterfowl and other waterbird species listed for Atwood Lake are possible here. Migratory shorebird potential is generally better here than at Atwood Lake, and good finds have included American Avocet, Whimbrel, Long-billed Curlew, and Marbled Godwit. Large flocks of Snow, Canada, and Cackling Geese occur from late fall through early spring.

If you have a decent map or navigation software, a drive along the River Road that follows Beaver Creek southwest from Atwood for over 20 miles is worth the time just for the appealing landscape. This road passes through varied habitats, including quality riparian woodland and some wetland areas. Listen for booming Greater Prairie-Chickens early on spring mornings. Near the Cheyenne County line is the largest of the roadside wetland areas.

4. Oberlin Area

County: Decatur

eBird Hotspot names: Sappa Park, Oberlin WTP

Hotspot type: Worth a stop

Oberlin is another town on US 36 near the Nebraska border that offers some good birding for the traveling birder with two or three hours to spare. One mile from the east edge of Oberlin is the 400-acre **Historic Sappa Park**. The lake was built in the 1930s as a Works Progress Administration (WPA) project and was acquired by the city of Oberlin in 1969. Due to excessive siltation, since the 1960s, there has been minimal surface water present except in a narrow area near the dam. The lake bed is mostly grown over with weeds and grasses. The surrounding areas of the park have woodlands and other wild habitat. There are approximately 5 miles of hiking and horseback riding trails. There has been a recent effort to revitalize the park that includes the development of wetland habitat in the old lake bed.

As you enter the park on 1300th Road from US 36, you will pass a nine-hole golf course where Eastern Bluebirds are present year-round. The road wraps around the east edge of the lake, passing through woodlands and eventually reaching the stone shelter house that was built during the WPA years. The woodlands near the shelter house and below the dam are among the most extensive anywhere in northwest Kansas. A good portion of the park is leased to Pheasants Forever and Quail Forever during hunting season, so restrict your birding to the developed park areas at that time of the year. Most birders park in the general vicinity of the shelter house and explore the park and along the trails as time allows. The birds here are generally like those described for St. Francis and Atwood. Because of the extensive timber, woodland species such as Yellow-billed Cuckoo, Great Crested Flycatcher, Red-eyed Vireo, White-breasted Nuthatch, Spotted Towhee, Black-headed Grosbeak, and Indigo Bunting can usually be found in the summer months. Orchard and Baltimore Orioles are abundant in summer, and Say's Phoebe is possible. Winter birds include Townsend's Solitaire. During migration, a good variety of transient species have been found, including Broad-winged Hawk, Blue-headed Vireo, Western Wood-Pewee, 12 species of warblers, and 16 species of sparrows. In general, this hotspot has not been visited all that often, and the potential for new discoveries is good.

While in Oberlin, most birders also include a stop at the **Oberlin WTP** at the southeastern corner of the town, where over 112 species of birds have been found, including 19 species of waterfowl; 18 species of shorebirds; and others, including Black-crowned Night Heron, Western Grebe, and Forster's and Black Terns. To reach the WTP, take Township Road south from US 36 about a quarter mile west of the Sappa Park entrance. After it reaches a "T" with South Sunflower Avenue, go south for another quarter mile to reach the WTP. There is a gravel approach to the locked gate, and from here, you can view most of the water impoundments. In summer, look for Mississippi Kite soaring overhead, especially over the town.

If you have time and a good county map, it can be worthwhile to wander northeast from Oberlin on county roads. Taking a zigzag route, you can cross heavily wooded Sappa Creek multiple times between Oberlin and the Nebraska state line. Stop at these bridges and spend a little time. It is hard to believe today, but in the initial era of European settlement of the area, species such as Wood Thrush and Louisiana Waterthrush had breeding populations along Sappa Creek, illustrating the fact that these riparian woodlands are among the oldest in western Kansas.

5. Soldiers Memorial Park

County: Sherman

eBird Hotspot names: Soldiers Memorial Co. Park, Sherman SFL, Rd. 20 & Rd. 62 Playa Lake

Hotspot type: Worth a stop

Located about 10 miles south of Goodland, **Soldiers Memorial Park** is always worth a stop. Formerly known as Smoky Gardens, the park is owned by Sherman County and was extensively refurbished in 2020. A 10-acre pond formed by a dam on the North Fork Smoky Hill River (the term "river" is highly relative this far west) is the centerpiece of the park. It usually holds water and is notable in the area by virtue of that alone. The pond is surrounded by stately mature cottonwood trees amid maintained campgrounds and picnic areas. The campgrounds are popular in the warmer seasons.

To reach the park, go south from the town of Goodland for 6 miles on K-27 and turn west onto Sherman County Road 57. Go 2 miles on Road 57, and then turn south onto Sherman County Road 17. Go south 2 miles to reach the entrance to the park. From spring through fall, watch for Lark Bunting on this drive, especially after leaving the highway. Within the park, there are roads on both sides of the pond that begin at the entrance and end at the dam. The tall trees around the lake attract many birds. The brushy and grassland areas outside the mowed areas of the park should also be checked carefully. The substantial stands of trees east of the park are on private lands that are clearly posted against trespassing.

Over 155 species of birds have been reported at this park. Hairy, Downy and, Red-bellied Woodpeckers and White-breasted Nuthatch are present year-round. Summer residents are typical of western Kansas: expect Red-headed Woodpecker; Eastern and Western Kingbirds; Great Crested Flycatcher; Say's Phoebe; Warbling Vireo; Cassin's, Grasshopper, and Lark Sparrows; Orchard and Baltimore Orioles; Dickcissel; and Yellow Warbler. In the winter months, look for a variety of raptors, sparrows, and finches. Merlin and Townsend's Solitaire are usually present in winter, and migrating Mountain Bluebirds are often seen in late fall and early spring. The park is a classic western Kansas oasis of trees and water, and a good variety of migratory species, including about a dozen warbler species, have been reported here in spring and fall.

Directly across from the Soldiers Memorial Park entrance is the entrance to the **Kansas Veterans WA** (formerly known as Sherman SFL) to the west. Like several other artificial lakes built in the early and mid-twentieth century in western Kansas, a combination of poor site selection, flawed engineering, and a steadily falling regional water table means that this lake is almost always without surface water. If there has been recent significant rain, it is worth checking to see whether it is holding water. On those rare occasions when surface water is present, waterfowl and other species such as Western Grebe and Common Loon have occurred. It will take some walking to reach them, but cottonwoods and other trees around the lake basin have matured over time and attract interesting migrants in spring and fall. Otherwise, shortgrass prairie is the dominant habitat, with limited species diversity.

On your way back to Goodland, there is a nearby playa that is worth checking. **The Rd. 20 & Rd. 62 Playa Lake** is 1 mile east of K-27 and 1.5 miles south of Goodland at the intersection it is named for. Like all playa lakes, this can be full of water or bone-dry. When it is holding water, waterfowl, shorebirds, grebes, and herons can be found, depending on the season. When water is present, the playa is best viewed from Road 62.

6. Colby Area

County: Thomas

eBird Hotspot names: Jack Kriss Natural Area, Northwest Research Extension Center

Hotspot type: Worth a stop

Birders traveling on I-70 can take a break to visit two hotspots in Colby that are worth the short amount of time required to explore them. Allow two or three hours to cover both locations. The **Northwest Research Extension Center** is located on the west edge of Colby. Take exit 53 on I-70 and go north on K-25 for 2.5 miles, then west for a half mile. This agricultural research station is affiliated with Kansas State University and has a variety of tree, shrub, and flower plantings. There are numerous mature coniferous and deciduous trees. You are free to wander the grounds. Most birder visits to this park have been in May and September, and about 80 species have been recorded. These include nine species of warblers (including Pine and Townsend's Warblers), Black-headed and Blue Grosbeaks, and other migrants. Red-breasted Nuthatch, Golden-crowned Kinglet, Townsend's Solitaire, and several species of sparrows are present through the winter months in most years.

Near the north edge of Colby is the 23-acre **Jack Kriss Natural Area** on K-25. As you head north out of town, just after you cross the railroad tracks is an inconspicuous grass parking area west of the highway. Park here to access 2 miles of trails that span the entire park area. Originally known as Ferguson Park, this hotspot was named for Jack Kriss, who devoted a great deal of time and work to preserve and enhance natural habitats in the park. There are a variety of habitats, including some mature timber along the intermittent creek, native grass, and shrubby thickets. At times the park can seem quiet in terms of bird numbers, but the quality of bird species found often makes up for the scarcity of birds. A small flock of Wild Turkey is resident along the creek, seemingly out of place given the location well out on the High Plains. Northern Cardinal is uncommon and local this far west, but a few pairs can be found year-round. Other permanent resident eastern species found here include Red-bellied and Hairy Woodpeckers and White-breasted Nuthatch. Summer residents include Mississippi Kite, Cooper's and Swainson's Hawks, Great Crested Flycatcher, Bell's Vireo, and Orchard and Baltimore Orioles. In winter, look for raptors such as Sharp-shinned Hawk and Merlin. The mature cedars and pines along the south edge of the park seem especially attractive to Townsend's Solitaire in the winter months, and several can be present at once in most years. During migration, numerous migratory passerine species have been found, including at least 10 species of warblers, such as MacGillivray's and Mourning, and Northern Waterthrush. Broad-winged Hawk is regular in migration, and on at least one occasion in early May, four were seen in one morning.

7. Sheridan State Fishing Lake

County: Sheridan

eBird Hotspot names: **Buffalo Bill Park, North Creek-110 E, Sheridan SFL**

Other hotspot: Hoxie WTP

Hotspot type: Worth a stop

Sheridan SFL is one of the larger bodies of water in northwestern Kansas and consequently attracts a good variety of birds. The lake is located 11 miles east of Hoxie on US 24. Look for the sign on the highway and go north a half mile on County Road 110, then east for another half mile on the entrance road to the park. The lake covers 60 acres, and there are 188 acres of developed parks, woodlands, and prairie around the lake. A perimeter road circles most of the lake. The most extensive and mature woodlands are on the west side of the lake and mostly must be birded on foot.

At least 159 bird species have been recorded at the lake and surrounding public land. At least 16 species of sparrows, including Brewer's, have been recorded here.

Numerous species of waterfowl, including most of the diving duck species, are possible in spring and fall. Other migratory waterbirds that have been seen here are Eared, Pied-billed, and Western Grebes; Black and Forster's Terns; White Pelican; five egret and heron species; and 10 shorebird species. Marsh Wren is somewhat reliable in the cattail stands in spring and fall. In late summer, migrating swallows of five species can usually be found. The grasslands and wooded areas around the lake attract passerine species and should be birded as time allows. In addition to typical summer western Kansas species, look for scarcer species, such as Bell's Vireo, Yellow Warbler, Yellow-breasted Chat, Blue Grosbeak, and Indigo Bunting. Migratory flycatchers, vireos, sparrows, and warblers occur in spring and fall. Olive-sided Flycatcher is frequently reported in spring and fall. As you leave the lake and return to Road 110, go north for about a half mile to a spot where the road curves slightly and crosses the creek (eBird hotspot **North Creek-110 E**), where there are many mature trees. This is still public land, allowing you to walk in if you choose to do so. Otherwise, just see what you can find from the road.

If you have time to spare, return to Hoxie and take K-23 south. At the south edge of town, **Buffalo Bill Park** is located within a corridor of substantial woodland along a small creek. Hike down the creek as time allows. Interesting migrants reported here have included Hermit Thrush, Summer Tanager, and Black-headed Grosbeak.

The **Hoxie WTP** are not shown as a hotspot in eBird but can easily be viewed from the road and have produced a good variety of shorebirds and waterfowl in the past. To reach them from US 24 from the east edge of Hoxie, go south for 2 miles on South Road 10 E and you will see the lagoons on the east side of the road.

8. Sheridan Wildlife Area

Counties: Sheridan and Gove

eBird Hotspot name: Sheridan WA, Quinter WTP & Cemetery

Hotspot type: Worth a stop

Sheridan WA is in the southeastern corner of Sheridan County and covers 458 acres, including 80 acres of timber along the Saline River. This is by no means a destination hotspot, but it does offer a birding diversion for the traveler on I-70. To reach this hotspot from the town of Quinter (take exit 107 on I-70), go east on Old Highway 40 for a half mile from the edge of town and turn north onto Gove County Road 74. In a half mile, you will reach the Quinter Cemetery. If you have a spotting scope, from the cemetery entrance, you can scan the **Quinter WTP** on the east side of the road, where a variety of waterfowl and shorebirds have been found over the years. Continue to the next intersection and turn east on Gove County AA Road, continue east for 2 miles, and turn north on Road 78 for 4 miles to Gove County EE Road. The Sheridan WA is north of EE Road for the next mile to the west. Stopping anywhere on this road can provide good birding. An entry road leading to a stone picnic shelter is located a quarter mile west on the north side. From the intersection of EE Road and Gove County Road 78, you can also continue north, and for the next half mile, the wildlife area is on your left. This road crosses the river, after which you are no longer on public land. The area receives little maintenance, oversight, or visitation, and there are no developed trails. You are free to explore as you wish on foot, but most birders settle for those birds that can be seen from the county roads. The Saline River this far west is more accurately described as a creek, but usually there is at least some water flow in the channel, and the trees are healthy. Summer species include Great Crested Flycatcher, Bell's Vireo, Yellow-breasted Chat, and Black-headed and Blue Grosbeaks. Winter species include Yellow-bellied Sapsucker, Ruby and Golden-crowned Kinglets, Brown Creeper, Townsend's Solitaire, and Spotted Towhee. Eastern Screech-Owl, Hairy Woodpecker, Red-bellied Woodpecker, and White-breasted Nuthatch are present year-round.

9. Historic Lake Scott State Park

Counties: Scott, Logan, and Gove

eBird Hotspot name: Scott SP

Hotspot type: Full day

Historic Lake Scott State Park is an oasis of water and woodlands on the High Plains of western Kansas, located about 10 miles north of Scott City. This is a true "destination" birding hotspot where you can easily spend an entire day. The centerpiece of the park is Lake Scott, created by a dam on Ladder Creek. The 100-acre lake lies in a large, rugged canyon. The surrounding park and wildlife area covers 1,000 acres. Habitats include rocky slopes, shortgrass prairie, cattail marsh, and woodlands. The maximum depth of the lake is 14 feet.

Historic Lake Scott State Park is located about 10 miles north of Scott City in Scott County. Take US 83 north from Scott City for 10 miles to the junction with K-95 and proceed west on K-95 to the park entrance. If you are coming from the north, the junction of US 83 and K-95 is about 35 miles south of Oakley.

A birding trip to this site at any time of the year is rewarding, and western species are well represented at all seasons. Two hundred seventy-six species of birds have been seen in Scott County, and the majority of those have been seen at Historic Lake Scott State Park. During the summer, the park is home to a diverse group of species, including Common Poorwill, Say's Phoebe, Bell's Vireo, Rock Wren, Yellow-breasted Chat, Black-headed Grosbeak, and Bullock's Oriole. During the winter months, waterfowl, raptors, and passerines are attracted to the park. The variety of raptors is especially good in the winter and usually includes Cooper's, Sharp-shinned, and Ferruginous Hawks; Merlin; and Prairie Falcon. Also in winter, be alert for western species such as Townsend's Solitaire, Red Crossbill, and roosting Long-eared Owls. An impressive number of migratory birds have been observed in the park during both spring and fall and have included a long list of exceptional rarities.

Following the shore on the west side of the lake is a paved road that allows you to access most of the habitats found in the park. Check all the stands of trees, but do not ignore the grassland and brushy habitats. Be sure to walk the Big Springs Nature Trail, where the Yellow-breasted Chat is present in summer. Rock Wren and Common Poorwill inhabit the rocky cliffs. The Elm Grove Campground at the south

Hotspot 9: The western Kansas oasis of Historic Lake Scott State Park is cradled in limestone bluffs and was listed by National Geographic as one of the country's 50 must-see state parks.

edge of the park has many mature trees and is a productive birding area. Near the north end of the state park is a mostly undeveloped and heavily wooded area known as Timber Canyon, which has produced numerous good birds over the years. A paved road goes westward along the length of Timber Canyon. This road is now closed to vehicles beyond the last campground area, but you can still walk beyond the gate for about a half mile as it climbs back up to the arid plains that surround the park. Because of the elimination of vehicle traffic, the road closure has made this stretch an improved birding location. Western Wood-Pewee nested here in 2021 for the first documented nesting record in Kansas.

Check the cattail stands below the dam and those adjacent to the El Cuartelejo historic site for species such as Virginia Rail, Marsh Wren, and Swamp Sparrow, which are wintering species in most years. The lake itself is the largest body of water in this region of Kansas and consequently attracts waterfowl and other water-dependent species, especially in winter, when park usage is minimal.

The roads on the east side of the lake are unpaved, and there are many brushy thickets that attract sparrows and other birds during the winter months. You have a good overlook of the lake here, and the road passes directly below the rocky bluffs.

A great nearby location for prairie dog–dependent raptors is located 1 mile south of the park on US 83, then 3 miles west on Road 270 near a large electric transmission line. Pull north into a recessed gate that allows two cars to pull off this sometimes busy road. The prairie dog town is located north and west and is a prime winter hunting ground for Golden Eagle and Ferruginous Hawk. A few Burrowing Owls occur in summer.

10. Little Jerusalem Badlands State Park and Smoky Valley Ranch

County: Logan

eBird Hotspot names: TNC—Little Jerusalem Badlands State Park, TNC—Smoky Valley Ranch, Haverfield Ranch, Russell Springs (town), Russell Springs Park

Hotspot type: Half day

Just a few miles north of Historic Lake Scott in Logan County is one of the newest Kansas state parks. **Little Jerusalem Badlands State Park** is owned by The Nature Conservancy in Kansas. Via a partnership, it is operated by the KDWP as a state park. This 330-acre property preserves a large formation of deeply dissected Niobrara Chalk that is about 1 mile long and up to 100 feet tall. Because of the fragile nature of the chalk formations, visitors are strictly limited to two hiking trails on the rim. The quarter-mile Overlook Trail ends at a scenic point overlooking the chalk formations. This trail allows you to sample the habitat with minimal effort. The 1.2-mile Life on the Rocks Trail goes farther along the rim and has two additional overlook points. There are scheduled hikes down into the formations a few times a year led by park staff that require advance reservations. The KDWP office at Historic Lake Scott State Park also administers this park.

A modest 75 species of birds have been reported from this hotspot to eBird. Western species are well represented. Species to expect in summer include Ferruginous and Swainson's Hawks; Say's Phoebe; Rock Wren; Cliff Swallow; Cassin's, Grasshopper, and Lark Sparrows; Lark Bunting (rare); and Blue Grosbeak. Species seen outside the breeding season have included Golden Eagle, Prairie Falcon, Mountain Bluebird, Townsend's Solitaire, and Chestnut-collared Longspur.

To reach Little Jerusalem from Lake Scott, return to US 83 and go north for 7 miles to Gold Road, then west for 3.5 miles on Gold Road to the intersection with County Road 400. Go north on 400, and you will reach the entrance and parking area after 1 mile. There are restrooms that are open seasonally.

Owned and operated by The Nature Conservancy in Kansas and conjoined with Little Jerusalem, the 18,600-acre **Smoky Valley Ranch** preserves a beautiful area of chalk bluffs and shortgrass prairie overlooking the Smoky Hill River valley. The trailhead for two hiking trails into the ranch property is located on 350 Road. To reach the trailhead from Little Jerusalem, return to Gold Road, which zigzags north and west for 8.7 miles (becoming Indian Road) to the intersection of Indian Road and 350 Road. Go north 5 miles to a parking area and trailhead kiosk. Alternatively, to reach the trailhead from the town of Oakley, take US 40 west from Oakley for 9 miles to Logan County Road 350. Go south on Road 350 for 15 miles. A 1-mile hiking trail and a 5-mile hiking trail both begin at the parking lot. A total of 134 species of birds have been recorded at this hotspot. Because the ranch preserves a major colony of Black-tailed Prairie Dogs, your chances of seeing Burrowing Owl, Ferruginous Hawk (has nested in the area multiple times), and Golden Eagle are good. Greater and Lesser Prairie-Chickens both occur on the ranch, and hybrids are possible. In summer, look for Say's Phoebe, Rock Wren, and Cassin's Sparrow. Sprague's Pipit and Chestnut-collared Longspur have been recorded during

Hotspot 10: Walking trails, with interpretive signs, allow visitors access to 220 acres of Little Jerusalem Badlands State Park and Kansas's most dramatic Niobrara chalk formations.

migration multiple times. Black-billed Magpie is possible year-round. Listen for their raucous call notes.

The Nature Conservancy in Kansas is currently developing an auto tour of the Little Jerusalem Badlands State Park and Smoky Valley Ranch areas, with planned completion in 2026. It will have several pullouts, which will include a hiking opportunity along the Smoky Hill River at the bridge on Road 350. Detailed information will be available online when it has been completed.

Several other good birding stops in Logan County are a relatively short distance from Smoky Valley Ranch. From the ranch entrance, go north on Road 350 to Plains Road. Turn left (west) and go 10 miles to reach the village of Russell Springs. The road curves south and merges into Road 270 at the edge of Russell Springs. Do not go into town yet. Continue south on 270 for about 1 mile to the low-water bridge over the Smoky Hill River. This is a good place to stop and check for birds in the trees and marshy habitats near the bridge. From here, continue south for about 7 more miles to the intersection with Gold Road and turn right (west) on Gold. You will see Lone Butte about 2 miles away. From this intersection for over a mile in all directions, you are driving through the heart of the **Haverfield Ranch**, which includes one of the largest remaining prairie dog colonies in Kansas. You can scan a huge area here from the road. Remember that this is privately owned land and all birding should be done from the roadside. Look for the same species mentioned above for Smoky Valley Ranch. Because of the density of prairie dogs, raptor numbers can be impressive, especially from fall through spring.

Return to **Russell Springs** and go into town on Broadway Street. You will see the city park on your right. From the old stone bridge on the edge of the park, you have a good vantage point into the thick stand of trees on the south side of the road. Warblers and other migrants are often seen from the bridge. Walking through the park can result in productive birding, especially during migration. Flocks of Wild Turkey roam around the town regularly. Eastern Screech-Owl is often seen and heard in the park.

11. Arkansas River Scenic Drive

Counties: Hamilton and Kearny

eBird Hotspot names: Beymer Park, Klotz Sandpits, River Rd-Kearny Co., River Rd-Hamilton Co., Arkansas River-Road Y, KS-27 Park, Sam's Pond

Hotspot type: All-day auto tour

This 60-mile drive along the **River Road** parallels the Arkansas River through most of Kearny and Hamilton Counties. Taking this road from the town of Lakin to the town of Coolidge offers a variety of birding opportunities and makes an excellent day trip. Before starting down the River Road, a brief visit to **Beymer Park**, an oasis of water and cottonwood trees located several miles south of Lakin on K-25 is usually worthwhile. Open water is at a premium in western Kansas, and the ponds in the park hold water year-round, attracting waterfowl and other water-dependent species. In summer, look for Say's Phoebe, Western Kingbird, Blue Grosbeak, Lark Sparrow, and Bullock's Oriole. Both Northern Bobwhite and Scaled Quail are possible at any season. A variety of migratory species have been seen here in spring and fall. Return north to Lakin, passing the extensive **Klotz Sandpits**, which are privately owned but can be viewed from the roadside. From fall through spring, these sandpits attract waterfowl, including various species of diving ducks, from late fall through spring.

The River Road begins at the southwestern corner of Lakin and proceeds west parallel to the Arkansas River, although not always close to the river channel. This road also follows the route of the old Santa Fe Trail, and there are marker signs for the trail along the road. The road traverses Kearney and Hamilton Counties to the town of Coolidge on the Colorado state line. There is little cultivated land on this drive, and the mostly undisturbed sand sage habitat is visually appealing as well as attractive to birds and other wildlife. Look for Scaled Quail and Black-billed Magpie at any season. Red-headed Woodpecker; Say's Phoebe; Eastern and Western Kingbirds; Cassin's, Lark, and Grasshopper Sparrows; and Bullock's and Orchard Orioles are common in summer. In winter, the raptor diversity can be excellent, with Bald and Golden Eagles and Rough-legged and Ferruginous Hawks possible, as well as Prairie Falcon and Northern Shrike. About 12 miles west of Lakin, the road converges with the river at the Lakin Diversion Dam, where water is diverted from the river channel for irrigation use. This is a good place to stop and carefully check for birds. From this point westward, there is usually at least some water in the river channel, and species diversity increases. Keep an eye out for cholla cactus shrubs, as even a single cholla may harbor a Curve-billed Thrasher. Reaching Kendall at the Hamilton County line,

Hotspot 11: Anywhere along the Arkansas River Scenic Drive, be alert for Scaled Quail near farms and ranches where old farm equipment is parked.

turn left at County Road Y and cross the river. On the south side of the river by the bridge, there is access to the river channel, where you can park and bird along the river and the adjacent thickets. Curve-billed Thrasher has been seen here. Just south of the bridge, Road Y merges into the River Road. From here, the River Road continues west to Coolidge at the state line. Be especially alert for Scaled Quail on this stretch, especially near farms and ranches where old farm equipment is parked. The road eventually reaches K-27 just south of the town of Syracuse. There are several worthwhile stops near Syracuse. **Sam's Pond** is a developed park area around a large sandpit lake on the south side of the river where interesting birds have been found over the years, including uncommon species such as Western Grebe, White Pelican, and Common Loon. On the north side of the river is the **K-27 Park**, where an impressive 184 species of birds have been recorded, including 11 flycatcher, 14 sparrow, and 17 warbler species; Summer and Western Tanagers; and a variety of other migrants. From the developed part of this park, you can follow well-worn dirt-bike and ATV trails along the river through the riparian corridor for as far as you care to go, although you will mostly need to do so on foot. Return to the River Road and continue west, eventually reaching the parking area for the **Cottonwood Flats WA**, where you can walk in to access more woodland habitat along the Arkansas River. Almost 100 species of birds have been recorded at this wildlife area. The River Road ends at K-27 south of Coolidge. Take the highway north. Just south of Coolidge is a bridge over the Arkansas River where you can stop to look for birds in the wooded corridor.

12. Hamilton County Wildlife Area

County: Hamilton

eBird Hotspot name: Hamilton SFL

Hotspot type: Worth a stop

Located 5 miles northwest of Syracuse, the **Hamilton County WA** covers over 500 acres. To reach it, take US 50 west from Syracuse for 3 miles. Turn north on Hamilton County Road M and go 2 miles to the lake entrance on your left. The entrance road crosses the top of the dam.

This hotspot was formerly known as Hamilton SFL. Western Kansas receives an average of only 18 inches of rain per year, and in drought years, there is sometimes no surface water at all at this hotspot. When there is water, it is often only near the dam and in a narrow channel going farther up the lake. The original lake bed was 94 acres in size, but the surface water now rarely exceeds 30 acres even in a good year, and there is considerable shrubby growth in most of the original lake bed. There are 12 acres of riparian timber near the upper end of the lake and an additional 338 acres of shortgrass prairie surrounding the lake. In most circumstances, you will need only an hour or two to bird this area, but it can be a worthwhile stop. Because of the lack of water, there is little usage of the area, and often you will likely be the only person(s) on the property.

Over the years, 19 species of waterfowl and 19 species of shorebirds have been observed at the lake. Cross the dam and follow the perimeter road that encircles the lake basin. At the northernmost point of the perimeter road is a small spring in the hillside on your left, and the largest trees at the lake are in this area. At the nearby low-water bridge is a small wetland area extending along a steep hillside that has dense thickets of aromatic sumac on the slope. The convergence of habitats at this spot can produce an interesting mix of bird species. Continue around the lake on the perimeter road. There are two dead-end side roads that go down to the lake bed and wooded areas. Birding along these can be worthwhile. During spring and fall migrations, this hotspot is one of the more reliable places in Kansas to see Brewer's Sparrow. Check flocks of Chipping and Clay-colored Sparrows carefully. Remember to look first for the clearly visible eye-ring of Brewer's.

13. Garden City Area

County: Finney

eBird Hotspot names: Concannon SFL, Finnup Park & Lee Richardson Zoo, Put-Em Back Pond, Valley View Cemetery

Hotspot type: Worth a stop

Finney County in southwestern Kansas has an all-time checklist of 355 species of birds, one of only seven Kansas counties with a list exceeding 350 species. The county seat of Garden City is the largest town in southwestern Kansas and acts as a major oasis for birds in an otherwise spartan environment. A dedicated group of ardent birders have lived in Garden City for many years. Many of the most impressive bird records from the county come from the yards of these birders, which typically have extensive feeder operations. An impressive eight species of hummingbirds have been observed at feeders and flower gardens in Garden City. The population of White-winged Dove in the city numbers in the hundreds.

Three public locations offering good birding in Garden City are **Valley View Cemetery** and the **Finnup Park/Lee Richardson Zoo** complex. Valley View Cemetery is located on the northern edge of town. The main entrance is on N 3rd Street roughly a half mile south of US 400. The large trees in the cemetery attract a variety of raptors, woodpeckers, flycatchers, warblers, sparrows, and finches during spring and fall migrations. Townsend's Solitaire is regular here in the winter months. Mississippi Kite is abundant in summer.

Finnup Park is a 110-acre park on the south edge of Garden City that includes the Lee Richardson Zoo. The main entrance is at 301 S 4th Street, south of Fulton Boulevard. Over 150 species of birds have been seen in the park. There are many mature trees and brushy areas that attract birds. The park gets a lot of recreational use, but an early morning walk usually provides quiet and rewarding birding.

Put-Em Back Pond on the southeast side of the city was formerly the Garden City WTP with restricted access. Now open to the public, it offers an excellent opportunity to view waterfowl, waders, and shorebirds, depending on the time of the year. A total of 130 species of birds have been reported at this modest-sized hotspot, including 21 waterfowl and 19 shorebird species. Arriving from the east on US 50/400, take the US 83 south exit (to Liberal) and head south across the highway and a railroad overpass. Then at 0.2 miles, take a left (east) onto an unnamed road off US 83 and then south on a paved road. At the next turn, exit south off the pavement when you see a sign, "Put-Em Back Pond." Continue on the sand road south and east about a half mile after making several turns. You will arrive on the north side of the pond.

If it has been a year with good rainfall, you might consider a trip to **Concannon SFL**, located about 17 miles east of Garden City on K-156. This state fishing lake is best thought of as a playa lake. As such, it is dry in most years and at those times is not of much interest to birders. However, when it does fill after major rains, the birding can be lively. Twenty species of waterfowl and 25 species of shorebirds have been recorded at the lake when it is holding water.

14. Manter Dam Recreation Area

County: Stanton

eBird Hotspot names: Manter Dam RA, Bear Creek—Rd. X

Hotspot type: Worth a stop

The **Manter Dam Recreation Area** is a seldom-visited location near the Colorado border. In an area dominated by shortgrass prairies and dryland farming, the timber and brushy habitats along Bear Creek offer an oasis of habitat for bird species that are generally absent elsewhere in the area. Allow two or three hours if you want to explore the area thoroughly. The name comes from a WPA project to create a dam and reservoir that began in 1935. It is hard to imagine today, but this was originally intended to be one of the largest reservoirs in Kansas. After two years of work on the dam, two major flood events caused substantial damage to the partially completed construction work. In 1937, the WPA withdrew funding, and the project was never finished. Today, the 662 acres of land that were purchased for the reservoir remain in the public domain. To reach the area, go 2.75 miles west on US 160 from the small community of Manter to Stanton County Road 17. Go 4.5 miles west on Road 17 (just past the crumbling walls of an old stone house on the south side of the road) and then 1.5 miles north on a gravel road that ends at a parking area. From the parking area, you can explore along the creek in either direction for over a mile. There are groundwater seeps in various locations along the creek, especially west of the parking area. These form pools that attract wildlife. There are numerous mature cottonwoods along the creek bed. If your time allows, explore the boulder-strewn grassland along the stream channel, the adjacent rocky bluffs, and the sumac thickets below them. West of the parking area is a prominent rocky bluff. Please be advised that no water, restrooms, camping, or any other amenities are available. The area is subject to limited access during hunting season.

This hotspot offers the most productive birding during spring and fall migrations, when a variety of passerine migrant species are possible. As with other western Kansas hotspots, these can be either eastern or western species. On an exceptional day, migrants can be abundant, but the birding can sometimes be decidedly slow. The area has not been visited much in the summer months, but expect species such as Western Kingbird, Warbling Vireo, Rock Wren, Cassin's Sparrow, and Bullock's Oriole. Scaled Quail and Curve-billed Thrasher are possible at any time of the year. In the winter months, there is little available food crop, and the area can be nearly devoid of birds.

Two nearby low-water bridge crossings of the creek are worth checking while you are in the area. These are both on private land, but birds can be observed from the roadside. From the Manter Recreation Area parking area, return to Road 17 and go west for about a half mile to reach the first crossing. Stop, look, and listen for birds. To reach the second crossing, go back east for about a mile to Stanton County **Road X** and go north for 1 mile to reach the low-water bridge. This crossing is an eBird hotspot and has produced interesting birds in past years.

15. Ulysses Area

County: Grant

eBird Hotspot: Frazier Park

Hotspot type: Half day

Frazier Park in Ulysses offers a variety of habitats and is one of the better birding hotspots in southwestern Kansas. As of 2023, 204 species of birds have been found at this park. Allow several hours to bird the park thoroughly. Look for the highway sign for the park on US 160 near the east edge of town. Proceed south on Stubbs Road for about a quarter mile to reach the entrance to the park on the left. As you enter the park, you will pass an impressive (at least by western Kansas standards) stand of trees. There is a parking area near the wooded area, or you can continue to Frazier Lake, where the road ends at the main parking lot. There are several miles of trails in the park that go through a variety of habitats, including wetlands, grasslands, and wooded areas. The lake is adjacent to the Bentwood Golf Course as well as the city of Ulysses water treatment plant. As at Arkalon Park near Liberal, wise use is made of the effluent water from the WTP, which is directed through a series of streams, waterfalls, and wetlands to filter the water before it enters Frazier Lake, which also has a substantial aerator fountain. As a result, the water quality of the lake is good. There is a modern campground in the park.

The woodland and brushy habitats in the park have attracted many interesting migratory flycatchers, thrushes, vireos, warblers, grosbeaks, sparrows, and other bird species in spring and fall. Seventeen species of warblers have been observed in Frazier Park, an excellent number for western Kansas. Follow the footpaths along the wooded creek channel and carefully check for birds. There is a large area of tall grasses and forbs near the inlet to the lake that typically hosts a good variety of sparrow species from late fall through spring. Closer to the lake, the weeds give way to an area of wetland plants and mudflats where shorebird, heron, and waterfowl species can be found according to the season. Twenty-two species of waterfowl and 17 species of shorebirds have been recorded on or near the lake. Waterfowl are usually abundant from fall through spring.

Lakin Draw is a nearby location that is worth the brief amount of time it requires to check for birds. As you exit the park, go back north on Stubbs Road. Cross the highway, go two more blocks, and turn right (east) onto E Hampton Avenue. After about a half mile, stop at the small bridge near a limestone marker for Wilson's Landing. Scan the wetland area south of the road. This can be dry or nearly so, but there is usually some open water. A respectable number of interesting birds have been found here, including one of the westernmost Kansas records of Yellow-crowned Night Heron. Continue a bit farther east on Hampton Avenue and turn right (south) onto Road L. Cross the highway again, stop, and scan the low area west of the road. In years with sufficient moisture, there is surface water here, and shorebirds, including Black-necked Stilt and other species, have been seen.

16. Cimarron National Grassland

County: Morton

eBird Hotspot locations: Elkhart Cemetery, Shelterbelt & WTP, Cimarron NG—Middle Spring, Cimarron NG—Recreation Area, Cimarron NG—Tunnerville Work Center, Cimarron NG—Turkey Trail, Cimarron NG—Point of Rocks, Cimarron NG—Cottonwood Picnic Area, Cimarron NG—Boy Scout Area, Cimarron NG—Western Crossing, Cimarron NG—Wilburton Crossing.

Hotspot type: Major destination

The Cimarron National Grassland (CNG) and the adjacent town of Elkhart are among the most popular western Kansas destinations for birders. Many birders spend two or three days exploring the area, which covers over 108,000 acres of land. Because of the location in the extreme southwestern corner of the state, many western species

Hotspot 16: In Cimarron National Grassland, the Sante Fe Trail passes just below Point of Rocks, which overlooks the wooded valley of the Cimarron River.

of birds that are rare elsewhere in Kansas are relatively easy to find. The CNG is managed by the US Forest Service. It was created after the dust bowl years of the 1930s, when the federal government bought up a substantial portion of the land in Morton County, removed it from cultivation, and began to restore native vegetation on the barren landscape. Today, healthy riparian woodlands are found along the Cimarron River, flanked by shortgrass prairie and sage habitats. Lying just to the south of the CNG, the town of Elkhart is an oasis of trees and water that attracts many interesting birds.

During both the spring and fall migrations, a variety of western species are possible on the CNG, including Rufous Hummingbird, Common Poorwill, Dusky Flycatcher, Townsend's Warbler, Western Tanager, and many others. The months of May and September are the most popular with birders, as the greatest variety of migratory western species can be found then. Also seen each year are a surprising number of migratory eastern species that have wandered westward from typical migration routes. In summer, look for abundant Cassin's Sparrow, Lark Bunting, and Orchard and Bullock's Orioles. Rock Wren is present in the rocky outcrops north of the Cimarron River, and large prairie dog towns attract Burrowing Owl. There was formerly a robust population of Lesser Prairie-Chicken on the CNG, but successive years of drought have drastically reduced their numbers, and they are no longer reliably found here. Scaled Quail are present year-round, usually near farms and abandoned farm equipment. During the winter months, a substantial population of Ferruginous Hawk is found on the CNG along with many other raptor species, such as Cooper's Hawk, Merlin, Prairie Falcon, and both Bald and Golden Eagles. Other winter specialties include Long-eared Owl, Townsend's Solitaire, and Chestnut-collared and Thick-billed Longspurs.

Begin a tour of this area in the **Elkhart Cemetery** at the eastern edge of town and the dense shelterbelt north of the cemetery, where many exciting birds have been found over the years. Pay special attention to the shelterbelt and the adjacent grove of locust trees. Many warblers, vireos, tanagers, flycatchers, and other species have been found in this small area during migration, including many outstanding rarities. American Barn Owls often roost in the dense cedars of the shelterbelt. The adjacent **Elkhart WTP** used to be open for birders, but recent regulatory action has closed access to the roads inside the fence. However, most of the surface area of the three pools can be viewed from the southwestern corner of the WTP at the dead end of Lagoon Lane. As the only significant body of water anywhere in the county, it attracts an impressive variety of shorebirds, gulls, terns, waterfowl, and other water-dependent species. Black-crowned Night Herons have nested in the trees around a small pool outside the southeastern corner of the fenced area. Red-necked Phalarope is seen during fall migration in most years, and Sabine's Gull has been recorded several times. An amazing 305 species of birds have been recorded at least once in the small area occupied by these two adjoined hotspots.

After checking this area, proceed north on K-27 for about 3 miles to the **Tunnerville Work Center**. There used to be extensive juniper and pine plantings here where many interesting birds were found, but many of the trees have died and been removed. However, it is still worth checking the remaining trees and brush. Scaled Quail are seen at or near the work center in some years. From the work center, continue north on K-27 to the Cimarron River. In Morton County, most of the land along the river is within the CNG, where you are free to explore as you wish. There are roads that follow both the north and south sides of the river and offer many access points. The **Cottonwood Picnic Area** just off K-27 on the south side of the river is always worth a stop. At the east end of the picnic area, a dirt road (called the **Turkey Trail** in eBird) begins and winds eastward for several miles through the open cottonwood and tamarisk woodlands on the south side of the river. The Turkey Trail road eventually returns to the South River Road. From there, you can continue east to the **Cimarron Recreation Area**, where there are campgrounds, restrooms, and several small excavated ponds. This is another location where many good birds have been found. Park and walk around for a while. A huge range fire burned through the CNG, including the recreation area, 20 years ago, and it is still easy to see the lingering evidence of the damage. Return to the South River Road and follow it east to the paved Wilburton Road. Turn north and stop at the river crossing. This spot is called the **Wilburton Crossing**. Like all the CNG birding locations, it can be devoid of birds or quite lively. Remember this is all public land, and you are free to roam as far down the river channel as you wish. After you are done here, you can proceed north to the North River Road and drive west for 3.2 miles to an unmarked dirt road that you can follow down the hill to one of the more legendary birding locations on the CNG: the **Boy Scout Area**. Park at the bottom of the hill and walk east through the cottonwoods and tamarisks below the bluffs. During spring and fall migrations, just about anything is possible here. The big holes in the cliffs usually have roosting American Barn Owls.

To visit two of the best-known sites on the CNG, go about a half mile north of the bridge over the Cimarron River on K-27 and turn left (west) onto the North River Road. Drive west for about 2 miles to reach **Middle Spring**, a watered oasis where many interesting birds have been found. One mile farther west, you reach a famous landmark called **Point of Rocks**, which offers an excellent view of the river valley. Rock Wrens are easy to find on its rocky outcrops. Check the scrubby brush below the rim for birds. The North River Road continues west to the state line. When driving this road after dark, drive slowly and be alert for Common Poorwills and Burrowing Owls standing in the road. About 2 miles from the Colorado line, you can take a county road south to conclude the "typical" birding tour of the CNG. This road crosses the river at a place called the **Western Crossing**. This location has fewer and smaller trees than the other river crossings but has produced many good birds. Stands of larger timber can be reached by hiking west along the river channel.

17. Hugoton Area

County: Stevens

eBird Hotspot names: Hugoton Storm Runoff Ponds, Hugoton WTP, Hugoton Cemetery

Hotspot type: Worth a stop

Located on US 56, the city of Hugoton is not a destination hotspot, but it does offer several birding locations that are worth checking if your travels take you through the area. On the north edge of Hugoton, turn south from US 56 at the Dirtona Raceway onto Road 13. A short distance south, turn in at the entrance sign for the city landfill to view the **Hugoton WTP** from the entrance road (closed on Sundays). Please be sure to park well to the side of the road, as there is traffic coming and going from the landfill farther down the road. These treatment impoundments are sizable and usually have a variety of shorebirds and waterfowl. During migration, American Avocet, Black-necked Stilt, Wilson's and Red-necked Phalaropes, and other shorebird species can be seen. Ducks and geese are usually present in season. Snow, Ross's, Cackling, and Canada Geese are often abundant from late fall through early spring. Look and listen for Scaled Quail around the perimeter of the treatment ponds and landfill.

Continue a short distance south on Road 13 to the intersection with E First Street/Road P and go east a few blocks to Cemetery Road. Most of the storm water for the city drains into the **Hugoton Storm Runoff Ponds**, located along Cemetery Road. Even in drought years, there is usually at least some water present. Numerous species of waterfowl, shorebirds, and herons have been seen here. On the west side of the pond is an area of brush and trees (accessible from Road P) where a notable number of migratory bird species have been found in spring and fall in past years. This undeveloped area is owned by the city and has trails for dirt bikes and ATVs that you are free to walk on. Many species of warblers and other migrant passerines have been seen in this small patch of habitat.

Continue south on Cemetery Road, cross K-51, and enter the **Hugoton Cemetery**. White-winged Doves sometimes join the abundant Eurasian Collared-Doves here. At the south edge of the cemetery, look for birds in the adjacent sand sage habitat. Curve-billed Thrashers can sometimes be seen in the cemetery perched on top of the manicured juniper bushes and probably have nested in the past.

18. Arkalon Park

County: Seward

eBird Hotspot names: Arkalon Park

Other Hotspot: West River Road

Hotspot type: Worth a stop

Located in the Cimarron River valley, **Arkalon Park** is an oasis of riparian woodland and wetland habitats offering rewarding birding. Two hundred nine species of birds have been reported at the park. This is not a stand-alone destination, but if a trip through the area takes you near the park, it can be a productive birding stop. Allow two or three hours to cover the area thoroughly. Named for a now-vanished town from the nineteenth-century settlement era, this large rural park is owned by the city of Liberal and is located in a corridor of riparian cottonwood forest along the Cimarron River. Formerly closed seasonally, the park is now open year-round. The park entrance is 10 miles northeast of Liberal on US 54. A small sign on the highway at the entrance marks the turn off the highway onto Arkalon Park Road. This road descends the hill to the campgrounds and woodlands along the river. Greater Roadrunner has been reported occasionally along the entrance road. In spring and summer, listen for Cassin's and Lark Sparrows singing in the sand sage habitat adjacent to the road. About halfway down the hill, you can turn left and take a park road to access a series of three terraced ponds that are fed by treated wastewater from the city of Liberal. The largest and deepest pond is at the top of the hill. These ponds e attract a variety of waterfowl, shorebirds, and other water-dependent species. Continue down the hill to reach the campground and developed park area. Several miles of maintained nature trails begin in the campground area and wind through the substantial undeveloped areas of the park. These trails pass through grassland, brushy, and wooded habitats and provide good birding opportunities. Some of these trails access the marshy areas adjacent to the river, where Black-crowned Night Heron, Green Heron, Marsh Wren, Common Yellowthroat, and other wetland species have been found. In the summer months, the park has a robust population of Red-headed Woodpecker, Yellow-billed Cuckoo, Mississippi Kite, Eastern and Western Kingbirds, Eastern Bluebird, Lark Sparrow, Blue Grosbeak, and Baltimore and Orchard Orioles. About 1 mile upstream from the campground area, Bald Eagles have nested, which is noteworthy in this dry landscape. During both fall and spring migration, many interesting migrants have been found at Arkalon, including numerous flycatcher, vireo, sparrow, and warbler species.

As you exit the park at US 54, turn left (northeast) and drive about 1 mile. Just after crossing the river near the Big Samson Bridge, turn right (west) onto the **West**

River Road, which follows the river channel for several miles until it reaches a dead end. There are only two or three homes on this stretch, and the sand sage habitat is largely intact. This stretch of road can produce a lot of birds. Listen for Common Poorwills calling at dusk from the bluffs to the north. Chuck-will's-widow and Whip-poor-will have been heard calling from the riparian timber in May, notably far west for both species.

19. Meade State Park

County: Meade

eBird Hotspot names: Meade SP, Meade SP—North Ponds Hatchery

Hotspot type: Worth a stop

An isolated oasis of trees and water on the cultivated western plains, **Meade State Park** is the westernmost outpost for several eastern breeding species and has a history of attracting unexpected birds. At least 217 species of birds have been seen at this hotspot. It is located about 13 miles southwest of the town of Meade. Due to its location in the semiarid High Plains, the lake acts as a magnet for waterfowl, rails, shorebirds, and other aquatic species. An extensive area of woodland at the upper end of the lake, cattail marshes, and shortgrass prairie provide diverse habitats for the birdlife.

To reach Meade State Park, go south from Meade on K-23 for 8 miles. The highway then curves to the west, and the entrance to the park is reached after another 5 miles. There are paved roads and campgrounds on both sides of the lake adjacent to the dam. The trailheads for several hiking trails through the extensive undeveloped woodland areas are in the campgrounds. Meade County V Road is a perimeter road on the north and west sides of the state park. On V Road at the west end of the state park is the Fish Hatchery and the Artesian Well Campground. This is a good place to park and spend a little time. You can hike a short distance north to reach the **North Fish Hatchery Ponds**, where shorebirds and other waterbirds are possible, depending on the season.

During the summer months, resident birds include Mississippi Kite, Chuck-will's-widow, Great Crested Flycatcher, and Baltimore and Orchard Orioles. In the winter, look for multiple species of waterfowl if the water is unfrozen; a variety of raptors, including Prairie Falcon; Townsend's Solitaire; and several winter sparrow species. A large fire killed many trees in the park during the 1990s, and there is still substantial standing dead timber that attracts multiple species of woodpeckers year-round. Barred Owl and Carolina Chickadee are year-round residents, and this is near the western limit of their range in Kansas. Be sure to check the large stands of cattails for rails and Marsh Wren. During spring and fall, birding can be exciting, as numerous migrants can occur when conditions are favorable. Bird-banding operations conducted at Meade State Park over many years have documented an exceptional variety of migratory birds.

20. Lakeview Playa

County: Meade

eBird Hotspot name: Lakeview Playa

Hotspot type: Worth a stop if there is water

Meade County and other areas of southwestern Kansas have numerous ephemeral wetlands known as "playas." When water is present, playas attract numerous waterfowl, herons, and shorebirds that can sometimes seem as though they appeared out of thin air. One of the largest and most accessible of these is **Lakeview Playa**. To reach this playa from the intersection of US 54 and US 160 (1 mile east of Meade), go east for 6 miles on US 160, and you will see the playa on the south side of the road. Although it is on private land, this playa can be easily viewed from the highway right-of-way and from Meade County Road 26 on the east end of the playa. In times of drought, this playa can be completely dry and birds are largely absent. When there has been sufficient precipitation, the surface water can cover several hundred acres and birding can be exceptional. The list of shorebirds, waterfowl, and other aquatic bird species that have been observed during migration at this playa is extensive. Decades ago, there were numerous dead trees standing in the water that hosted a nesting colony of Black-crowned Night Herons but the trees are gone now. During the winter months, large numbers of geese use the playa, and Short-eared Owls are often seen at dawn and dusk.

North-Central Region

North-central Kansas represents a gradual but steady transition from eastern to western landscapes, plant communities, and wildlife. Most of the north-central region as we have defined it in this book lies in the Smoky Hills physiographic region. The High Plains physiographic region extends eastward to parts of Graham, Norton, Phillips, Rooks, and Trego Counties. The Smoky Hills region as a whole has less land area devoted to cultivation than other parts of the state. Extensive rolling hills covered with grasslands and scattered rock outcrops create a visually appealing landscape.

The principal rivers in the region are the Republican, Saline, Smoky Hill, and Solomon, all of which flow in a generally eastern direction. The cumulative watershed area of these rivers and their tributaries drains a huge area of Kansas. Within the Republican and Solomon River systems, riparian oak forest extends to the western edge of the region and beyond. The tree community on the Saline and Smoky Hill Rivers is dominated by cottonwood and hackberry.

During the 1950s and 1960s, the Bureau of Reclamation and the US Army Corps of Engineers constructed a series of major reservoirs on these rivers for flood control and irrigation. These reservoirs and their surrounding lands constitute most of the public lands in north-central Kansas and consequently represent many of the hotspots in this area of the state. These are Cedar Bluff, Glen Elder, Kanopolis, Kirwin, Keith Sebelius, Lovewell, Webster, and Wilson Reservoirs. Kirwin is the only national wild-

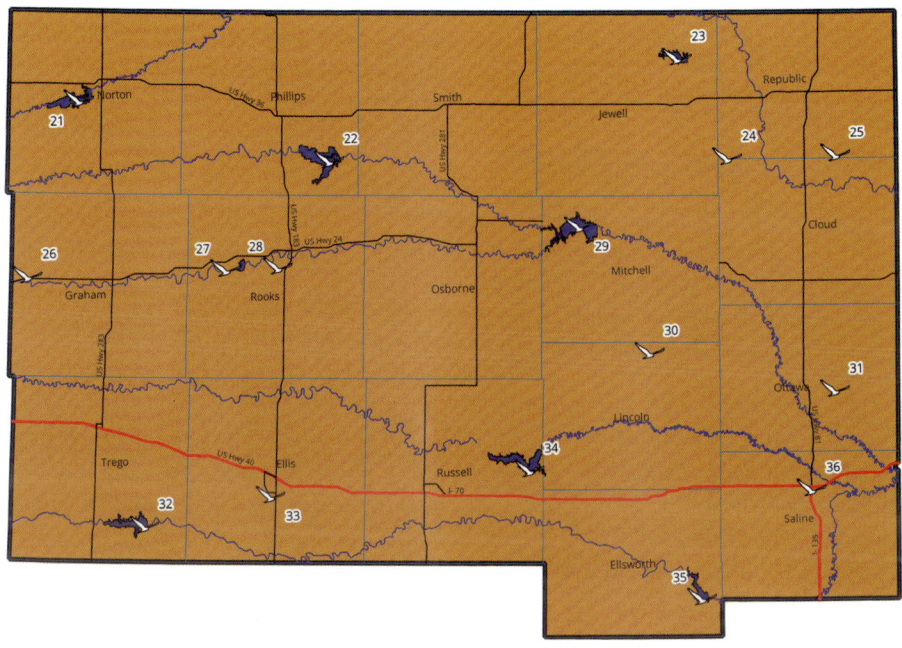

North-Central Region Hotspots

21. Keith Sebelius Reservoir

22. Kirwin National Wildlife Refuge

23. Lovewell Reservoir

24. Jamestown Wildlife Area

25. Talmo Wildlife Area

26. Antelope Lake

27. Webster Reservoir

28. Rooks State Fishing Lake

29. Waconda/Glen Elder Reservoir

30. Gurley Salt Marsh

31. Ottawa State Fishing Lake

32. Cedar Bluff Reservoir

33. Hays Area

34. Wilson Reservoir

35. Kanopolis Reservoir

36. Salina Area

life refuge in the region. As detailed below, these reservoirs attract huge numbers of birds and a diversity of species.

There are several important surviving wetlands in the region that predate European settlement of Kansas. These are the Gurley Salt Marsh, Jamestown Wetlands, and Talmo Marsh. All of these are owned and managed by the KDWP and preserve vital habitat.

Eastern bird species that are near the western limit of their range in the region are largely the same as those in the south-central region and tend to be found along wooded stream corridors. In general, the range of these species is gradually pushing westward as the woodland plant and tree community matures and expands. These species include Red-shouldered Hawk, Barred Owl, Pileated Woodpecker, Eastern

Wood-Pewee, Carolina Wren, Red-eyed Vireo, Eastern Towhee, and Rose-breasted Grosbeak. Western species such as Ferruginous Hawk, Burrowing Owl, Say's Phoebe, Black-billed Magpie, Rock Wren, Mountain Bluebird, Spotted Towhee, and Black-headed Grosbeak are all found in the north-central area, depending on the time of year.

21. Keith Sebelius Reservoir

County: Norton

eBird Hotspot names: Keith Sebelius Res., Norton Rest Area, Norton WA, Norton WA—Bluff Overlook, Norton WA—Darlington Access, Norton WA—Horseshoe, Norton WA—Longspur Area, Norton WA—Sand Pit, Norton WA—Schoens Cove, Norton WA—Three Jakes, Prairie Dog SP

Hotspot type: Half day

Keith Sebelius Reservoir and the adjoining **Norton WA** are located in Norton County, just west of the town of Norton. The reservoir was created in 1964 by the construction of Norton Dam on Prairie Dog Creek. The lake and wildlife area are owned by the Bureau of Reclamation and managed by the KDWP. The reservoir covers 2,200 acres when full and has a maximum depth of 42 feet. Prairie Dog State Park on the north shore of the lake is 1,150 acres in size, and Norton WA covers an additional 6,500 acres of land surrounding the lake. While most of the major reservoirs in north-central Kansas are within the Solomon River watershed, Prairie Dog Creek lies within the Republican River watershed.

A counterclockwise birding tour around the entire lake and wildlife area can be accomplished in a half day. Begin at **Prairie Dog State Park**, which extends along 2 miles of shoreline on the north shore close to the dam and offers good views of most of the lake. The park entrance is via K-261 at the intersection with US 36 about 4 miles west of the town of Norton. The park office is 1 mile south of the highway on K-261. Check the pine trees around the office for a few minutes. Just south of the office is a small prairie dog town where you might see a Ferruginous Hawk or Burrowing Owl. The various park roads have multiple spots where you can scan the lake for birds. When water levels in the lake are low, (which is often the case), good numbers of shorebirds, gulls, and terns can be seen loafing on the exposed flats and points. Return to US 36 and go 2 more miles west to the junction with K-383. Go south for 1.5 miles to the **Norton Rest Area**, where there are numerous mature pines. Species such as Townsend's Solitaire and Red-breasted Nuthatch have been found at the rest area in winter. Cross K-383 at the rest-area entrance and take the gravel road west and then south to the **Darlington Access**, which has some of the better woodlands in the wildlife area. This is a good area for summer songbirds.

To access the south side of the lake, continue south on K-383 (crossing Prairie Dog Creek). Immediately south of the bridge, access roads lead eastward along the south shoreline for several miles, providing access to the **Three Jakes** and **Sandpit** campgrounds. Three Jakes has extensive stands of sumac and other shrubs and has been a reliable location for Yellow-breasted Chat and Spotted Towhee during the breeding season. This area of the lake still has many dead trees and stumps standing in the water that date back to the construction of the lake and so has a bit of an odd feel. From the Sandpit campground and eastward, O Road winds east to the dam with turnouts for several other access points to the lake, including **Schoens Cove** about a mile to the east. Closer to the dam is a small parking area for the rock formation known as the **Bluff**, which provides a good vantage point for the lake and the surrounding landscape. Continue east for another 2 miles to the dam. Adjacent to the dam on the south shore is the **Longspur** camping area, from which you can scan the lake close to the dam. The Longspur area has a good stand of trees and brush that has produced a decent list of songbirds during both spring and fall migrations. At the north end of the dam is a gravel road leading down to the shore next to the dam. This offers one final vantage point of the lake and along the face of the dam.

A total of 249 species of birds have been observed in Norton County, and the vast majority of these have been seen at Keith Sebelius Reservoir. Illustrating the birding potential of the area, during the 2013 Kansas Ornithological Society (KOS) spring meeting based in Norton, 203 species of birds were identified over a three-day period. This still stands as the highest number of species found at any spring KOS gathering anywhere in the state. While some of these sightings were from adjacent counties, it does show the exceptional potential of Norton as a birding hotspot. Typical of other north-central reservoirs, birds reported from Sebelius/Norton include over 25 waterfowl, 27 shorebird, 6 gull, 5 grebe, 12 hawk, and 19 sparrow species. Eastern woodland species such as Red-bellied Woodpecker, Eastern Wood-Pewee, Eastern Phoebe, Red-eyed Vireo, and Indigo Bunting are present in the appropriate habitats. Western species seen regularly according to the time of year include Ferruginous Hawk, Burrowing Owl, Say's Phoebe, Black-billed Magpie, Mountain Bluebird, Townsend's Solitaire, and Chestnut-collared Longspur. Both Black-headed and Rose-breasted Grosbeaks are seen in migration and in the breeding season. Spotted Towhee is another breeding species that should be looked for in summer. Some notable finds have included Surf Scoter, Red-breasted Merganser, Clark's Grebe, Piping Plover, Whimbrel, Ruddy Turnstone, Sabine's Gull, Lesser Black-backed Gull, and Scarlet Tanager.

22. Kirwin National Wildlife Refuge

County: Phillips

eBird Hotspot names: Kirwin NWR, Kirwin NWR—Bluegill Point, Kirwin NWR—Cottonwood Grove, Kirwin NWR—Crappie Point, Kirwin NWR—Dam & Outlet Park, Kirwin NWR—Knob Hill, Kirwin NWR—North Dam, Kirwin NWR—Prairie-Dog Town, Kirwin NWR—Scout Cove, Kirwin NWR—Silver Bridge, Kirwin NWR—Solomon Bend, Kirwin NWR—South Dam, Kirwin NWR—South Shore Boat Ramp

Hotspot type: Half day

Kirwin NWR covers 10,778 acres of land in Phillips County, including Kirwin Lake. The lake covers about 5,000 surface acres when full, but the reservoir is frequently below conservation-pool water level. Kirwin was the first national wildlife refuge in

Kansas, established in 1954. The reservoir was created by the Bureau of Reclamation in 1955 with the construction of Kirwin Dam on the North Fork of the Solomon River. The national wildlife refuge is an overlay on the Bureau's Kirwin Reservoir project, which holds the actual water rights to the lake. The land area within the refuge is a mix of about 2,000 acres of cropland and 4,000 acres of mixed-grass prairie. The tributary streams for the reservoir are Bow Creek and the North Fork of the Solomon River. There are well-established woodlands along both streams. The refuge is an area of rolling hills, with the tops of some of the hills 200 feet higher than the bottoms of the wooded stream corridors. Due to its primary purpose as a refuge, there are no developed park or campground areas at Kirwin, although there are several boat ramps and restroom facilities.

The refuge headquarters is a good place to begin a tour of Kirwin. From Phillipsburg, go south on US 183 to the small town of Glade, then take K-9 east for 6 miles. Look for the refuge sign and turn south on County Road 700. The refuge headquarters is 1 mile south of the highway on Xavier Road. There is a modest visitor center and interpretive display at the headquarters. From the headquarters, go west on Xavier Road. The first mile is often one of the best shorebird spots at Kirwin. There is an extensive wetland area north of the road where shorebirds are often found during spring and fall migrations. At the 1-mile mark is a slight curve in the road, and there is a good vantage point for the lake here marked on eBird as the **Solomon Bend**. If it is not gated, Xavier Road follows the Solomon River channel for several more miles, eventually returning to K-9. Return to the headquarters and go 1 mile east, turning right onto the North Refuge Road. The road forks almost immediately. Take the left fork. For the next mile, there are parking lots at several good vantage points where you can scan the lake, including the **North Shore Boat Ramp** and **Cottonwood Grove**. The refuge road reaches the intersection of Xavier and Warrior Lane. Remain on N Refuge Road for another 2 miles. The **Knob Hill** and **North Dam** parking lots are on this stretch, offering additional good vantage points where you can scan the lake. There is not a road across Kirwin Dam, so to reach the south side of the lake, you will need to go into the town of Kirwin and take 1st Street south for about 2 miles to the intersection with S Kirwin Lake Road. Between the town and the intersection, you will pass the **Outlet Park**. You can stop here and check the extensive cattails in the channel for rails, herons, Marsh Wren, and wetland sparrow species. At the end of the dam is a short access road to the **South Dam** parking area, where you can scope for waterbirds along the face of the dam. Continue west on the South Refuge Road for about 7 miles. On this stretch are parking areas where you can check the lake for birds at the **South Shore Boat Ramp**, **Bluegill Point**, **Crappie Point**, **Scout Cove**, **Kiln Access**, and **Silver Bridge**. Silver Bridge is in a wooded area along Bow Creek and can be good for woodpeckers, songbirds, and other woodland species. Just after Silver Bridge, turn right onto E 700 Road and go north for 3 miles to the intersection with Yankee Road. From the corner, go east for 1 mile to the **prairie dog town** and check for Burrowing Owl in the warm season and other raptors year-round. Go back west on Yankee Road for 5 miles. You will pass the **Willow Flats** and **Quillback** parking areas, which have shoreline access. When you reach E 300 Road, turn north and go

2 miles to return to K-9. If you are there near dawn or dusk during the colder months, Short-eared Owls can be seen on Yankee Road between Quillback Cove and 300 Road from late fall through early spring. This stretch is also good for other raptors. Listen for booming Greater Prairie-Chickens at dawn from March through May.

Kirwin is most famous as a migration and winter staging area for immense flocks of waterfowl. Flocks of Snow Geese, Canada Geese, Mallards, and other species sometimes number in the hundreds of thousands. Kirwin is also one of the critical stops for migrating Whooping Cranes in spring and fall. Shorebird numbers vary substantially as water levels in the lake rise and fall but can be impressive when conditions are ideal. Thirty shorebird species have been reported at Kirwin, including noteworthy species such as Piping Plover, Long-billed Curlew, Short-billed Dowitcher, and Red-necked Phalarope. In early June of 2015, an experienced birder found 30 Snowy Plovers on flats at Kirwin, which is one of the largest concentrations ever reported in Kansas away from Quivira NWR and Cheyenne Bottoms, strongly suggesting potential breeding. Sabine's Gull has been reported in fall in multiple years. Numbers of Double-crested Cormorants are large during migration, and in some years, they have a nesting colony on the refuge. Large flocks of American White Pelicans are also seen during migration, and a few nonbreeding individuals linger through the summer. Western species including Burrowing Owl (summer), Say's Phoebe (summer), Black-billed Magpie (year-round), Mountain Bluebird (winter), and Townsend's Solitaire (winter) are seen in most years. Because of the scarcity of woodlands, the number of sparrow and warbler species that have been reported at Kirwin is modest in comparison to those reported at other north-central Kansas hotspots. There are quite a few summer records for Yellow-headed Blackbird, and this species likely nests in suitable habitat in some years.

23. Lovewell Reservoir

County: Jewell

eBird Hotspot names: Lovewell Res.—Dam Overlook, Lovewell Res., Lovewell Res.—John's Creek, Lovewell Res.—Oak Hill RA, Lovewell Res.—Pawnee Point, Lovewell Res.—Two Mile Access, Lovewell SP, Lovewell WA

Other hotspots: Montana Creek Access

Hotspot type: Half day

Lovewell Reservoir is located about 18 miles north of Mankato in Jewell County. The 2,900-acre lake was created by the construction of a dam on White Rock Creek in 1955 and has a maximum depth of 35 feet. Surrounding the lake is the 2,100-acre Lovewell WA. On the north shore of the lake is the 1,160-acre Lovewell State Park. The geography of the area is interesting and gives this hotspot exceptional visual appeal. The deeply dissected Niobrara escarpment crosses much of Jewell County. It is composed of Fort Hays Limestone and clearly marks the geological division between the High Plains and the Smoky Hills. It is one of the most prominent formations of its kind in Kansas, and its northern edge borders White Rock Creek, including the entire southern shore of Lovewell Reservoir, reaching its eastern extremity

© Pete Janzen

Hotspot 23: Oak woodland habitat along John's Creek at Lovewell Wildlife Area provides a western outpost for eastern species such as Red-shouldered Hawk, Chuck-will's-widow, Barred Owl, and Pileated Woodpecker.

near the reservoir dam. These bluffs are 150–200 feet above the lake surface, giving Lovewell an overall visual appearance that is distinct from that of any other central Kansas reservoir. Another unique aspect of Lovewell is the extent of upland deciduous woodland dominated by oaks, especially on the south and west sides of the lake. As noted elsewhere in this book, oak woodlands can be found much farther west in Kansas in the counties along US 36 but are generally confined to riparian corridors. At Lovewell, the oak forest covers the slopes of the escarpment for many miles, further contributing to the visual appeal.

More than any other Kansas reservoir, Lovewell is used as an irrigation source. There is a sophisticated water control structure at the reservoir outlet. The outlet channel connects directly to a network of irrigation canals and related infrastructure that provides water to farmland in Jewell and Republic Counties. The Courtland Canal provides water to the reservoir from the Republican River north of the lake. Especially in late summer, there are dramatic drawdowns of water from Lovewell into the irrigation canals. At these times, there can be substantial areas of shallows and mudflats that attract shorebirds, gulls, and herons.

A typical birding visit to Lovewell begins with a stop at the **Dam Overlook**, where there is a panoramic view of most of the lake. The small area of woodland at the **Outlet Park** is usually worth a few minutes to check for birds. From the dam area, go north on 260 Road to Y Road (also called North Shore Road) and turn west. Just after crossing the adjacent railroad tracks, there is an access road to the Courtland Canal inlet. This road is sometimes closed, but if it is open, it is worth checking the inlet area. Continue west on Y Road to 250 Road, then south on 250 Road for 1 mile to the entrance and offices for **Lovewell State Park**. The state park covers over 1,100 acres and includes a lot of shorelines. The park is extensive and well-kept. It receives heavy usage for much of the year. There are lots of mature trees and unmowed areas, and the birding can be quite good in the park. Excellent viewing points for the lake within the state park are at the marina as well as the Cottonwood, Walleye Point, and Willow Campgrounds.

Return to Y Road/North Shore Road and continue west for 4 miles until you reach K-14. There are several wildlife area roads that lead from North Shore Road to the lake. These include the **Montana Creek**, **Pawnee Point**, and **Two Mile** access areas. These are sometimes gated, but foot traffic is permitted unless posted otherwise. Montana Creek is closest to the upper end of the lake and perhaps the most interesting of these. Here there is older hardwood forest, especially across the river channel. The "woodland trio" of Red-shouldered Hawk, Barred Owl, and Pileated Woodpecker have colonized Lovewell in recent years, and all have been seen at the Montana Creek area.

The south shore of the lake is far less developed and receives much less public usage than the north shore. It offers equal or better birding potential than the north shore areas. To reach it from the North Shore Road/K-14 intersection, go south on K-14 for 4 miles to U Road and turn east. After 1 mile, turn north on 220 Road. The road goes north for 3 miles along **John's Creek** and then turns east, following the shoreline for another mile to a dead end. Thick oak forests grow along the creek parallel to the road and on the adjacent bluffs for much of this drive. The last 2 miles of this road are within the Lovewell WA. The grasslands and wooded areas are mostly undisturbed and attract many species of birds. The final mile along the lakeshore provides an unobstructed view of the upper end of the reservoir and is a good area to find shorebirds, gulls, and herons when lake levels are low. Return to U Road and go east for 2 more miles to 240 Road. Go north on 240 Road for 2 miles. The road gradually curves to the left and enters the **Oak Hill Recreation Area**, where there is a boat ramp, campground, and several spots where you can view the lake easily. As at the John's Creek drainage, there is a lot of oak woodland. If you have birded all the areas described above, you have essentially covered the entire reservoir and surrounding public lands.

Well over 200 species have occurred at Lovewell. Twenty-five waterfowl species have been seen, and geese sometimes numbers in the tens of thousands in late fall and early winter. Expect large numbers of Common Goldeneye and Common Merganser in winter when the water is open. Common Loon and four species of grebe, including Western Grebe, have been seen during migration. Twenty shorebird species have been found, and the potential for additional species is good. Nine heron species have included Yellow-crowned Night Heron. Seven gull species have been seen, including Glaucous and Sabine's. Eleven hawk species and all four falcon species have occurred. As at the other major reservoirs, Bald Eagles concentrate at Lovewell in the winter, when large numbers of waterfowl are present and the lake is partially frozen. Chuck-will's-widow and Summer Tanager are summer residents of the wooded slopes on the south shore and are at or near the western edge of their breeding range in northern Kansas. In summer, listen for Common Poorwills calling along V Road south of the dam. Ten warbler and 20 sparrow species have been recorded, largely during migration. As at some of the other north-central hotspots, western breeding species such as Say's Phoebe and Black-billed Magpie nest in Jewell County, as do eastern species such as Eastern Wood-Pewee, Red-eyed Vireo, and Blue-gray Gnatcatcher.

One of the more interesting aspects of the birdlife at Lovewell is that the breeding ranges of Eastern and Spotted Towhee appear to overlap here. There are multiple records of both species from the summer breeding season. Eastern Towhee is a breeding species of eastern Kansas, especially in the Flint Hills, while Spotted Towhee is a breeding species of the Nebraska border counties in the western half of Kansas. All sightings of both species should be uploaded to eBird, particularly during the summer breeding season. This is also in the east/west contact zone for Black-headed and Rose-breasted Grosbeaks, as nesting by both species has been documented in Jewell County.

24. Jamestown Wildlife Area

Counties: Cloud, Jewell, and Republic

eBird Hotspot names: Jamestown WTP, Jamestown WA, Jamestown WA—Greenwing Marsh, Jamestown WA—Gun Club Marsh, Jamestown WA—Mallard Marsh, Jamestown WA—Marsh Trail Bridge, Jamestown WA—North Buffalo Marsh, South Buffalo Marsh, Jamestown WA—Puddler Marsh

Other hotspot: Game Keeper Marsh

Hotspot type: Half day

Jamestown WA consists of 5,124 acres of land in Cloud, Jewell, and Republic Counties. Over 1,900 of these acres are classified as wetlands. This is one of the historical wetland complexes that was present in Kansas prior to European settlement. It has been under state management since 1932 and is managed by the KDWP. The entire area is within the watershed of Marsh Creek. Most of the property is in a contiguous corridor of land that stretches along Marsh Creek from near the town of Jamestown northwest to K-148. The separate Jewell County tract slightly to the northwest is known as Puddler Marsh and covers several hundred acres. There has been considerable habitat manipulation of the area over the years to conserve and manage water and wetland habitat. There are several water control structures on Marsh Creek that allow water to be conserved and allocated as needed. These wetlands are shallow and have experienced significant accumulation of silt over many decades. At the time of this publication, the KDWP has a series of long-range wetland renovation projects in progress to combat siltation, remove invasive plants, and create additional wetland areas. Jamestown receives a substantial amount of hunting pressure in the fall and winter, and some areas are closed to all activities from October through March. Despite its large size and importance as one of the principal wetlands in the state, Jamestown receives far less attention from birders than comparable wetlands elsewhere in Kansas.

The tour outlined below will allow you to view a lot of the marsh. More than at some other wetland hotspots in Kansas, you will need to hike in to reach some of the best habitat areas.

Begin a visit to Jamestown WA just north of the town of Jamestown with a brief stop at the **Jamestown WTP** on 40th Road. These are easily viewable from the road and sometimes have a variety of waterfowl during migration and in winter. Proceed north on 40th for 1 mile to Vale Road and turn west, entering the wildlife area. In this first mile, the **North Buffalo Marsh** is on the north side of the road and the **South Buffalo Marsh** on the south side. After 1 mile, you will reach the southernmost water control structure, where there is a small parking lot adjacent to the **Marsh Creek Marsh** to the north. Continue west on Vale for another mile. The private farmland south of the road in this mile sometimes attracts migratory shorebirds in wet years and after rain events. When you reach 20th Road, go north for about three-quarters of a mile to the parking area for the **Greenwing Marsh**. The road beyond the parking area is gated, but you can walk into this area of well-developed wetland habitat for over a mile. In some drier years, the most productive mudflats at Jamestown are north of this parking area. You may need to walk in for up to a mile to reach them. Return to Vale Road, which merges with K-28. Go west for 1 mile and north for 1 mile on K-28. Where the highway turns west, continue north on 70 Road for 1.5 miles to the intersection with Marsh Trail Road and turn east. After a half mile, you will reach the **Marsh Trail Bridge** over Marsh Creek, with good views of the wetlands on both sides of the road. Another gated road goes south from the bathroom parking area at the bridge, and you can walk it to reach (at times) extensive wetland and mudflat habitat, depending on water levels. Continue east for another half mile to the intersection with 30 Road. Turn south onto 30 Road, and after about a half mile, turn into the wildlife area to access excellent wetland habitat at the **Gun Club Marsh** and the **Mallard Marsh**. This stretch of road usually provides excellent birding opportunities for shorebirds, waders, and other wetland bird species. Return to the intersection of 30 Road and Marsh Trail and continue north on 30 Road for 1.5 miles to Xavier Road, then west on Xavier for 1 mile until you reach a dead end where there is a small parking area with a vehicle gate. From here, you can hike in to reach the eastern edge of the **Game Keeper Marsh**, which is usually the largest single wetland impoundment on the refuge. This impoundment is quite shallow, and the surface water area is highly variable. From the eastern shore, you can see most of the open water and mudflats. In drought years, you may need to hike some distance across dry flats to reach the mudflats and shallows that attract birds. There is an additional walk-in access to the Game Keeper Marsh from the west side from a parking area on K-148 at the northwestern corner of the wildlife area.

Puddler Marsh is a separate unit located 1 mile west and 1 mile north of the primary Jamestown property. Access is via Jewell County 300 Road from the north. The road dead-ends at West Marsh Creek, and from there, you need to hike in for about a quarter mile along a levee to reach the wetland habitat. In dry years, this is not a rewarding stop, but during wet years, the wetland pool undoubtedly attracts good numbers of waterfowl and shorebirds. The grasslands surrounding the pool are thick. Check these for Marsh Wren and migratory sparrows.

Jamestown attracts species similar to those you would expect at other Kansas wetlands such as Quivira NWR. Over 25 waterfowl and 30 shorebird species have been recorded. Because of the generally shallow water, numbers of diving ducks are usually modest in comparison with those of dabbling ducks. Raptors are numerous, especially during migration and winter. Look for Peregrine Falcons stooping on shorebird flocks in spring and fall. In the colder seasons, be alert for Short-eared Owls as they appear near dusk. In years when water conditions are right, large numbers of Wilson's Snipe, Virginia Rail, and Sora have been reported during migration. Yellow-headed Blackbird nests in some years. Song Sparrow is rare as a nesting species in Kansas but has been confirmed to breed at least occasionally at Jamestown. Some of the more noteworthy species reported from Jamestown have been Western and Clark's Grebes, Whooping Crane, Piping Plover, Long-billed Curlew, Buff-breasted Sandpiper, Ruddy Turnstone, Caspian Tern, and Least Bittern.

25. Talmo Wildlife Area

County: Republic

eBird Hotspot names: Talmo Marsh

Hotspot type: Worth a stop

The **Talmo WA** is located in southeastern Republic County. It is a state wildlife area that is managed by the KDWP. It covers about 950 acres, of which 290 acres are wetlands. Talmo Marsh was a historic alkaline salt marsh basin covering 1,400 acres at the time of European settlement. In the 1960s, much of the wetland was modified in an attempt to convert it to cultivated land. In recent years, a wetland restoration was initiated that was completed in 2017. As at Jamestown, this included the creation of water management units and control structures to allow managers more habitat control. Expect substantial numbers of hunters when seasons are open.

From the intersection of K-9 and US 81 in Concordia, drive north 8.5 miles to K-148, then 4.5 miles east. A half mile east of Talmo Road and the few buildings known as Talmo is a road to the south. The sign calls this County Road 20; maps might call it County Road 200. This is at the northwestern corner of the wildlife area. From this corner, go south for 2 miles. The wetlands are on both sides of the road in much of this 2-mile stretch. There are several parking areas from which you can walk in. The wetland is divided into named impoundments: the Borchardt, Nutter, Trost, and Warren Marshes. You need to hike in on the levees around these pools to view them well. Continue south on 200 Road to the intersection with Young Road. In wet years, there can be shallow water or a mudflat on the private land south of Young Road near this intersection that attracts shorebirds.

Since the renovation of the wetlands was done relatively recently, birders have not visited Talmo often, but the quality of the birding has been excellent. Thus far, 126 species of birds have been observed, and the potential for adding additional species is quite high. Sixteen waterfowl species have included Cinnamon Teal and Hooded Merganser. Twenty-four shorebird species observed have included American Golden-Plover, Hudsonian and Marbled Godwits, Ruddy Turnstone, Dunlin, and

Short-billed Dowitcher. Other wetland species have included Black Tern, American White Pelican, Sora, Virginia Rail, American Bittern, White-faced Ibis, and Short-eared Owl.

26. Antelope Lake

County: Graham

eBird Hotspot names: Antelope Lake

Hotspot type: Worth a stop

Statistically, Graham County is the least birded county in Kansas, with fewer than 350 eBird checklists having ever been submitted from the entire county as of 2023. **Antelope Lake** is by far the location in the county most visited by birders and is worth a stop regardless of the season. It is located 14 miles west and 1 mile north of Hill City. The entrance road on 125th Avenue is marked by highway signs on US 24. The lake is encircled by Antelope Lake Road. The lake covers 70 acres and is 11 feet deep at its deepest point. The total area of the park is approximately 100 acres. The lake was created in 1935 by the construction of a dam on Antelope Creek. There are stone shelter houses and restrooms.

Birding this lake is straightforward. Drive the entire perimeter road to check the lake for waterbirds. These have included 24 waterfowl, 4 grebe, and 12 shorebird species, with notables such as Greater Scaup, all three merganser species, Western Grebe, Marbled Godwit, and Willet. Wood Duck is a nesting species. On the north and west sides of the lake are areas of mature trees and brush. In summer, expect Great-crested Flycatcher; Eastern and Western Kingbirds; Bell's, Red-eyed, and Warbling Vireos; Orchard and Baltimore Orioles; and Yellow Warbler. Yellow-breasted Chat and Black-headed Grosbeak have been found here during the breeding season. Shoreline areas of cattails and wetland plants should be checked for Green Heron, Black-crowned Night Heron, Marsh Wren, and Song Sparrow. Ten sparrow species have been found, including some that are scarce in this part of the state, such as Field, Fox, and White-throated Sparrows and Eastern Towhee. Nine warbler species have been reported, and additional warbler species may well be found in the future.

27. Webster Reservoir

County: Rooks

eBird Hotspot names: Webster Res., Webster SP—Coyote Trail, Webster SP—Oldtown Area, Webster WA, Webster WA—Morel Campground

Hotspot type: Half day

Webster Reservoir in Rooks County is one of several federal reservoirs in north-central Kansas. The lake covers 3,767 surface acres when full and has a maximum depth of 42 feet. There are 880 acres of state park land near the dam. Webster WA covers an additional 8,000 acres. The reservoir was created in the 1950s by the construction of the dam on the South Fork of the Solomon River.

The **Oldtown Area** of Webster State Park extends west from the dam along several miles of shoreline on the north shore of the lake. The Oldtown Area can be reached from US 24 via 9 Road and 10 Road. There are several locations within the park where you can view the lake for waterfowl, gulls, and other waterbirds. K-258 crosses the dam, and near the south end of the dam is the much smaller **Goose Flat Area**, which offers additional vantage points.

Within the **Webster WA** farther up the lake is good access to the lake and land habitat. On the south side of the lake, Road N follows the shore for 6 miles, eventually ending in woodland along the Solomon River. In some years, there is good wetland habitat south of the road adjacent to **Morel Campground** on N Road where shorebirds and herons can be found. On the north side of the lake, M Road extends westward from the state park for several miles along the lake shoreline, with multiple lake viewing points, including the **Coyote Trail Area**. Webster WA extends west of the reservoir to 3 Road, and there are various access roads. Use a map or phone app to explore these if you are inclined. The woodlands at the upper end of the lake are well-developed and worth exploring. There is also extensive grassland and brushy habitat.

Despite the relative scarcity of birding reports from Webster, 28 waterfowl, 31 shorebird, 10 gull, 5 tern, 18 sparrow, and 10 warbler species have been found here. Both Tundra and Trumpeter Swans have been reported in multiple years, most often in December. Common Merganser, Common Goldeneye, and other waterfowl are abundant in winter. Especially in the shallow waters within the wildlife area, flocks of geese can be impressive in late fall and early spring. Rarer shorebirds that have been seen at Webster include Whimbrel, Long-billed Curlew, both godwits, Buff-breasted Sandpiper, and Red-necked Phalarope. Piping Plover has been documented as a breeding species at least once, and there are summer records for American Avocet, suggesting that it could breed here as well. Northern Shrike is reported on most visits during the winter months. Species such as Say's Phoebe, Black-billed Magpie, and Mountain Bluebird are seen in most years and impart a western flavor to the mix of bird species.

28. Rooks State Fishing Lake

County: Rooks

eBird Hotspot names: Rooks SFL

Hotspot type: Worth a stop

Built by the WPA in the 1930s, **Rooks SFL** is not large. The lake covers 67 acres when full, and there are an additional 245 acres of public land surrounding the lake. This hotspot is only 5 miles from Webster, so both can be visited on the same day. The lake is located 2 miles west and 2 miles south of Stockton. The main entrance on 16 Terrace (by the stone picnic shelter) is always open. The water source of the lake is intermittent, and in some years, the lake can go completely dry. There are modest-sized wooded areas on the east shore of the lake and below the dam. Where Boxelder Creek feeds into the lake, there is some shallow wetland habitat, again depending on the overall water level in the lake.

Despite the small size of this hotspot, over 200 species of birds have been found here. Twenty-five waterfowl species have been seen, including Cinnamon Teal, White-winged Scoter, and Hooded Merganser. The erratic water levels sometimes create mudflat and shoreline habitat that is attractive to shorebirds, which have included notables such as Piping Plover, Hudsonian and Marbled Godwits, Ruddy Turnstone, Short-billed Dowitcher, and Red-necked Phalarope. Ten species of wading birds have been reported, including uncommon species such as Least Bittern, Yellow-crowned Night Heron, and Glossy Ibis. Other waterbirds have included Common Loon, Neotropic Cormorant, and Osprey, and Bald Eagle has also been seen. Greater Prairie-Chickens have been heard booming near the lake property in spring. Despite the limited amount of suitable habitat, 15 sparrow species and 12 warbler species have been found at the lake.

29. Waconda/Glen Elder Reservoir

County: Mitchell

eBird Hotspot names: Glen Elder SP, Glen Elder WA, Glen Elder WA—Carr Creek, Glen Elder WA—Fisherman's Bridge, Glen Elder WA—Granite Creek Access, Glen Elder WA—North Fork, Waconda Lake, Waconda Lake—Lake Dr. Causeway

Hotspot type: Half day

Waconda Reservoir, also called **Glen Elder Reservoir**, is one of the largest Kansas reservoirs and one of the better birding hotspots in the north-central part of the state. The surface area of the lake is 12,500 acres, and the surrounding **Glen Elder WA** covers an additional 13,200 acres. The KDWP operates the 1,450-acre Glen Elder State Park on the north shore of the lake adjacent to the dam. The lake was created in the 1960s by the construction of Glen Elder Dam on the Solomon River. The North and South Forks of the Solomon River are the principal tributaries of the lake.

The dam is located on the east end of the lake. Great Spirit Lane is a paved road that crosses the dam. Parking on the dam is not permitted, but there are gravel roads on both ends from which you can scan the face for waterfowl, loons, grebes, and gulls, depending on the time of year.

The main entrance to **Glen Elder State Park** is about 2 miles west of the town of Glen Elder on US 24. The park stretches for about 4 miles from the dam westward and has a network of paved roads. Within the park are multiple vantage points where you can view the lake for aquatic birds. Returning to US 24, continue west for 4 miles from the state park entrance to the **Granite Creek Boat Launch**, where there is a large parking lot. From the parking lot, you can scan the lake and the inlet cove of Granite Creek. Outside the state park, this is the best lake vantage point on the north shore. There are often good numbers of waterfowl, shorebirds, and gulls at this spot. Continue west into Cawker City, turning south onto Lake Drive in the center of town near the famous "World's Largest Ball of Twine." As you leave town, the **Cawker City WTP** are on your right. These are easily viewed from the road and have interesting birds on occasion. Continue south on Lake Road. The road crosses the levee and then for the next 2 miles crosses the lake via **Lake Drive Causeway**. The causeway is unique. The road predated the construction of the lake and was considered important enough that the causeway was constructed to allow it to remain in use. There are several pullouts on the causeway that allow you to scan a substantial area of the lake on both sides of the road. If you have time for only a few stops at Glen Elder, make sure the causeway is one of them. The area of the lake closest to the causeway is shallow and consequently can produce a good variety of shorebirds and wading birds depending on the overall water levels in the reservoir. The trees and brush scattered along the length of the causeway often have good numbers of land birds.

If you have birded all the areas described above, you will likely have used a half day of time. If it seems like the birding is good and you want to continue, the extensive Glen Elder WA has additional productive birding areas. For reasons of space, detailed directions to many of them are not included here, but you can navigate to all of them using maps, eBird map links, or other internet resources. Favored birding hotspots west of the causeway are **Carr Creek**, **Fisherman's Bridge**, and **North Fork**. Especially at Carr Creek and North Fork, there are often extensive mudflats that can attract shorebirds and waders. These are also good locations for summer-breeding songbirds such as kingbirds, vireos, and orioles. The public wildlife area extends westward all the way to the town of Downs. From US 24, there are several access roads into this western portion of the wildlife area, including one just south of Downs via a gravel-road entrance on K-181 south of the river bridge. On the south side of the lake and east of the causeway, there is far less public usage and only a few access points to the lake. The best south-shore lake vantage points are **Harrison Point** and **Schoen's Cove**.

Nearly 300 species of birds have been recorded in Mitchell County, and the vast majority of these have been observed at or near Glen Elder. Thirty-one waterfowl species have been reported, including multiple records of all three scoter species and Long-tailed Duck. Waterfowl numbers can number in the hundreds of thousands from late fall through early spring, depending on prevailing weather and water conditions. These flocks are dominated by Cackling, Canada, and Snow Geese and Mallard. Winter flocks of Common Merganser and Common Goldeneye sometimes number in the thousands. Six species of grebes, including Western and Clark's, have been reported. An impressive 34 shorebird species have been reported at Waconda, a diversity of shorebirds rivaled in Kansas only by Cheyenne Bottoms, Quivira NWR, and a few other Kansas hotspots. These have included Black-bellied, Piping, and Snowy Plovers; Hudsonian and Marbled Godwits; Ruddy Turnstone; Buff-breasted Sandpiper; and Red-necked Phalarope. Waconda is also one of the better Kansas hotspots for finding rare species of gulls during migration and in winter. These have included Sabine's, Iceland, Glaucous, and Lesser Black-backed Gulls. Migrating flocks of Sandhill Crane are often seen in early spring and late fall, and Whooping Cranes have been observed with them more than once. Other notable waterbird species observed at Waconda have included Neotropic Cormorant, Glossy Ibis, and Caspian Tern. A good variety of raptors occur during migration and the winter months. Osprey is a common migrant. Bald Eagle can be abundant in winter when conditions are favorable, and one or two pairs remain to nest. Northern Shrike, Townsend's Solitaire, and Mountain Bluebird are all winter possibilities.

30. Gurley Salt Marsh

County: Lincoln

eBird Hotspot name: Gurley Salt Marsh

Hotspot type: Worth a stop

Gurley Salt Marsh is a 160-acre refuge in Lincoln County that is owned and managed by the KDWP. The primary wetland basin covers about 45 acres when full. There are 50 acres of high-quality mixed-grass prairie south and east of the wetland. Gurley is one of only a few salt marshes in Kansas. It is a recent addition to Kansas public lands, having been acquired from owner Jim Gurley only in 2012. It is located on K-14 about 11 miles north of Lincoln. The wetland is open to the public from March 2 through August 31. For the rest of the year, all public access is managed through the KDWP Special Hunts program. When the marsh is closed to public access, you can use a spotting scope to spot birds on the wetland basin from the parking area without entering the closed refuge area. This interesting wetland is not visited frequently by birders and deserves more attention than it gets.

All access is on foot from the parking area on K-14. Follow the vehicle tracks leading northeast from the parking area. This road curves around the north edge of the wetland for about a half mile to the water outlet control structure. From that point, you can walk south and back to the west around the perimeter of the wetland for another half mile. Except when the marsh is nearly dry, you will not be able to completely circle the wetland without knee-high wading boots.

The complete absence of trees at Gurley means that the expected species are restricted largely to grassland birds, waterfowl, shorebirds, waders, and other wetland species. Diversity and overall numbers of birds are dependent primarily on water conditions, which fluctuate substantially from year to year. Early-spring trips frequently produce a variety of duck species. During the peak of spring shorebird migration in early May, it is possible to see 15 or more plover and sandpiper species on a good day. There are very few bird records of any kind from the southbound shorebird migration period that begins in July, but when the right water conditions exist, the fall birding should be equally interesting. In late October and November, try walking along the edge of the wetlands next to K-14 south of the parking area. Don't cross into the closed refuge area, and stay in the grass well off the shoulder for safety. LeConte's Sparrow is possible here along with other sparrows, such as Swamp and Song Sparrows, and Marsh Wren. Some of the better finds at Gurley have been Cinnamon Teal, Horned and Eared Grebes, Common Gallinule, Virginia Rail, Sora, Black-necked Stilt, Snowy Plover, Hudsonian and Marbled Godwits, and Willet.

There are other good birding opportunities nearby, although they are on private land and can be viewed only from the roadside. Just north of the Gurley parking lot, turn west onto X-Ray Road. The first 2 miles of road west of the highway have additional salt marsh wetland areas along Rattlesnake Creek that can be in prime condition or completely dry, depending on prevailing weather conditions. Most of these are easy to view from the road and can produce good birding when water conditions are favorable. Return to K-14 and go north a short distance to Yarrow Lane, then proceed east on Yarrow for 1 mile until it curves south into N 200th Road. Just after the intersection with X-Ray Road, 200th Road crosses Rattlesnake Creek. Park near this bridge and walk around for a few minutes. At this point, you are below the outlet channel from Gurley and the habitat consists of dense thickets amid tall prairie grasses. In late fall, this can be an exceptionally good spot for sparrows, including noteworthy numbers of LeConte's. After you have finished birding here, continue south on 200th for 3.5 miles to where the road crosses Battle Creek. Here there is a mature riparian oak forest. Summer Tanager is probably a nesting species at this location, near the western edge of its breeding range. Look for other breeding species of mature woodland, such as Eastern Wood-Pewee, Blue-gray Gnatcatcher, and Red-eyed Vireo.

31. Ottawa State Fishing Lake

County: Ottawa

eBird Hotspot names: Markley Grove Park, Minneapolis WTP, Ottawa SF, Rock City Park

Hotspot type: Worth a stop

Ottawa SFL is a 728-acre property managed by the KDWP that is located 16 miles north and 4 miles east of Salina. The lake is among the oldest in Kansas. The dam was completed in 1929. The surface area of the lake is 110 acres, and it is 11 feet deep at the deepest point. Water levels are more stable than at some other Kansas lakes and reservoirs. The main entrance to the state lake is at the southwestern corner of the lake near the intersection of Justice Road and N 190th Road. From there, Lake Drive goes north along the western shore of the lake, eventually crossing a low-water bridge and exiting the park via Kiowa Road. The general area near the low-water bridge is a good place to spend some time walking along the road. There is a second entrance at the southeastern corner of the lake at the intersection of Justice Road and Goodwin Drive. Goodwin Drive follows most of the eastern lakeshore. The shoreline areas on the southern half of the lake consist largely of primitive campgrounds and fishing jetties. Below the dam is an area of mature woodland. There is a short hiking trail on top of the dam. The northern and eastern quadrants of the property are also heavily wooded. The lake is fed by two different streams. At the upper end of the lake where these streams enter are areas of wetland vegetation, including water lilies and cattails. The southern half of the property is open year-round; the northern components are closed from October through March. A substantial amount of invasive Asian Bush Honeysuckle is present throughout the woodlands, but healthy stands of native shrubs and prairie survive in places.

Over 190 species of birds have been reported at Ottawa SFL. These include 24 waterfowl, 10 hawk, and 19 sparrow species. In summer, expect good numbers of Great Crested Flycatcher, Warbling and Red-eyed Vireo, Blue-gray Gnatcatcher, Baltimore Oriole, and other birds typical of the eastern woodlands. Winter residents are also typical of eastern woodlands. In late fall and early spring, waterfowl diversity and numbers can be excellent. Bald Eagle is usually present in winter. Pileated

Woodpecker and Barred Owl are rare but possible in any season. The lake is not noted as a place that attracts an impressive diversity of migrants. However, there are only a handful of eBird reports from early May and early September, when migration is at its peaks in Kansas. As an example, only eight warbler species have been reported to eBird from Ottawa SFL. The lack of shorebird habitat is evident, with only seven species reported here.

If you have time, two locations in the nearby town of Minneapolis are worth a visit. The **Minneapolis WTP** on the east edge of town are expansive and can be easily viewed from the entrance road. As you come into town on K-106, turn south onto an unmarked road on the west edge of the golf course and go south for a quarter mile until the road ends at the treatment ponds and the Solomon River. One hundred three species have been reported here, mostly waterfowl and other waterbirds but also several warbler species seen in the adjacent woodlands along the river. **Markley Grove Park** is located at the southwestern corner of town along the Solomon River. The entrance is on K-106. The park has numerous mature trees in the developed areas and a short hiking trail that goes through the woodlands along the river. A brief visit, especially during spring and fall migrations, is usually worth the short amount of time required.

From Markely Grove Park, you can continue south for 3.5 miles to Ivy Road, then follow the signs for **Rock City Park**. Covering 5 acres, this is more of a geological point of interest than a birding stop. This National Natural Landmark preserves 200 boulders called calcite concretions that are up to 9 meters in diameter. The grassland and wooded habitats adjoining these unique formations have produced 70 species of birds. There are restrooms and picnic shelters and even a gift shop. There is a modest visitor fee.

32. Cedar Bluff Reservoir

County: Trego

eBird Hotspot names: Cedar Bluff SP—Agave Ridge Nature Trail, Cedar Bluff SP—Bluffton Area, Cedar Bluff Res.—Dam and Outlet Park, Cedar Bluff Res.—Cedar Bluff Overlook, Cedar Bluff Res., Cedar Bluff SP, Cedar Bluff WA, Cedar Bluff SP—Page Creek Area

Hotspot type: Half day

Lying southwest of Hays on the Smoky Hill River, **Cedar Bluff Reservoir** is the westernmost large reservoir in Kansas. It was created by the construction of a dam that was completed in 1963. This reservoir experiences exceptional fluctuations in water levels. When it is full, the reservoir covers 6,869 surface acres, but in most years, the surface water area is substantially less than that. **Cedar Bluff State Park** consists of the 350-acre Bluffton Area on the north shore and the 500-acre Page Creek Area on the south shore. The **Cedar Bluff WA** is 9,000 acres in size. The semiarid shortgrass prairie habitat surrounding the lake has a substantial amount of sage and yucca. This plant community, combined with the numerous rocky cliffs and rock outcrops in the area, gives Cedar Bluff a decidedly western feel.

Hotspot 32: The view from the Cedar Bluff Overlook is atop the most prominent cliff overlooking Cedar Bluff Reservoir.

The dam and the entrance to the **Bluffton Area** of the state park are on K-147 about 15 miles south of I-70. If your time is limited, you can find a good variety of birds by scanning the lake and shorelines from the scenic overlook at the south end of the dam and the various park roads in the Bluffton Area.

You can access the **Outlet Park** by turning east on Y Road a short distance south of the dam. There is a sizable wetland area here that is worth checking, although you will have to proceed on foot to reach the best habitat. Multiple Marsh Wrens are likely here in late fall and remain during mild winters.

If you have additional time to explore, there are other good stops, especially on the south shore of the lake. From the dam, continue south for 4 miles to the intersection with CC Road and go west. CC Road curves south to join AA Road in 3.5 miles at the entry to the Page Creek Area of Cedar Bluff State Park, where there are campgrounds, restrooms, and multiple vantage points from which to view the lake. Over 150 species of birds have been seen in the Page Creek Area. At the west end of the Page Creek Area is the **Agave Ridge Nature Trail**, which traverses productive birding habitat.

From the Page Creek entrance, continue west on AA Road for 2 miles to 290 Road. Go north 1 mile on 290 Road, where you can continue north for another half mile to the lake, or go east and then north for 1 mile to reach the **Cedar Bluff Overlook**. At the overlook, you are on top of the most prominent bluffs at the reservoir and you can scan most of the lake for birds, especially if you have a spotting scope. This overlook offers an incredible view and is a scenic landscape that is rare in Kansas.

For your last stop, return to AA Road and go another 4 miles west to the intersection with US 283. Go north 2 miles to the entrance road for the **Cedar Bluff WA**. This road follows the river channel eastward through grasslands, open woodlands, and brushy slopes for several miles. The woodlands here are the most mature and extensive at Cedar Bluff and offer the best opportunity to see eastern woodland species such as Red-bellied Woodpecker, Great Crested Flycatcher, Red-eyed Vireo, Black-capped Chickadee, and White-breasted Nuthatch. Yellow-breasted Chat is often found in summer in the brushy thickets along with Bell's Vireo, Field and Lark Sparrows, and Blue Grosbeak.

The reservoir attracts good numbers of waterfowl, grebes, shorebirds, gulls, and other waterbirds according to the season and water conditions. The sighting list from Cedar Bluff includes 30 waterfowl, 30 shorebird, and seven gull species. There is at least one confirmed breeding record for Piping Plover, and Snowy Plovers with downy young were observed in the summer of 2024. In addition to the waterfowl and shorebird species typically found at Kansas reservoirs, several western bird species can be encountered in Cedar Bluff WA. These include Black-billed Magpie, Say's Phoebe, and Rock Wren. In the winter, expect multiple species of hawks and falcons, Northern Shrike, Townsend's Solitaire, and Mountain Bluebird.

In some winters, several of the large plantings of cedar and pine trees within the wildlife area have attracted large roosts of Long-eared, Short-eared, and American Barn Owls that have occasionally numbered in the hundreds of birds. Both Greater and Lesser Prairie-Chickens have populations in Trego County, and both species have been observed at Cedar Bluff or nearby. Listen for them booming early on spring mornings.

33. Hays Area

County: Ellis

eBird Hotspot names: Frontier Historical Park, KSU Agricultural Research Center—Hays, Sternberg Natural Area

Hotspot type: Worth a stop

Located in Ellis County, Hays is the largest city in north-central Kansas and is a major commercial hub. It is also home to Fort Hays State University, where much important field biological research has been conducted over the years. Due largely to decades of bird-banding efforts and other ornithological studies, the all-time checklist of species that have been seen in Ellis County is currently at 329, ranking among the highest of all Kansas counties.

Several locations in the city provide good opportunities for the visiting birder. These can be a pleasant break for the traveling birder passing through on I-70. The Dr. Howard Reynolds Nature Trails in the **Sternberg Natural Area** are adjacent to the Sternberg Museum of Natural History in northeast Hays. This hotspot is always worth a stop. Two miles of trails wind through 22 acres of habitat adjacent to the museum parking lot, mostly along the wooded riparian corridor with several good-sized pools. Despite the small size of the property, at least 144 species have been identified here. These are predominantly songbirds, including 15 sparrow and 15 warbler species seen during spring and fall migrations, with notables such as LeConte's Sparrow, Black-throated Blue and Townsend's Warblers, and Lazuli Bunting. The Sternberg Museum has excellent natural history exhibits and is worth a visit if your time allows. At the time of this book's publication, the museum was also serving as the headquarters for the Audubon of Kansas Lek Trek, centered around viewing opportunities for both species of prairie-chicken.

Another productive hotspot is **Frontier Historical Park** on the south edge of Hays. The entire park covers 177 acres and includes golf, disc golf, ball fields, and other athletic facilities. On the south edge of the park adjacent to the US 183 Bypass, 3 miles of trails wind along both banks of Big Creek. There is a considerable amount of riparian woodland, including numerous mature trees. The list of bird species that have been seen here stands at about 130 species.

Located directly across the highway from Frontier Park is the **KSU Agricultural Research Center**. The entrance to the research center is on 240th Avenue one block south of US 283. The primary draw for birders at the Ag Center are the numerous mature pine trees scattered across the grounds. When irruptions of Pine Siskin and Red Crossbill move into Kansas, those pines are one of their favored locations. In most winters, one or two Red-breasted Nuthatches and Townsend's Solitaires can also be found.

34. Wilson Reservoir

Counties: Russell and Lincoln

eBird Hotspot names: Wilson Lake, Wilson WA, Wilson Lake—Dam, Wilson Lake—Sylvan Park, Wilson SP—Hell Creek Area, Wilson Lake—Minooka Park RA, Wilson SP—Otoe Area, Wilson Lake—Lucas Park WA, Wilson WA—Cedar Creek Boat Ramp

Hotspot type: Full day

Wilson Reservoir is one of the most popular birding destinations in Kansas. A full day can easily be spent birding the lake and adjacent public lands. The reservoir is large and deep, surrounded by the scenic rolling hills and rock formations that typify the Smoky Hills region. Formed by a huge dam on the Saline River, the lake covers 9,000 surface acres when full and is one of the deepest impoundments in Kansas, with a maximum depth of 65 feet. It is surrounded by 13,000 acres of land managed by the Corps of Engineers and the KDWP. The Wilson WA covers over 8,000 acres. Several major park areas occupy the remainder of the public land at Wilson.

Take the Wilson exit on I-70 and then K-232 north for 5 miles to reach the dam, where you can begin a tour of this hotspot. **Sylvan Park** below the dam is managed by the Corps and has mature timber along the river channel and a wetland area. This is one of the best areas to find woodland bird species at Wilson. A variety of eastern woodland bird species can be expected according to the season.

The locations at Wilson that are visited most frequently by birders are on the south shore of the reservoir. **Wilson State Park** at the east end of the reservoir is divided into the **Otoe Area** and the **Hell Creek Area**. Farther west is the **Minooka Park Recreation Area**, managed by the Corps. On the north shore, **Lucas Park Recreation Area** stretches for some distance along the shoreline. All these parks and recreation

Hotspot 34: Carved by the Saline River, Wilson Reservoir is surrounded by red rock formations of the scenic Smoky Hills region.

areas offer multiple vantage points from which to view the lake and provide good opportunities for land birds.

Farther up the lake are additional access points to the north and south shores within **Wilson WA** via county and park roads. Use a map or Google Maps to explore the sprawling wildlife area as your time allows. In addition to lake access points, the wildlife area preserves a substantial area of midgrass prairie and a variety of brush and woodland habitats. Among the better birding locations in the wildlife area are the **Cedar Creek Boat Ramp** and nearby **Horseshoe Bend**, which are excellent shorebird spots in some years.

Overall in Russell County, 341 species of birds have been recorded, and the majority of these have been recorded at Wilson. Only seven other counties in Kansas have recorded more species. Over 30 waterfowl species have been seen at Wilson, including all three scoter species and Long-tailed Duck, all of which are recorded in most years. Common Goldeneye and Common Merganser number in the hundreds or thousands in the winter months. Six species of grebes, four species of loons, 12 species of gulls, and 34 species of shorebirds have been seen in Russell County, most of them at Wilson. There are numerous records of the rarer gull species from the winter months. In spring and fall, migrating flocks of Sandhill Cranes can frequently be heard and seen overhead. Whooping Crane has been seen with these flocks several times. Raptor numbers and diversity are also impressive, especially during migration and in winter. Mountain Bluebird is usually present during the winter in the big stands of cedar scattered around the lake. Wilson is reliable for Northern Shrike in the winter months. Be sure to check rocky outcrops for Rock Wren from spring through fall, and listen for

Common Poorwill in this habitat at dusk. Notable for a location this far west, over 20 species of warblers have been reported at Wilson during migration. Twenty species of sparrows have been found at Wilson, and during migration, you can expect a good variety of them.

35. Kanopolis Reservoir

County: Ellsworth

eBird Hotspot names: Kanopolis Lake, Kanopolis Lake—Dam and Outlet Park, Kanopolis SP—Horsethief Area, Kanopolis SP—Horsethief Canyon Trails, Kanopolis SP—Langely Point Area, Kanopolis Lake—Venango PUA

Hotspot type: Half day

Kanopolis Reservoir is the oldest Corps of Engineers reservoir in Kansas. It was created by the construction of a dam on the Smoky Hill River in 1948 and covers 3,400 surface acres. This is an exceptionally large public land area. The Corps and KDWP own or manage 16,000 acres of land around the lake, stretching almost to the town of Kanopolis, over 12 miles from the dam. The rolling hills, midgrass prairies, and scattered rock formations at Kanopolis typify the Smoky Hills region. Because the public lands are now well over 70 years old, the woodland and grassland habitats are mature and well-established.

Kanopolis Dam is 7 miles west and 3 miles north of the town of Marquette. K-141 crosses the top of the dam. The entrance for **Riverside Park** is at the south end of the dam. On this south side of the outlet channel at Riverside are campgrounds set amid open woodlands. Take Sand Creek Road from the north end of the dam to reach the less developed north side of the outlet channel, where there is an area of substantial woodland with foot trails that usually provides good birding. On the south shore near the dam is the entrance to the **Langley Point Area** (spelled "Langely" by eBird). In the off-season when recreational use is minimal, the birding can be worthwhile in the campground area, notably at the **South Swimming Beach**. This is a good place to view the lake, especially along the dam, where interesting waterfowl and gulls are found in the colder months. These can include uncommon species such as Black and Surf Scoters and Western Grebe. Continue past the campground area on Langley Point Road. As you near the lake, the woodlands along this stretch of the road are mature and extensive. At the end of Langley Point Road is the Sandplum Campground, where you can view a large part of the lake

Starting from the north end of the dam and continuing for several miles up the east shore of the lake are several birding hotspots connected by a series of paved roads. These are all accessed from K-141 from the main entrance at Venango Road. As you enter, bear left on Venango Road to access Venango Park or right on Horsethief Road to reach Kanopolis State Park and Horsethief Canyon.

Remain on Venango Road to reach **Venango Park** and the **Venango Public Use Area**, which are close to the dam and have multiple viewing points of the lake. Loder Point is the southernmost of these points and offers the most unobstructed view of the reservoir. It also has a short nature trail and a small man-made wetland that

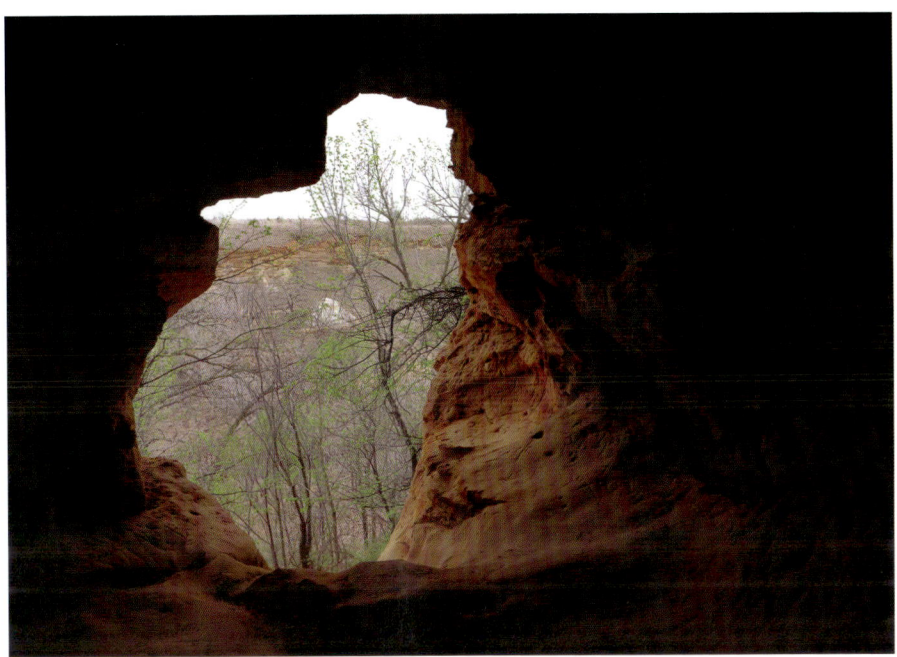

Hotspot 35: A shallow cave is a popular feature for hikers along the Buffalo Track Canyon Nature Trail at Kanopolis State Park.

sometimes attracts wetland birds. The public swimming beach at the end of the dam can produce shorebirds and roosting gulls when swimmers are not present. There is a considerable amount of mostly undisturbed woodland in the Venango area, and this is a good place to explore on foot.

Farther north along the east shore, **Kanopolis State Park** and **Horsethief Canyon** are reached by taking Horsethief Road from the entrance area and following it as it wanders north and west following the shoreline. As you proceed north, several secondary park roads lead from Horsethief Road to the lakeshore, all of which can provide good birding. In drought years, the receding water provides mudflats for foraging shorebirds in this area. Horsethief Road ends at the trailhead for the **Buffalo Track Canyon Nature Trail**. This nature trail follows the lakeshore and shoreline bluffs. Even if the birding is slow, this trail is a great hike. In the warm months, be alert for Rock Wren in the taller rock formations on this trail, about as far east as this species regularly occurs in Kansas. Common Poorwill has also been reported multiple times calling from these bluffs at dawn and dusk. Upland stands of cedars within the state park have hosted small roosts of Long-eared Owl in some winters.

Well over 200 species of birds have been reported at Kanopolis Lake. These include 26 waterfowl, 23 shorebird, 9 gull, and 4 grebe species. Especially in the winter months, this is an excellent location for raptors.

36. Salina Area

County: Saline

eBird Hotspot names: Lakewood Park, Salina, Saline SFL, Smoky Hills Audubon Sanctuary

Hotspot type: Worth a stop

Salina is one of the largest cities in north-central Kansas. Because of its location at the intersection of two interstate highways, your travels may well take you there. Several hotspots in or near the city offer good birding.

The **Smoky Hills Audubon Sanctuary** is managed by the Smoky Hills Audubon Society, based in Salina. It is located close to the I-135/I-70 interchange and is easily reached from the Halstead Road exit on I-70. From the exit, go south a short distance to the intersection with West Stimmel Road and go east for 1 mile until you see refuge signs on your left. There is parking available near the west end of the lake, and the entire lake can be scoped from the parking area. The lake was created as a borrow pit during the construction of the highway interchange. In the late 1970s, the Saline County Conservation District granted the Smoky Hills Audubon Society a 50-year lease to manage the property as a wildlife refuge and education property. The property covers 67 acres and has 2.5 miles of mowed hiking trails passing through open woodland, grassland, and shoreline habitats. The Smoky Hills Audubon Society has done an excellent job of keeping the site well maintained and wildlife friendly. There is a handicap-accessible photo blind at the southwestern corner of the lake, and an adjacent bird-feeder array is usually stocked with food. Despite its relatively small size and its location adjacent to a major highway, at least 205 species of birds have been observed at the sanctuary. Twenty-five species of waterfowl have been recorded, including less common species such as Cinnamon Teal, Greater Scaup, and Red-breasted Merganser. Nineteen shorebird species have included Hudsonian and Marbled Godwits, Semipalmated Plover, and Willet. Herons are usually common in the warmer months, and nine species of them have been seen here. Raptors and songbirds are typical of eastern Kansas and have included 18 sparrow, 12 warbler, 9 flycatcher, and 6 vireo species.

Lakewood Park is a 100-acre city park located in northeast Salina. The park is located east of Ohio Street between North Street and Iron Avenue. There are park entrances on both streets to North Lakewood Park Drive, which wraps around the west edge of the park.

The park is located on the site of an old sandpit and has a lake covering 13 surface acres, with a maximum depth of 6 feet. Much of the dry land in the park lies within the old sandpit excavations, which are covered with dense woodland and shrub habitat. The Smoky Hill River channel borders the eastern edge of the park and has a corridor of mature riparian timber on its banks. There is a small area of restored native grassland on the east side of the park. The Lakewood Discovery Center, which offers interpretive exhibits and bird feeders with a nearby viewing blind, is located a half mile from the park entrance. There are numerous hiking trails in the park.

Nearly 200 species of birds have been identified at Lakewood Park. Despite the relatively small size of the lake, these include over 20 waterfowl, 14 shorebird, and 8 heron species. Passerine birds are the focus for birders at this hotspot. Sixteen warbler, 15 sparrow, and 7 vireo species have been reported. These include uncommon migrants such as White-eyed, Yellow-throated, and Philadelphia Vireos and Mourning, Magnolia, Chestnut-sided, and Canada Warblers. Other noteworthy migrants have included American Woodcock, Black-billed Cuckoo, and Summer Tanager. During the winter, the park provides good birding, and species including Yellow-bellied Sapsucker, Brown Creeper, Golden-crowned Kinglet, Winter Wren, Hermit Thrush, and White-throated and Fox Sparrows are usually present. Raptors including Cooper's, Sharp-shinned, and Red-shouldered Hawks and Merlin are often seen in the winter as well. Cooper's Hawk undoubtedly nests in or near the park.

Saline SFL is another local hotspot worth checking. The lake is 2 miles north of the I-135/I-70 interchange. From the Halstead Road exit on I-70, go north 1.5 miles to Watkins Road, east 1 mile to Gerard Road, and then north a short distance to the entrance. The vehicle gate for the entrance road is usually locked, and you will need to walk in. A spotting scope will be useful at this hotspot. The lake covers 38 surface acres when full, but it is not usually full. In drought years, the lake goes completely dry, and in these conditions, the birding is of minimal interest. Because of the dramatic changes in water level, this hotspot is best thought of as an ephemeral playa lake. There are 39 acres of surrounding land, most of which is grassland. There are cottonwoods around most of the perimeter of the lake.

Over 150 species of birds have been reported here. Waterfowl diversity and numbers can be good when sufficient water is present, and 25 species of ducks and geese have been reported, including Cinnamon Teal, Greater Scaup, and all three merganser species. Other swimming birds seen here have included Horned Grebe, Double-crested Cormorant, and White Pelican. When water levels are low, as is often the case, a good diversity of shorebirds can be seen during spring and fall migrations. Other wetland species reported here have included Virginia Rail, Sandhill Crane, American Pipit, Sedge Wren, and LeConte's Sparrow.

A discussion of Saline County hotspots is not complete without mention of the thriving tallgrass prairies west of Salina, especially those along W Crawford Street between 6 and 9 miles west of the city limits. This area is well-known as one of the most reliable places in Kansas to find Short-eared Owl from fall through spring. Look for these enigmatic crepuscular owls at dusk as they start to forage. If you are lucky, you can observe their acrobatic territorial and mating displays, which feature dramatic aerial plunges and unique vocalizations. In some years, up to 10 can be seen by driving several miles on Crawford and adjacent roads. This entire part of the county near Brookville and the Smoky Hill Air National Guard Range has a stable population of Greater Prairie-Chickens. Listen for their booming at dawn on calm spring mornings

South-Central Region

South-central Kansas epitomizes the diversity of habitats and birdlife in the state. The High Plains physiographic region extends from Ford and Hodgeman Counties east to Kingman and Reno Counties, giving way to the Wellington-McPherson Lowlands of Harvey, McPherson, Sedgwick, and Sumner Counties. Within the Wellington-McPherson Lowland region are two major wetlands: the McPherson Valley Wetlands and the Slate Creek Wetlands. The Arkansas River Lowlands bisect the region in a broad swath following the course of the river. Within the Arkansas River Lowlands and south of the river are several extensive areas of stabilized sand dunes known as the sandhills that have a unique plant and animal community. These are most prevalent in Edwards, Harvey, Reno, and Stafford Counties. Pratt Sandhills WA and Sandhills State Park are excellent places to explore this unique habitat. The well-known wetlands of Cheyenne Bottoms and Quivira NWR are also within the Arkansas River Lowlands. Along the Oklahoma border, covering most of Barber, Clark, Comanche, and Harper Counties, is the unique Red Hills physiographic region, typified by rugged hills and canyons, making this one of the most scenic regions of Kansas. Because of the rugged topography of the Red Hills, there is less cultivated land than in other areas of the state.

Several eastern songbird species reach the western limit of their Kansas distribution in the south-central region. These include Red-shouldered Hawk, Barred Owl, Pileated Woodpecker, Eastern Wood-Pewee, Red-eyed Vireo, Blue-gray Gnatcatcher, Tufted Titmouse, Louisiana Waterthrush, Northern Parula, and Summer Tanager. Conversely, several western species are near the eastern limits of their range. These include Common Poorwill, Say's Phoebe, Rock Wren, Cassin's Sparrow, and Lark Bunting in summer and Golden Eagle, Ferruginous Hawk, Prairie Falcon, and Townsend's Solitaire in winter.

Several Arkansas River tributaries flow to the east and southeast across the region. These include the Chikaskia, Little Arkansas, Medicine, Ninnescah, Pawnee, and Salt Fork Rivers as well as Walnut Creek. As the woodlands bordering these streams have matured, eastern bird species of the eastern woodlands have gradually expanded westward along these corridors.

Cheney Reservoir is the only large federal reservoir in the south-central region, but there are numerous small state, county, and city reservoirs, many of which are discussed in the following hotspot accounts. All these impoundments attract waterfowl, grebes, shorebirds, gulls, and terns. Most of the largest and most important wetland complexes in Kansas are located in the south-central region. These are Cheyenne Bottoms, McPherson Wetlands, Quivira NWR, and Slate Creek Wetlands. Many Kansas wetlands were drained and converted to agricultural use after European settlement, making those that remain vital habitats for migrating waterfowl, shorebirds, and other wetland species. They are critically important for breeding species such as King Rail, Common Gallinule, Least Bittern, Snowy Plover, and Least Tern. In addition to the large wetlands, there are playa lakes in the western half of the region and various smaller wetland areas in the eastern half. These provide additional important islands of habitat for a variety of wetland-dependent species.

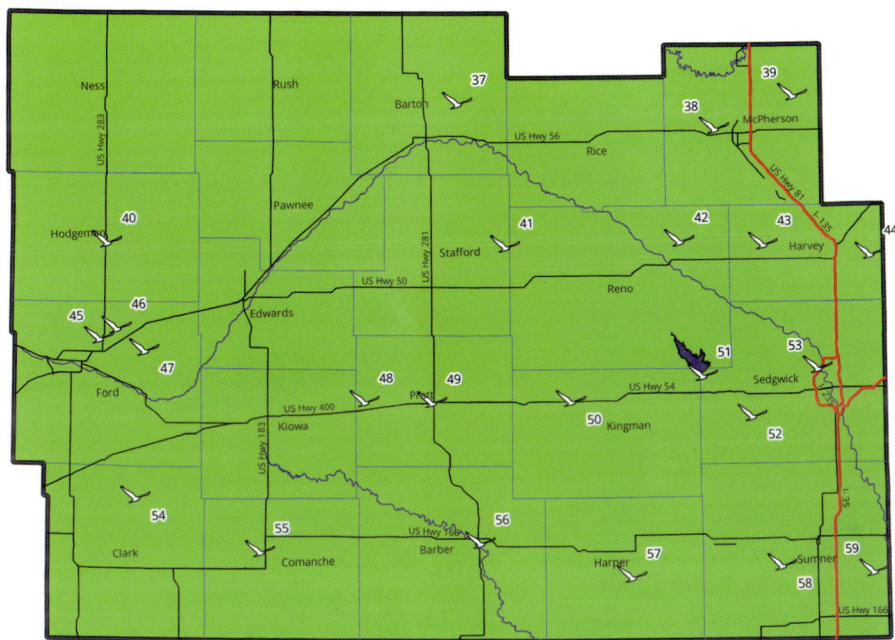

South-Central Region Hotspots

37. Cheyenne Bottoms
38. McPherson Valley Wetlands
39. McPherson State Fishing Lake and Maxwell Wildlife Refuge
40. Jetmore Area
41. Quivira National Wildlife Refuge
42. Sandhills State Park and Dillon Nature Center
43. Harvey County West Park
44. Harvey County East Park
45. Ford State Fishing Lake
46. Hain Wildlife Area
47. Herron Playa Wildlife Area

48. Texas Lake Wildlife Area and Pratt Sandhills Wildlife Area
49. Pratt Area
50. Kingman State Fishing Lake and Byron Walker Wildlife Area
51. Cheney Reservoir
52. Lake Afton
53. Wichita Area
54. Clark State Fishing Lake
55. Coldwater Lake
56. Red Hills Scenic Drive
57. Anthony City Lake
58. Wellington Lake
59. Slate Creek Wetlands

37. Cheyenne Bottoms

County: Barton

eBird Hotspot names: Cheyenne Bottoms WA, Cheyenne Bottoms WA—Campground, Cheyenne Bottoms WA—Inlet Canal, Kansas Wetlands Education Center, TNC—Cheyenne Bottoms Preserve

Hotspot type: Major destination

Cheyenne Bottoms is the largest remaining wetland in Kansas. It lies in a 60-square-mile basin northeast of Great Bend. Over 320 species of birds have been recorded on the refuge. These include 34 waterfowl, 39 shorebird, 13 gull, 15 heron and ibis, 14 hawk, and 21 sparrow species. Many of these cumulative totals eclipse those from any other birding hotspot in Kansas. The entire wetland covers about 41,000 acres of public and private land. The Cheyenne Bottoms WA consists of 20,000 acres managed by the KDWP and 7,300 adjacent acres owned and managed by The Nature Conservancy in Kansas.

To reach Cheyenne Bottoms from Great Bend, take K-156 northeast for 6 miles to the south entrance to the Cheyenne Bottoms WA.

The **Kansas Wetlands Education Center** is located on the southeast edge of the wetland off K-156. It is an interagency partnership of Fort Hays State University, the KDWP, and The Nature Conservancy in Kansas. The center offers interpretive programs, displays, and a gift shop. Current information on recent bird sightings is posted near the building entrance.

The south entrance to the **Cheyenne Bottoms WA** is adjacent to the education center. The extensive dike system divides the area into pools that allow water management. Most of the dikes have roads on them, and most are open to the public. This road system provides the best viewing opportunities for birders visiting the Bottoms, and most birders take the time to drive all these roads. The wetland complex is characterized by dramatic fluctuations in water supply. It is difficult to predict from year to year where the best birding will be. Concentrate your search on areas with extensive areas of mudflats or shallow water. Pool 1A has an observation tower, and there is frequently a long, exposed spit into Pool 1A near the tower. The spit is a popular roosting area for shorebirds, gulls, and terns. On windy days, the shallow water is forced to the windward portions of the pools, and shorebirds move onto the exposed mudflats in search of food. In many years, the best birding is along the east and south sides of the center Pool 1A. Continue driving around Pool 1 on the dike road, and eventually you will reach an entrance road from the west along the **Inlet Canal**. This canal is also a good birding area. If you want to supplement your trip list with some passerine species, visit the **Cheyenne Bottoms WA Campground** along the inlet canal, located about a mile west of the KDWP Headquarters.

To reach The Nature Conservancy in Kansas's **Cheyenne Bottoms Preserve**, go north from the KDWP headquarters building for 4 miles. After crossing a low-water bridge, turn right (east) on NW 100 Road and follow the road past an extensive wetland area. This road should be avoided after heavy rains. The property does not have a system of dikes and levees and more closely resembles the presettlement appearance and dynamics of these wetlands.

In years when water conditions are favorable, the massive population of larval insects inhabiting the mudflats is a food source that makes this the single most important stopping point for migratory shorebirds in the interior US. Up to 45 percent of the entire population of some North American shorebird species, including Hudsonian Godwit, Wilson's Phalarope, and White-rumped Sandpiper, pass through the Bottoms in spring. Shorebird migration peaks during April and May in the spring and July through October in the fall. On an exceptional day, you can find 25 or more shorebird species as well as grebes, waterfowl, herons, cranes, rails, gulls, terns, and sparrows. Large flocks of Double-crested Cormorant and American White Pelican are seen in spring and fall. In summer, look for nesting American and Least Bitterns, Black-crowned Night Heron, all three egret species, American Avocet, Black-necked Stilt, Forster's and Black Terns, and Yellow-headed Blackbird. In areas of dense cattails, carefully watch the edge of the vegetation for Least Bittern, Virginia Rail, and Common Gallinule. In late summer, postbreeding egrets and herons can number in the hundreds or thousands of birds. During the prolonged fall migration, numbers of shorebirds may not be as large as in the spring, but they tend to linger longer. Early spring is highlighted with large flocks of waterfowl and Sandhill Crane. Whooping Crane is usually reported in early April, late October, and early November. In winter, look for Rough-legged and Ferruginous Hawks, Northern Harrier, and Bald and Golden Eagles.

38. McPherson Valley Wetlands

County: McPherson

eBird Hotspot names: McPherson Valley Wetlands—Big Basin Marshes, McPherson Valley Wetlands—Chain of Lakes Marshes, McPherson Valley Wetlands—Little Sinkhole Marshes, McPherson Valley Wetlands—Kubin Marshes, McPherson Valley Wetlands WA, Lake Inman

Hotspot type: Half day

The **McPherson Valley Wetlands** were historically equal in size and importance to Cheyenne Bottoms, originally covering more than 9,000 surface acres of water in 52 wetland areas. In the early twentieth century, they were drained by farmers in an effort to convert them into farmland. In the late twentieth century, the KDWP began buying some of the original wetlands and restoring the wetland habitat. These efforts are ongoing today. The acquired wetlands are not contiguous but are in separate tracts of land that total 4,455 acres, including 1,760 acres of wetlands in 51 managed units.

The **Kubin Marshes** and the **Big Basin Marshes** are conjoined and are located west of McPherson. To reach the Big Basin, take US 56 west from McPherson to 11th Avenue, then head north on 11th to Moccasin Road, then west for 1.5 miles on Moccasin to reach the Big Basin. This road can be impassable after rainfall. To reach the Kubin Marshes, continue north on 11th Avenue 1 mile to Mohawk Road and then west on Mohawk for 1.5 miles to reach the parking area. These two areas cover several square miles and have 18 managed wetland units. The **Chain of Lakes Marshes** are

2.5 miles south of US 56 on 11th Avenue and another half mile west on Frontier Road, where there are six managed wetland areas. The **Little Sinkhole Marshes** are 3 miles east and 2 miles south of Inman west of the intersection of Arrowhead Road and 10th Avenue. This tract has 17 managed wetland areas and is the most favored by birders. To reach the best birding area here, park near the intersection and walk west along the levees, starting at the north side of the pond. Just east of 10th Avenue, there is often productive habitat as well.

Privately owned **Lake Inman** is not part of the wildlife area, but most birders visiting the area make a stop there. Lake Inman tends to attract more diving ducks than the other wetland tracts. Prior to European settlement, this was the largest natural lake in Kansas. Lake Inman is located about 3 miles east and 1 mile north of Inman at the intersection of 11th Avenue and Cheyenne Road.

To assist with navigation directions for the McPherson Wetlands, you will find it useful to download or obtain a hard copy of the KDWP brochure for McPherson Valley Wetlands. Success with birding these wetland tracts is greatly influenced by water availability and levels.

When conditions are favorable, these wetlands attract waterfowl, shorebirds, herons, sparrows, and a variety of other birds, and over 200 species of birds have been identified. These have included 30 shorebird, 12 heron and ibis, and over 15 sparrow species. Wetland specialists such as Least Bittern and Yellow-headed Blackbird have been found during the breeding season in multiple years. Notable finds here have included Black-bellied Whistling Duck, Cinnamon Teal, Whooping Crane, Piping Plover, Ruddy Turnstone, and Dunlin.

39. McPherson State Fishing Lake and Maxwell Wildlife Refuge

County: McPherson

eBird Hotspot names: McPherson SFL & Maxwell Wildlife Refuge

Hotspot type: Half day

Located 6 miles north of Canton in McPherson County, the **Maxwell Wildlife Refuge** preserves 4.5 square miles of mixed-grass prairie. Small herds of bison and elk are managed on the refuge and can often be seen from the road. McPherson SFL is located on the eastern edge of the wildlife refuge. The lake covers 29 surface acres and has a maximum depth of 30 feet. The lake property includes 46 acres of land surrounding the lake. To reach the refuge from Canton, go north for 6 miles on 27th Avenue to Pueblo Road and turn west. The refuge is on the south side of the road for 1 mile and then on both sides of the road until reaching the lake. Shortly before the lake, there is an observation tower south of the road that often provides views of the bison and sometimes of the elk. Please note that other than the observation tower and the state lake property, visitors are not allowed to leave the roadway at Maxwell.

At the lake, there are access roads on both sides, and visitors are free to wander on foot. At the southwestern corner of the lake is the trailhead for the Gypsum Creek Nature Trail, which traverses a tract of mature woodland. At the south end of the lake is some wetland habitat. In addition to the large mammals and birds, the refuge is well-known for its variety of wildflowers and prairie grasses.

Twenty-one species of waterfowl have been recorded at the lake. Because of its depth, Horned Grebe and Common Loon are found regularly during migration. In early spring, displaying American Woodcock has been heard near the upper end of the lake, and it has been confirmed to nest here. Nine heron and ibis species have been seen, and Osprey visit in spring and fall. Red-shouldered Hawk, Barred Owl, and Pileated Woodpecker have been found in recent years. Prairie birds are abundant in summer and include Upland Sandpiper, Common Poorwill, Common Nighthawk, Scissor-tailed Flycatcher, Eastern and Western Kingbirds, Bell's Vireo, Grasshopper Sparrow, Lark Sparrow, Yellow-breasted Chat, Baltimore and Orchard Orioles, Eastern and Western Meadowlarks, and Blue Grosbeak. Sedge Wren and LeConte's Sparrow are possible during fall migration. During migration, the wooded areas on the west side of the lake and on the nature trail have produced 13 warbler species, Summer Tanager, Rose-breasted Grosbeak, and other neotropical migrants.

40. Jetmore Area

County: Hodgeman

eBird Hotspot names: Buckner Valley Park, Horsethief Reservoir, Jetmore Cemetery, Jetmore City Lake

Hotspot type: Half day

There are several productive birding hotspots in Hodgeman County within the Buckner Creek watershed. **Horsethief Reservoir** covers 450 acres surrounded by 1,600 acres of developed park and shortgrass prairies. The entrance is 10 miles west of Jetmore on K-156. A vehicle permit is required to enter the property. The property surrounding the lake is nearly devoid of trees. There is a nice campground here with lots of campsites. Waterbirds, including waterfowl, shorebirds, grebes, gull, and herons, are among the 170 species of birds that have been seen here. Notable sightings have included Trumpeter Swan, Cinnamon Teal, Red-breasted Merganser, Western and Clark's Grebes, Black-bellied Plover, Long-billed Curlew, Hudsonian Godwit, and Red-necked Phalarope. Cassin's, Grasshopper, and Lark Sparrows are breeding birds in summer.

From the Horsethief entrance, drive east 4 miles to the entrance of **Buckner Valley Park**, which is an old picnic and camping property managed and maintained by the city of Jetmore. The park is located on Buckner Creek where a small dam creates a modest reservoir. By western Kansas standards, the woodlands along the creek are substantial. Even in drought years, there is at least some surface water. Below the dam is an area of thick brush. Over 150 species of birds have been found here. Species that are scarce in the west, such as Green Heron, Cooper's Hawk, Red-bellied Woodpecker, and Indigo Bunting, probably nest here. During spring and fall, migration species of interest have included Black-throated Green and Townsend's Warblers, Summer Tanager, Rose-breasted and Black-headed Grosbeaks, and Painted Bunting.

Return to K-156 and proceed east for another 2.5 miles to the intersection with 215 Road and go south for another 2.5 miles to reach **Jetmore City Lake**. This reservoir is much older than Horsethief Reservoir. It covers 106 surface acres with about 100 acres of public land. The access road forks, with the left fork ending near the west end of the dam. The right fork continues to the upper end of the lake. Because the lake is much older than Horsethief, the aquatic vegetation, invertebrates, and baitfish populations are well established. Probably for this reason, species such as loons, grebes, and diving ducks can often be more numerous here than at Horsethief. The flooded brush on the east shore may conceal Hooded Merganser, Pied-billed Grebe, and American Coot lurking in the cover.

Located a half mile south of Jetmore on US 283 is the **Jetmore Cemetery**. The cemetery has many mature juniper trees and is a reliable location for Townsend's Solitaire in winter. Some observers have reported up to 20 of them here in a single visit. Merlin is also reported here in the winter months. During spring and fall migrations, a few sparrows and warblers are usually present.

41. Quivira National Wildlife Refuge

County: Stafford

eBird Hotspot names: Quivira NWR—Big Salt Marsh—Wildlife Drive, Quivira NWR—Little Salt Marsh, Quivira NWR—Migrant's Mile and Park Smith Lake, Quivira NWR—Visitor's Center and Headquarters

Hotspot type: Major destination

Quivira NWR reigns supreme among Kansas birding destinations. Located in Stafford County about 30 miles west of Hutchinson, its 21,000 acres encompass a variety of wetland habitats, extensive sand prairie, and upland woods. The two major wetland impoundments at Quivira are the **Little Salt Marsh** at the south end of the refuge and the **Big Salt Marsh** on the north end. There are numerous smaller wetlands.

Three hundred forty species of birds have been seen in Stafford County, and most of these have been observed at the refuge. Quivira is one of the foremost birding destinations in Kansas regardless of the season. It is renowned for its variety of waterfowl, cranes, herons, shorebirds, and other wetland species, especially during spring and fall migrations. An amazing 34 waterfowl, 40 shorebird, 10 gull, 7 tern, 16 heron and ibis, 13 hawk, all 4 falcon, and 23 sparrow species have been observed at Quivira. The greatest variety of birds can be found during migration from mid-March through late May in the spring and again from mid-July through mid-October in the fall. During the summer months, many wetland specialists remain to nest, including American and Least Bitterns; Black, King, and Virginia Rails; Common Gallinule; Snowy Plover; Black-necked Stilt; American Avocet; and Least Tern. During the summer, the upland areas have numerous nesting Eastern and Western Kingbirds, Eastern Bluebird, Bell's Vireo, Field Sparrow, and Baltimore Oriole. In late fall, species diversity declines significantly, but huge flocks of geese, ducks, and Sandhill Crane are present, remaining into the winter months while open water remains. During the winter months, Bald Eagle, Red-tailed and Rough-legged Hawks, Northern Harrier, Prairie Falcon, and

Hotspot 41: In late fall, tens of thousands of Sandhill Cranes roost at Quivira National Wildlife Refuge. The best viewing is along the Wildlife Drive at dawn and dusk.

other raptors are conspicuous. Flocks of Harris's, White-crowned, and American Tree Sparrows; Dark-eyed Juncos; and other winter songbirds enliven the brushy sandhills. Small groups of Short-eared Owl are often seen foraging over prairies and marshes at dawn and dusk from late fall through early spring.

To reach Quivira from US 50, go east from Stafford for 6 miles to the small town of Zenith, where you will see a sign for Quivira. Turn north on the paved county road and proceed 8 miles to the headquarters. The route outlined here will take you past many of the most popular birding spots on the refuge. The **Visitor's Center and Headquarters** is located at the entrance on the south side of the Little Salt Marsh. The headquarters has a small interpretive display, and there you can obtain maps, a bird checklist, and other useful information. From the headquarters, proceed north on the gravel road that follows the eastern shore of the Little Salt Marsh. Near the start of this road, a large observation platform allows an expansive view of the Little Salt Marsh. A short trail leads to an observation blind. The refuge road continues north and east for several miles to **Park Smith Lake**, which often attracts good numbers of diving ducks. A 1-mile-loop walking trail, the **Migrant's Mile**, begins and ends here and includes a long boardwalk across a cattail marsh. Across from Park Smith Lake are the only restrooms on the refuge. Continue north, cross paved NE 140th Street, and continue north, passing several smaller wetlands, until you reach a T intersection with **NE 170th Street**. Turn left (west) onto 170th and proceed 1 mile to the beginning of the **Big Salt Marsh Wildlife Drive**. This 5-mile loop road offers the best wildlife viewing on the refuge. Wetlands border the drive on both sides for most of its length. In summer, look for Common Gallinule, Least and American Bitterns, King and

Virginia Rails, and Sora lurking at the edges of the cattails and sedges at the water's edge. Shorebird and waterfowl numbers can be impressive, depending on water conditions in the various pools and impoundments along the drive. Late-summer concentrations of herons and egrets can number into the hundreds. In late fall and winter, small numbers of both Trumpeter and Tundra Swans are often present.

The Wildlife Drive returns to 170th Street. Here you can turn right (east) and drive about 1 mile to a large mudflat area north of the road, which is often the best shorebird site at Quivira, although it can be totally dry. From here, turn around and go back to the west for 2 miles to an extensive wet sedge meadow on the north side of the road. In late fall, LeConte's and Nelson's Sparrows are possible in this sedge grass habitat. This is the best place on the refuge to hear Black Rail in late spring. Bobolink nests in this field in some years. At the corner with NE 105th Avenue, turn north for a drive through sandhill prairie habitat with dogwood, plum, and sumac thickets.

In late fall, hundreds of thousands of geese and tens of thousands of Sandhill Crane roost on the refuge. The best viewing is from the Wildlife Drive at dawn and dusk, when massive flocks fly to and from grainfields to forage. The sight of these incredible numbers of birds with the setting sun as the backdrop is one of the greatest wildlife spectacles in Kansas or anywhere in the US. Quivira is also one of the most frequently used stopping points in the interior US for migrating Whooping Crane in early April, late October, and early November. Northbound Whooping Cranes in spring rarely linger, but southbound migrants in the fall may remain at Quivira for several days. Occasionally 10 or more can be seen at once.

42. Sandhills State Park and Dillon Nature Center

County: Reno

eBird Hotspot names: Dillon Nature Center, Sandhills SP

Hotspot type: Half day

Sandhills State Park preserves a substantial tract of the unique sand prairie habitat and is 1,123 acres in size. Rolling sand dunes up to 40 feet tall give the area visual appeal. The park includes sand prairie as well as brushy thickets, upland woodland, and wetlands. There are 14 miles of interpretive, hiking, and horseback trails. The park is located just north of Hutchinson. The main entrance is located on E 56th Avenue 1 mile east of K-61 on the north side of the road. South of this entrance is a developed camping area with RV sites, restrooms, and an artificial pond. Over 150 species of birds have been recorded at this state park.

In summer, the brushy prairies have impressive numbers of Bell's Vireo, Field Sparrow, Indigo Bunting, and Blue Grosbeak. Yellow-breasted Chat is generally scarce, with a scattered distribution in Kansas. The extensive brushy thickets at Sandhills State Park are one of its Kansas strongholds, and several can be seen during a summer visit. Woodlands support Red-bellied Woodpecker, Eastern Wood-Pewee, and Red-eyed Vireo as summer residents. Other summer residents include Red-headed Woodpecker, Scissor-tailed Flycatcher, Eastern and Western Kingbirds, Baltimore and Orchard Orioles, and Mississippi Kites. During late fall migration, look for LeConte's

Sparrow in the wet meadows. In winter, the extensive thickets harbor flocks of sparrows, and early spring may produce displaying American Woodcock.

Many birders combine a trip to Sandhills State Park with a visit to **Dillon Nature Center**, located on E 30th Avenue about 2 miles south of the state park. The entrance is two blocks east of K-61 on the north side of the street. This 100-acre sanctuary has a visitor center with nature exhibits, a gift shop, and restrooms. The mission of the nature center is to act as a demonstration area for attracting wildlife to urban landscapes. There are woodlands, prairie, wildflower gardens, and two small ponds accessible via 3 miles of marked nature trails. More than 200 species of birds have been observed here. These include some waterbirds as well as migratory woodland and prairie passerines.

43. Harvey County West Park

County: Harvey

eBird Hotspot names: Harvey Co. West Park

Hotspot type: Worth a stop

This 310-acre park west of Newton is 14 miles west and 3 miles north of Newton. Take US 50 west from Newton to River Park Road, then drive north for 3 miles to NW 24th Street, then head east on 24th for a half mile to the park entrance. There are numerous picnic and camping areas, including an RV area, along a 1-mile stretch of the Little Arkansas River. The area includes mature riparian woodlands and a 16-acre lake with an old suspended metal swinging bridge that crosses the lake to an area of over 100 acres of sandhill habitat. Three miles of mowed trails traverse the sandhills. Most birders split their time between the riparian woodlands along the main park roads and the sandhill trail area.

At least 220 species of birds have been observed at this park. Woodland birds are numerous, and the park is home to Red-shouldered Hawk, Barred Owl, Pileated Woodpecker, and Summer Tanager in summer. The lake is not large but has attracted 21 species of waterfowl and other aquatic species such as Eared Grebe and American Bittern. Eastern Whip-poor-will nested here historically, and a small breeding population may still persist. Chuck-will's-widow is also a summer resident. During migration, the park is an excellent birding location with an assortment of vireos, warblers, and sparrows. Both Rose-breasted and Black-headed Grosbeaks have nested here. Other good finds have included Black-billed Cuckoo, White-eyed Vireo, Northern Shrike, Mountain Bluebird, Wood Thrush, and Scarlet Tanager.

44. Harvey County East Park

County: Harvey

eBird Hotspot names: Harvey County East Park, Osage Nature Trail

Hotspot type: Worth a stop

Located east of Newton is the 1,338-acre **Harvey County East Park**. Within the park is a 314-acre lake with camping and picnic areas. To reach the lake, go east on E 1st

Street from Newton for 6 miles. Look for the sign to the park, and turn north on East Lake Road. This road bisects the park and the lake and has several viewing locations. Below the dam, with parking access on 1st Street, is the short Steve Harper Trail, which goes through the woodlands near the outlet channel.

Nearly 250 species of birds have been found at this hotspot. Waterfowl are present from fall through spring. In drought years, shorebirds can be present in good numbers. Some of the better birds that have been found here are Surf Scoter, White-winged Scoter, both godwits, Ruddy Turnstone, and Smith's Longspur.

From the intersection of E 1st Street and East Lake Road, drive 2 miles north to NE 24th Street and then west on NE 24th 1 mile to the parking area and trailhead for the 1.8-mile **Osage Nature Trail**. The primary half-mile loop trail is mostly paved and winds through prairie and upland woodland habitats. The species list for the trail stands at nearly 150.

45. Ford State Fishing Lake

County: Ford

eBird Hotspot name: Ford SFL

Hotspot type: Half day

Located 3 miles north and 5 miles east of Dodge City, Ford SFL and the surrounding Ford County Park is the largest tract of publicly owned wildlife habitat in the Dodge City area. The entrance is on Garnett Road, 1 mile east of US 283. This hotspot is not far from Hain SFL, and the two locations can easily be visited on the same day. The property consists of 314 acres of land, including the lake of 48 surface acres and 10 acres of wetlands at the upper end of the lake. There are also 50 acres of mature woodland south of the lake. A 2.9-mile hiking trail begins and ends near the south end of the lake on the west shore. This trail can also be accessed from a small parking area on 116 Road on the west edge of the park property. The loop trail meanders along the lakeshore through grasslands and woodlands. This lake does not attract as many waterfowl as Hain. During drought years, shorebirds can be numerous on the mudflats.

Over 220 species of birds have been observed at Ford County Lake. Waterfowl and other waterbirds, shorebirds, and wading birds are well represented on the list. Fifteen species of hawks and falcons have been seen here, including Red-shouldered and Broad-winged Hawks and all four falcon species. During spring and fall migrations, an excellent variety of passerines has also been found, including 6 vireo, 6 wren, 12 sparrow, 18 warbler, and 6 grosbeak and bunting species.

46. Hain Wildlife Area

County: Ford

eBird Hotspot Name: Hain SFL

Hotspot type: Worth a stop

Hain WA is located northeast of Dodge City. Six miles east of Dodge City, at Wright, take US 283 5 miles north of US 50, then drive 2 miles east on Eagle Road. Because of its location on the dry High Plains and KDWP management, good numbers of waterfowl, shorebirds, and waders are often found here. The surface area of the lake is 35 acres, and 20 acres of upland habitat surround the lake The maximum depth of the lake is 5 feet. In drought years, the lake can go dry. A few mature cottonwood trees are near the shoreline and dam, but most of the land around the lake is shortgrass prairie. At the south end of the lake is an area of brush and taller vegetation where various passerine species are possible.

At this hotspot, 167 species of birds have been reported. Aquatic species are the main attraction. Thirty species of waterfowl and other swimming birds have been seen here, mostly from late fall through early spring. In the winter months, large flocks of geese are often present. Shorebird numbers fluctuate along with the water levels. Nineteen species of shorebirds have been reported, including Black-necked Stilt, American Avocet, Whimbrel, and Willet. Eight species of wading birds have been observed, including American Bittern and Glossy Ibis, both of which are noteworthy in the western half of the state. Cassin's, Grasshopper, and Lark Sparrows nest in the grassland habitat. In the lakeside cottonwoods, look for Eastern and Western Kingbirds, Bell's and Warbling Vireos, and Baltimore and Orchard Orioles.

47. Herron Playa Wildlife Area

County: Ford

eBird Hotspot name: Herron Playa WA

Hotspot type: Worth a stop

Herron Playa WA is located 4 miles south of Spearville in Ford County and covers 700 acres. There are 160 acres of wetlands, 360 acres of native grass, and 180 acres of cropland. This was historically a playa wetland but was largely drained in the 1950s. Since it was acquired by the KDWP, dikes have created two separate wetlands, and an irrigation well maintains water levels in times of drought. Consequently, there is water here more reliably than at most other playas in western Kansas. The area is open to hunting by posted notice. Park at the intersection of Iron Road and 125 Road, located 4 miles south of Spearville. From here, walk west for a half mile to 1 mile on what remains of Iron Road. In wet years, you will find good shorebird mudflats on the south side of the road and deeper marsh habitat on the north side. At the half-mile mark, there are trees around the small pond where Black-crowned Night Heron nests in some years.

About 140 species of birds have been identified at Herron Playa. Waterfowl, shorebirds, wading birds, and other wetland species represent most of the birds of interest at this hotspot. Geese are often on the area in the colder months. Most of the ducks are dabbling ducks, but some diving ducks can be expected as well. Look for Cinnamon Teal in spring and early summer. Twenty-four species of shorebirds have been seen here, including Black-bellied Plover, Piping Plover, Long-billed Curlew, Whimbrel, and Marbled Godwit. Ten species of wading birds have been reported, including regionally scarce American Bittern, Yellow-crowned Night Heron, and Glossy Ibis. Because of the relatively stable water levels, Herron Playa attracts wetland breeding species, such as Black-crowned Night Heron, Black-necked Stilt, American Coot, and Pied-billed Grebe. In favorable years, some of their nests can be observed from the road. July and August can be the best months to visit, as family groups of wetland species are joined by southbound migratory shorebirds and wandering postbreeding herons and egrets.

48. Texas Lake Wildlife Area and Pratt Sandhills Wildlife Area

County: Pratt

eBird Hotspot names: Texas Lake WA, Pratt Sandhills WA

Hotspot type: Worth a stop

Two public areas managed by the KDWP in western Pratt County are conveniently close to US 54/400 and offer good birding opportunities. These areas are managed primarily for hunting, and both receive significant hunting pressure, especially during waterfowl and upland game bird seasons.

Texas Lake WA covers a total of 1,232 acres and is located 4 miles west and 2 miles north of Cullison. At the intersection with Pratt County NW 130th Avenue, there is a sign for Texas Lake on US 400. From the highway, go north for 1.5 miles to the intersection with W 1st Street. From this intersection, county roads traverse the wildlife area to the east, north, and west. Depending on water conditions, good numbers of shorebirds, waterfowl, and other wetland species can be observed from these roads. Walking the fringes of the wetland areas into the prairie and along the old shelterbelts can lead to productive birding in all seasons. This natural wetland is managed today with the assistance of dikes and a pump system. The native grasslands are dotted with plum thickets, Eastern Red Cedar, and old shelterbelts. Over 140 species of birds have been reported at Texas Lake. Waterfowl and shorebirds are abundant when conditions are favorable. Other wetland species, including Virginia Rail, Sora, Marsh Wren, and Yellow-headed Blackbird, are likely. The upland and wetland habitats are especially good for sparrows. In summer, there are nesting Dickcissel along with Field, Grasshopper, and Lark Sparrows. Other nesting birds include Bell's and Warbling Vireos, Eastern and Western Kingbirds, Scissor-tailed Flycatcher, Orchard and Baltimore Orioles, and Blue Grosbeak. In winter, look for Dark-eyed Junco and American Tree, Harris's, and White-crowned Sparrows around the thickets of Sandhill Plum. Raptors include Rough-legged Hawk, Prairie Falcon, and Merlin. Look for migrating flocks of Sandhill Crane in early spring and late fall.

Pratt Sandhills WA covers 5,708 acres of land and is located a few miles north of Texas Lake. To reach it from Texas Lake WA, go west for 1 mile to NW 140th and then north for 4 miles (or 7 miles north of US 54) to reach the southern boundary of the area. For the next 2 miles, the wildlife area is on both sides of the road. The public land extends as far as 7 miles to the east of 140th, but there are no roads within the area. You can hike as far in as you care to, but the habitat can be sampled reasonably well without venturing too far on foot. Birds may not be as abundant as at Texas Lake WA, but a visit here is worth the time just to take in the appealing and largely undisturbed sandhill habitat. Nonwetland birds are mostly as described for Texas Lake.

49. Pratt Area

County: Pratt

eBird Hotspot names: KDWP Headquarters, Lemon Park, Ninnescah River Trail, Pratt Co. Lake

Hotspot type: Worth a stop

There are several productive birding spots around Pratt adjacent to the South Fork Ninnescah River. **Lemon Park** is a 117-acre park located on the river at the south edge of town. This large, multiuse park receives substantial visitation. Signs for the park are on both US 400 and US 281. The main entrance is from the north via S Pine Street, which becomes a loop road within the park. Several nature trails pass through undeveloped woodlands. Nearly 150 species of birds have been seen in the park, mostly in these woods. In summer, look for Eastern Wood-Pewee, Red-eyed Vireo, and Tufted Titmouse, all of which start to become scarcer west of Pratt. Red-shouldered Hawk, Barred Owl, and Pileated Woodpecker are all expanding their range westward along the Ninnescah River, and you have a decent chance of finding them in the woodlands of Lemon Park. Winter birds include Yellow-bellied Sapsucker, Hermit Thrush, Brown Creeper, Ruby-crowned Kinglet, Golden-crowned Kinglet, and Winter Wren. Both Carolina and Black-capped Chickadees have been reported in the park. Spring and fall migrants have included 6 vireo, 13 sparrow, 14 warbler, and 4 grosbeak and bunting species.

A trio of hotspots are adjacent to each other along the Ninnescah River just west of town. These are the Ninnescah River Trail, the KDWP headquarters, and Pratt County Lake. To reach the **KDWP headquarters**, take SE 25th Avenue for 1 mile south from US 54/400 or 3 miles east from US 281. There is public parking north of the headquarters building. There are small ponds and woodlands north and east of the headquarters, which can be explored on foot. Take K-64 north for a short distance across the bridge to reach the parking area and trailhead for the **Ninnescah River Trail** on the east side of the highway. The trail is roughly a half mile in length and passes through riparian woodland along the river. **Pratt County Memorial Veterans Lake** is located 1 mile east of the KDWP headquarters on SE 10th Street. Over 200 species of birds have been reported at the lake, including many waterfowl and shorebird species. This 51-acre lake has a heavily developed shoreline and grounds. Notable sightings from the lake have included Sabine's Gull, Neotropic Cormorant, and Common Loon. West of the parking lot is a large cattail marsh, which during migration has produced Sora, Virginia Rail, Marsh Wren, and Swamp Sparrow. On the north side of the lake, the perimeter road runs parallel to the river with good woodland habitat along the road.

50. Kingman State Fishing Lake and Byron Walker Wildlife Area

County: Kingman

eBird Hotspot names: Byron Walker WA, Kingman SFL

Hotspot type: Worth a stop

Located in western Kingman County, **Byron Walker WA** is one of the larger wildlife areas in Kansas, with a total area of over 4,500 acres extending for over 5 miles on both sides of US 400. Almost 8 miles of the South Fork Ninnescah River are within the boundaries of the wildlife area. Because most of the river watershed is uncultivated prairie habitat, the water runs clear and the quality is superior to that of most other Kansas streams. There are wetlands along the banks of the river and elsewhere. There are also several ponds scattered across the wildlife area.

At the eastern edge of the wildlife area is **Kingman SFL**, which covers about 140 surface acres. NW 70th is the entrance road from US 400 and goes north along the east side of the lake. The lake is only 4 feet deep and was formed by an earth dam that was built in 1936 around a historical wetland known as the Callahan Marsh. Over 210 species of birds have been recorded at Kingman SFL. Waterbirds include 27 species of waterfowl, including both Trumpeter and Tundra Swans as well as 4 grebe, 15 shorebird, 8 heron and ibis, and 11 hawk species. Waterfowl numbers peak in late fall and early spring. Common Loon, Osprey, and large flocks of both American White Pelican and Double-crested Cormorant occur in spring and fall. The wetland areas can produce rails and Marsh Wren.

The dam is located near the southeastern corner of the lake. From the south end of the dam, a dirt road goes west between the lake levee and the river and is a great place to search for birds such as vireos, warblers, sparrows, buntings, and finches. Barred Owl, Red-shouldered Hawk, and Pileated Woodpecker have all moved into this area over the past 20 years. Louisiana Waterthrush, Northern Parula, Painted Bunting, and Summer Tanager have all been reported in recent years in spring and early summer. If you have time for only a quick stop at this hotspot, make this 1-mile road the focus of your visit. From the end of the road, you can continue on foot for some distance around the west end of the lake, where there is considerable wetland and woodland habitat. From the county roads on the north and west edges of the state lake property, there are additional walk-in access points to productive birding habitat.

There are several miles of county roads that traverse the Byron Walker WA to the north and west of the state fishing lake, and a lot of the birdlife can be sampled with a leisurely tour by car. If you want to explore on foot, you can do so anywhere on these public lands. During at least two recent drought years, displaying Cassin's Sparrow was present in this part of the wildlife area.

51. Cheney Reservoir

Counties: Kingman, Reno, and Sedgwick

eBird Hotspot names: Cheney Lake, Cheney Dam, Cheney WA, Cheney WA—DeWeese Park, Cheney WA—Yoder Cove, Cheney SP—East Shore Area, Cheney SP—West Shore Area

Hotspot type: Full day

Cheney Reservoir covers 9,500 surface acres and has a maximum depth of 49 feet. The reservoir is surrounded by 5,240 acres of public wildlife area and an additional 1,913 acres of state park. It is the largest body of water and one of the largest tracts of public land anywhere in south-central Kansas. Because of its size and varied habitats, it has long been a major birding destination. The lake is popular and is heavily used for boating, fishing, hunting, camping, and picnicking. To reach Cheney Reservoir, travel west on US 54 to the Cheney Reservoir exit, then go north for 5 miles to the intersection with 21st Street. From this intersection, you can either turn left and go 3 miles to reach the entrance to the **West Shore Area** of Cheney State Park or continue north for 1.5 miles to reach the entrance of the **East Shore Area** of Cheney State Park. Vehicle permits are required to enter Kansas state parks. In both state park areas, there are paved roads giving access to multiple points from which the lake can be scanned for birds and sheltered coves that frequently attract birds. Most birders check all the points and coves to be sure they are covering the lake thoroughly. The reservoir is large, and a spotting scope is needed when trying to identify birds far out on the lake. Many birders park below Cheney Dam and walk to the top for a panoramic view of the lake; sometimes many birds can be seen from the dam.

In the **West Shore Area** of the state park, there is an excellent nature trail within the Spring Creek area adjacent to the Smarsh Creek Campground. This trail has a long boardwalk that passes over a flooded area created by beaver dams in an attractive stand of woodland. Tree Swallow nests in the dead trees near the trail, and in winter, this is a good place to look for species such as Winter Wren and Fox Sparrow. About a mile west of the nature trail, you reach the Ninnescah Yacht Club, where you can scan a broad area of the lake.

Cheney WA includes all the lakeshore area beyond the state park and has several good birding stops, especially on the east shore. At the intersection of Silver Lake Road and Mayfield Road, turn south on Mayfield Road at the sign for the Red Bluff Area, where there are several miles of roads, lots of good habitat, and multiple access points to the shore. From the intersection of Yoder Road and Sun City Road, go west for a mile on Sun City Road and follow the road as it curves south out to **Yoder Point**. The woodlands and brushy prairies along this stretch nearly always have a great variety of birds. From Yoder Point, you can scan most of the upper end of the lake. Return to Yoder Road and continue north to Parallel Road and then west for 1.5 miles to the entrance to **DeWeese Park**, located in a substantial area of mature woodland along the Ninnescah River. Red-shouldered Hawk, Pileated Woodpecker, Barred Owl, and Prothonotary Warbler (summer) are possible at DeWeese along with other eastern woodland species that are otherwise scarce at Cheney. Rose-breasted Grosbeak nested here in at least two years.

Cheney has produced nearly 300 species of birds, making it one of the better hotspots in this book. These include over 30 waterfowl, 30 shorebird, 13 gull, 20 sparrow, and 25 warbler species. In October, concentrations of Franklin's Gull on the lake total hundreds of thousands of birds. Large flocks of waterfowl, White Pelican, Double-crested Cormorant, gulls, and terns are present during migration. Common Merganser, Common Goldeneye, and Bald Eagle are common to abundant in winter. Numerous gulls are found in the winter months, including several of the rarer gull species in most years. During drought years, substantial mudflats appear around the lake and are most widespread at DeWeese Park and Yoder Point. Many shorebirds can be seen at these times and in the past have included rarities such as Piping Plover, Whimbrel, and Red-necked Phalarope as well as many others. In 2013, an especially severe drought year, several pairs of Snowy Plover nested on the exposed flats near Yoder Point, probably the easternmost inland nesting record of this species ever documented. Both Piping Plovers and Snowy Plovers with precocial young were observed at Cheney in the summer of 2024, another severe drought year.

52. Lake Afton

County: Sedgwick

eBird Hotspot name: Lake Afton

Hotspot type: Worth a stop

Covering 230 surface acres of water within the 738-acre Lake Afton Park, this is the largest body of water in Sedgwick County except for Cheney Reservoir. This area can be productive for waterfowl, gulls, and shorebirds, and over 220 species of birds have been found. To reach the lake from Wichita, take K-42 southwest from Wichita for 4 miles to the little community of Schulte. Turn right at the traffic light, and proceed west on MacArthur Road for 8 miles, where you will see the lake on both sides of the road. The main body of the lake on the south side of MacArthur is encircled by a park road. Gulls often congregate near the dam spillway, and shorebirds can be found in the spillway during migration. On the north side of MacArthur is a much shallower portion of the lake with large areas of cattails and other wetland vegetation bordered by woodland habitat. A short nature trail through this undeveloped area is worth walking at any season. In fall and winter, when the cedar trees have a good berry crop, American Robin and Cedar Waxwing can be abundant. Some of the better birds that have been reported from Lake Afton are Cinnamon Teal, Western Grebe, Neotropic Cormorant, Piping and Snowy Plovers, Marbled Godwit, Great Black-backed Gull, Bonaparte's Gull, and Mountain Bluebird.

53. Wichita Area

County: Sedgwick

eBird Hotspot names: Arkansas River—71st St. Access, Chisholm Creek Park & Great Plains Nature Center, Crane Park—Derby, Oak Park—Wichita, Pawnee Prairie Park, Pracht Wetlands, Sedgwick County Park, Sim Park

Hotspot type: Various

There are many excellent birding locations in the Wichita area that offer a variety of habitats. Described here are some of the better local birding hotspots, but there are many other worthwhile birding hotspots in the area.

Chisholm Creek Park is an island of 240 acres of natural habitat surrounded by developed areas. Despite a high volume of visitors and considerable highway noise, this park is one of the better birding locations in Wichita. The **Great Plains Nature Center**, located at 6232 E 29th Street N, provides access to the east end of the park and has excellent interpretive exhibits relating to Kansas wildlife. There are additional parking lots on the west end of the park, reached by traveling south from K-96 on Oliver Street for a quarter mile. Over 2 miles of paved trails in the park begin at the parking areas. Habitats in the park include tallgrass prairie, riparian woodlands, hedgerows, ponds, and a wetland area. The park is an excellent place to find migrants in both spring and fall. Over 200 species of birds have been found here, including 28 warbler and 20 sparrow species. The boardwalk over the wetland is a good place to look for wetland birds, including rails and Marsh Wren. In the fall, the dogwood thickets attract migratory warblers, vireos, and other songbirds.

Oak Park is 37 acres in size and preserves the oldest mature hardwood forest in the city. The site was an isolated stand of oak forest surrounded by prairies when J. R. Mead first described the Arkansas River valley in the 1860s and has survived to the present day. This park is known as the best single place to find migratory songbirds in the Wichita area. The park is located north of downtown in the Riverside neighborhood. From the corner of 13th Street N and Waco Street, go south two blocks to 11th Street, then proceed eight blocks west to the park. For birds, the best part of Oak Park is the woodland area north of 11th Street. A complicated network of footpaths provides easy access to search for a variety of birds. An impressive 35 species of warblers have been seen in the park, mostly in the first two weeks of May, although overall numbers of warblers are noticeably diminished in comparison to previous decades. Additional migratory flycatchers, vireos, thrushes, tanagers, sparrows, and finches are observed each spring. During fall migration, smaller numbers of species are found over a more prolonged period. In summer and winter, expect typical woodland bird species for this habitat. In recent years, Cooper's Hawk, Red-shouldered Hawk, and Barred Owl have all successfully nested in the park.

Sim Park is a 181-acre park located in the central museum district and is adjacent to Botanica, Cowtown Museum, and the Sim Park Golf Course. A broad swath of open woodlands extends for almost a mile between the golf course and the Arkansas River. To reach this park from Oak Park, continue west on 11th Street to Amidon Street, then go south on Amidon until you reach the entrance to Botanica. You can park at Botanica or go west along the river to any of the parking lots along the park drive. Trails access brushy and wooded areas on stabilized sand dunes.

Sedgwick County Park contains 400 acres and is located just west of the Sedgwick County Zoo. There are entrances at each end of the park on 21st Street N and 13th Street N, about a quarter mile east of Ridge Road. The park consists of extensive mowed areas intermingled with native habitats and several small lakes. The area is good for waterfowl, wading birds, and songbirds. Nearly 9 miles of multiuse trails crisscross the park. The best area for birding is the trail on the west bank of Big Slough Creek at the western edge of the park. Stands of cattails along the creek often have Black-crowned Night Herons, egrets, and Common Yellowthroat in the summer months. In the winter, this is a good place for Virginia Rail, Marsh Wren, and Swamp Sparrow. Look for Osprey in the trees around the lakes in April and October. In winter, the lakes attract a variety of waterfowl.

Pawnee Prairie Park contains 640 acres extending along 1.5 miles of riparian woodland along Cowskin Creek. It also includes grasslands and old shelterbelts dating from its farmland past. Habitat succession has transformed some native grasslands into upland woods. To reach the park, travel south on Tyler Road from the exit on Kellogg (US 54) for 1.5 miles to a parking lot south of the public golf course. There are restrooms here and a trailhead that accesses the several miles of multiuse trails. The most productive birding is often on the trails that are closest to the creek, but don't ignore the grassland areas. In the woods are Red-shouldered Hawk, Pileated Woodpecker, Blue-gray Gnatcatcher, and Red-eyed Vireo throughout the summer.

Pracht Wetlands Park is a 90-acre park preserving what survives of a natural wetland that predates the European settlement period. The entrance and parking area are located on 29th Street N about a half mile east of Maize Road. There is a half-mile paved trail along the north edge of the wetland area with two metal boardwalks ending at viewing blinds. The southern 40 acres of the wetland are inaccessible, with plans for additional trails in the future. Surface water varies with recent rainfall and can be productive for a variety of shorebirds, waterfowl, herons, and sparrows. Bell's Vireo nests in the dense thickets, and this is one of the last remaining places in Wichita where Ring-necked Pheasant can be found. Some of the more notable finds at Pracht have included Cinnamon Teal, Black-necked Stilt, Piping Plover, Hudsonian and Marbled Godwits, Whimbrel, Dunlin, and Glossy Ibis

Several parks in the suburb of Derby preserve tracts of mature oak forest that, like Oak Park, predate European settlement of the area. **Crane Park** is productive for warblers and other passerines, especially during the month of May. This park has a history of producing rare warblers. From the intersection of James and Woodlawn in Derby, proceed east on James to the next intersection and turn right on Marguerite Parkway. Crane Park will be on your left. The park is dominated by large oak and hackberry trees with dense understory running along the creek. There is a well-worn foot trail along the west bank of the creek. Yellow-crowned Night Heron has nested along the creek in several years. Summer and winter birds are typical of the area. Other parks worth exploring in Derby including **Zollinger Park**, **High Park**, and **Riley Park**, all found farther south along Dry Creek and Spring Creek. Just west of Derby, the **Arkansas River—71st St. Access** on the west bank of the river is known as the "Canoe Launch" and has produced over 200 species of birds. This is a less used and somewhat isolated location but is usually reliable for Fish Crow, Summer Tanager, Painted Bunting, and many others in summer as well as Pileated Woodpecker year-round. From the intersection of 63rd Street S and Grove, go south to the end of the road and turn left; in less than a half mile, you will reach the entrance and parking lot.

54. Clark State Fishing Lake

County: Clark

eBird Hotspot name: Clark SFL

Hotspot type: Half day

Clark SFL is an oasis of water and woodlands located within the Red Hills region. Lying in an impressive canyon created by Bluff Creek, it is one of the jewels of Kansas public lands. To reach Clark SFL from US 54/400, turn south at the town of Kingsdown onto K-94 and proceed 10 miles to the lake entrance where the road abruptly descends into the deep canyon.

The 300-acre lake is surrounded by 900 acres of public land. More than 250 species of birds have been observed at this hotspot. Twenty-seven species of waterfowl plus cormorants, pelicans, loons, grebes, gulls, and terns have been found at the lake. Likely because of the 40-foot depth of the lake, diving ducks are often numerous during migration. Uncommon species such as Clark's Grebe, Western Grebe, and Surf Scoter have been observed. In winter, waterfowl remain if the lake is unfrozen. Osprey is typically present at the lake during spring and fall. During drought years, migratory shorebirds are attracted to the exposed mudflats.

Cattails at the upper end of the lake attract herons, egrets, rails (including Black Rail), Marsh Wren, Song and Swamp Sparrows, and Common Yellowthroat, depending on the season. Below the dam along the outlet channel is an area of woodland dominated by cottonwoods. There is usually an Eastern Screech-Owl somewhere in this woodland tract. Shorebirds are sometimes found in the spillway channel below the dam. Mountain Bluebird and Townsend's Solitaire are possible from late fall through early spring.

At the north end of the lake is a stand of mature woodland along Bluff Creek accessed by the park road on the east side of the lake. A hiking trail allows access to search for woodland species such as Barred Owl, Red-bellied Woodpecker, Carolina Wren, Carolina Chickadee, and White-breasted Nuthatch. In summer, Red-eyed and Warbling Vireos and Blue-gray Gnatcatcher can be found, and in winter, look for species such as Winter Wren; both kinglet species; and Fox, Harris's, White-crowned, and White-throated Sparrows. During spring and fall migrations, at least 14 species of warblers have been seen at the lake, most of them in this woodland area.

In summer, Cassin's, Grasshopper, and Lark Sparrows are abundant in the prairies surrounding the lake, and in some summers, Lark Bunting is also present. During migration, Savannah and Vesper Sparrows are common in this prairie and Chestnut-collared Longspur is possible. Summer residents in the brushy habitats on the slopes

Hotspot 54: Yucca dots the bluffs overlooking one of the jewels of Kansas public lands at Clark State Fishing Lake in the Red Hills region.

around the lake include Say's Phoebe, Bell's Vireo, Blue Grosbeak, Painted Bunting, and Orchard and Baltimore Orioles. Rock Wrens sing from the canyon rims, and Common Poorwills call just after dusk. Eastern Bluebird and Bewick's Wren are year-round residents.

55. Coldwater Lake

County: Comanche

eBird Hotspot names: Lake Coldwater, US Rte. 160 Prairie-Dog Town

Hotspot type: Worth a stop

Located in Comanche County, Coldwater Lake covers 250 surface acres and is surrounded by an additional 930 acres of public land. **Lake Coldwater** is owned and operated by the city of Coldwater. There is a modest fee for a daily vehicle permit. The lake entrance is a half mile south of Coldwater on the west side of US 183. A park road encircles the lake, and there are several campgrounds and picnic areas. Over 200 species of birds have been observed at the lake and surrounding property. Twenty-four waterfowl, 4 grebes, 7 heron, and 17 shorebird species have been observed at the lake. American White Pelican, Osprey, Bald Eagle, and other water-dependent species occur regularly. In the scattered cottonwoods, Eastern and Western Kingbirds, Warbling Vireo, and Baltimore and Orchard Orioles are abundant in summer.

At the north end of the lake, Cavalry Creek feeds a wetland with well-established sedges. Several rail species, including Black Rail, have been found here in the past, but habitat manipulation in recent years has led to a decline in records. In late fall, this wetland habitat is good for Marsh Wren, Swamp Sparrow, and Song Sparrow. North of the wetland, in the brush and mature trees near the old railroad bed, Black-billed Cuckoo has been seen during the breeding season. In summer, expect to find Bell's Vireo, Yellow-breasted Chat, and Blue Grosbeak.

Most of the park is mowed, but there are two undeveloped wooded areas worth checking. On the west end of the dam is a parking area near a stand of trees with a short nature trail. Below the dam outlet is an area of riparian woodland where interesting bird species can sometimes be found during migration. Twelve species of warblers have been reported at Coldwater Lake, and most of them have been found in these two wooded areas. On summer nights, listen for Chuck-will's-widow calling at dusk in the woods along the creek.

A prairie dog town is found a few miles north of the lake on US 160 north of Coldwater. The **US 160 prairie dog town** is located on the south side of the road about a quarter mile east of the US 183 and US 160 junction. Be sure to park off the highway. Usually one or two Burrowing Owls can be found here in the warm season, and Ferruginous Hawk is possible in fall and winter.

56. Red Hills Scenic Drive

Counties: Barber and Kiowa

eBird Hotspot names: Barber SFL, Medicine Lodge River—Clawson Memorial Bridge, Mule Creek—Painted Post, Salt Fork Arkansas River—Aetna Road, Thompson Creek Rd. (M St.-73rd Rd.)

Hotspot type: Full-day auto tour

The Red Hills are a rugged, scenic physiographic region located in Barber, Clark, Comanche, and Kiowa Counties in south-central Kansas. The habitat consists of extensive mixed-grass prairie interspersed with picturesque buttes and ravines that shelter extensive stands of Eastern Red Cedar and other trees. The Salt Fork and Medicine Rivers and their tributaries have mature woodlands along their banks. Despite the absence of public land, the mostly uncultivated countryside and limited traffic in this sparsely populated country make a drive through the area appealing. Regardless of the season, a diverse variety of birds can be expected on a trip to this beautiful area.

An interesting mixture of eastern and western birds is found in the Red Hills. Eastern woodland species such as Carolina Wren, Tufted Titmouse, Blue-gray Gnatcatcher, and Louisiana Waterthrush reach the western edge of their range in Kansas. In contrast, several western species reach their extreme eastern range limit. These include Lesser Prairie-Chicken, Rock Wren, and Cassin's Sparrow. The birdlife of the region also has a southern influence, best characterized by the Greater Roadrunner, which is scarce elsewhere in Kansas, and a robust population of Painted Bunting during the summer. The Red Hills are renowned for attracting large flocks of Mountain Bluebird in winters when the abundant Red Cedar trees have a bumper crop of berries. Thousands can be seen in a single day during an auto tour taken during major invasion years, but totals in the hundreds are more typical. In years when berries are few, Mountain Bluebird is scarce or absent.

Hotspot 56: The Red Hills near Medicine Lodge are renowned for attracting flocks of Mountain Bluebirds in winter when Red Cedar trees have a bumper crop of berries.

To begin an auto tour, start at the intersection of US 281 and US 160 on the south edge of Medicine Lodge. Give attention to the steep and narrow draws you encounter on this drive. These are most effectively checked by getting out and observing on foot. Many of the roads are gravel and, because of the soil composition, are treacherous after recent rain or melting snow. From Medicine Lodge, drive west for 4 miles to the intersection with Gypsum Hill Road (paved). Proceed south on this road for 18 miles, eventually reaching the intersection with Hackberry Road near the Oklahoma state line. Turn right (west) and take Hackberry Road west for 16 miles until you reach the intersection with Aetna Road. Proceed north on Aetna Road. After about 1 mile, stop where there is a steep ravine on the east side of the road. This is usually a reliable spot for Painted Bunting in summer. Another mile to the north is the **Salt Fork Arkansas River—Aetna Road** bridge, a good place to stop and look for birds. Continue north to the intersection with Painted Post Road and go west for 2 miles to the **Mule Creek—Painted Post** bridge. In this area, you are in healthy sagebrush habitat, so be alert for Cassin's Sparrow in summer. Return to Aetna Road and continue north for approximately 15 miles until you reach US 160. Go east for 1 mile to Sun City Road. Turn north on Sun City Road and proceed another 8 miles to the small town of Sun City, where you might want to stop for a sandwich or a beverage at Buster's, a renowned local institution. Just before you reach Sun City, the road crosses the Medicine River at the **Clawson Memorial Bridge**, where there are thick woods along the river. Red-shouldered Hawk, Barred Owl, and Pileated Woodpecker are all possible near the bridge. From Sun City, you can return to Medicine Lodge on the River Road (paved).

If you want to sample even more of this beautiful country, take the River Road northwest for another 10 miles to the intersection with Wilmore Road and go west for 1 mile to the tiny town of Belvidere. From Belvidere, go northwest on 76th Road, cross the river, and at the fork with 73rd Road, bear left to stay on 76th Road, also known as **Thompson Creek Road**. Over 150 species of birds have been found along this stretch of road. Golden Eagle, Ferruginous Hawk, and Greater Roadrunner are all possible on the River Road and Thompson Creek Road. Northern Shrike is possible in winter. Listen for Louisiana Waterthrush along the creek in spring. After 8 miles, you reach the intersection of Thompson Creek and M Street. If you are in this area in the early morning in spring, listen for Lesser Prairie-Chickens booming in the sandhills to the east along M Road.

While in this area, you may want to stop in Medicine Lodge at **Barber SFL** on the north edge of town, where over 150 species have been observed. The main attraction here is the lake, where many species of waterfowl and other water-dependent species have been recorded. The upper end of the lake is shallow and may be dry in drought years. It attracts waders and shorebirds when water is present.

57. Anthony City Lake

County: Harper

eBird Hotspot names: Anthony Municipal Lake

Hotspot type: Worth a stop

Covering approximately 150 surface acres, Anthony City Lake is the largest body of water in Harper County. The main entrance is located 1 mile north of Anthony on K-2. The property covers a total of 450 acres, including the lake. The lake has produced 25 waterfowl, 3 grebe, 8 heron and ibis, and 12 shorebird species as well as American White Pelican, Double-crested Cormorant, and other aquatic birds. There is always at least some sandy or muddy shoreline.

While you can expect reasonably good birding in the developed areas around the lake, the best birding is typically in the undeveloped woodland accessed from NE 30 Road and in the wooded areas below the dam near the dog park. Local birders have developed a modest birding trail at the lake with signs at designated stops. The chickadees here are all Carolinas. Summer species often include Bewick's Wren and Painted Bunting. In winter, a variety of sparrows are present in the weedy areas and Rusty Blackbird can sometimes be found. During migration, at least 12 species of warblers have been found here, mostly in May.

58. Wellington Lake

County: Sumner

eBird Hotspot names: Wellington Lake, Wellington City Lake.

Hotspot type: Half day

Nearly 3 miles in length with 674 surface acres of water and 10 miles of shoreline, **Wellington Lake** is one of the larger lakes in south-central Kansas. It is owned and operated by the city of Wellington. To reach the lake, take US 160 west from the town of Wellington to Anson Road, then go south 4 miles on Anson Road to the lake entrance. Park roads encircle the lake and provide access to nearly 200 campsites. At the north end of the lake, W 30th Street S crosses the upper end of the lake. At this bridge, there is good birding in the wetlands on both sides of the road. Below the spillway is another wetland dominated by cattails, where Virginia Rail and Marsh Wren can be found. In the lake is a partial levee about a half mile north of the dam. The levee is a roosting area for waterfowl, wading birds, shorebirds, gulls, and terns.

Over 220 species of birds have been found at Wellington Lake. The lake is an excellent location for waterfowl from late fall through early spring. Twenty-six species of waterfowl have been seen at the lake, including Black and Surf Scoters. Small flocks of Eared and Horned Grebes are often seen during migration and occasionally Western Grebe. Twenty-six species of shorebirds have been seen, including notables such as Black-bellied and American Golden-Plovers, Whimbrel, Long-billed Curlew, Hudsonian and Marbled Godwits, Dunlin, and American Woodcock. Seven gull, 7 owl, and 16 hawk and falcon species have been observed. Osprey is frequently reported in April and October, and Bald Eagle is common in winter. Woodlands around the lake are not substantial; consequently, the diversity of passerine species is somewhat limited. Seventeen sparrow and nine warbler species have been recorded here.

On the east side of Wellington is **Wellington City Lake**, which you can check if time allows. It is smaller and more developed than Wellington Lake. To reach the lake from the intersection of US 160 and North A Street in the center of Wellington, go north for 10 blocks on A Street to the lake entrance on the right. Park roads provide three access points on the west side of the lake from which you can scan the lake.

59. Slate Creek Wetland

County: Sumner

eBird Hotspot name: Slate Creek Wetland WA

Hotspot type: Full day

Slate Creek Wetland covers 945 acres of land owned and managed by the KDWP. This wetland complex is located 7.5 miles south and 1.5 miles west of the small town of Oxford. Exploring Slate Creek requires some hiking, as many of the wetland areas are not visible from roads.

After Cheyenne Bottoms and Quivira NWR, Slate Creek is one of the most significant wetlands in Kansas. Although it is smaller in size, the bird species are much the same as you would expect to find at the larger Kansas wetlands. Over 250 species of birds have been found here. These include 25 waterfowl, an impressive 34 shorebird, 7 gull, 13 heron and ibis, 5 rail and gallinule, 15 hawk and falcon, 21 sparrow, and 15 warbler species. Slate Creek is well-known for its variety of sparrow species, particularly during October and November. During winter, this is one of the most reliable locations for Short-eared Owl in south-central Kansas.

To reach the best location on the state land, travel south from Oxford for 5 miles on S Oxford Road. At the intersection with 60th Street, turn right, go west for 2 miles to Greenwich Road, and then head south for 1 more mile. Here you will see an observation tower and a small parking area. You probably won't see many birds from the tower, but it does give you a good feel for the area. Cross the gate and walk west down the dirt road, which eventually curves around one of the larger water pools fed by a diversion from Slate Creek. When water conditions are good during migration, shorebirds can be numerous on this pool. When you reach the west side of the pool, and if you are here in October or November, be alert for Nelson's Sparrow in the taller patches of prairie cordgrass near the water's edge. You will usually find LeConte's, Savannah, Song, and Swamp Sparrows as well as Sedge and Marsh Wrens on this hike in late fall.

The wetlands 2 miles due south of the state wildlife area are privately owned but can be birded from County Road E 90th Street, which bisects this part of the wetland from east to west. This is a minimum-maintenance road, and at times it is necessary to proceed on foot for part or much of the distance. The mudflats here are often the largest at Slate Creek and often have many shorebirds. Remember that all the land on both sides of 90th is private and leased for hunting, so you need to restrict your birding to what can be seen from the road and at the appropriate times of the year.

East-Central Region

The east-central region of Kansas, as we have defined it in this book, includes most of the Flint Hills physiographic region. While the Flint Hills seem relatively modest in appearance as a geographical feature, they still act as a distinct natural boundary for plant and animal communities. The ranges of nearly all western bird species in Kansas end when they reach the Flint Hills. From the Flint Hills eastward, the birdlife is entirely eastern in composition, and the few western birds that occur are extralimital strays.

The Flint Hills are dominated by the tallgrass prairie and its special community of grassland birds, which include species such as Greater Prairie-Chicken, Upland Sandpiper, Common Poorwill, Common Nighthawk, Eastern and Western Meadowlarks, Henslow's Sparrow, Grasshopper Sparrow, and Dickcissel. Large areas of the Flint Hills are burned each spring by ranchers, and the burned prairies attract large flocks of migrating American Golden-Plover, Buff-breasted Sandpiper, and other shorebirds. Corridors of gallery forest follow rivers and streams through the Flint Hills. These ribbons of riparian woodland are inhabited by a diverse group of eastern woodland bird species, such as Red-shouldered Hawk, Pileated Woodpecker, Summer Tanager, Louisiana Waterthrush, and Northern Parula.

East of the Flint Hills are the Osage Cuestas, one of the predominant physiographic regions of eastern Kansas. This region is typified by east-facing ridges alternating with gentle slopes that have a western gradient. These geographical features are often subtle and not evident. In our east-central region, nearly all of Chautauqua, Elk, Greenwood, Lyon, Wabaunsee, and Woodson Counties are within the Osage Cuestas.

As in the Flint Hills, gallery forests bordering rivers and streams within the Osage Cuestas act as corridors of habitat for woodland species, contrasting with upland grasslands and farmland. Several rivers, including the Caney, Cottonwood, Elk, Fall, Neosho, Walnut, and Verdigris, are tributaries of the Arkansas River and flow south or southeast through the southern Flint Hills and Osage Cuestas regions.

There are many city lakes, state fishing lakes, and large federal reservoirs in the region, many of which are discussed in the following hotspot accounts. These attract waterfowl, shorebirds, gulls, and other water-dependent bird species. Additionally, nearly all of these impoundments have public lands surrounding them, offering many birding opportunities.

Within the Osage Cuestas, the Cross Timbers is a separate, unique physiographic region of rocky hills and scattered upland oak forests interspersed with expanses of prairie. The predominant trees are blackjack and post oak, which are small in comparison with other species of oak. Because they grow in shallow, rocky soils, even trees that are hundreds of years old have trunks less than 18 inches in diameter. This fragile, threatened habitat occurs in a patchy distribution that extends from southeastern Kansas through northeastern Texas. In Kansas, the Cross Timbers habitat is found in a narrow corridor from Chautauqua County north to Woodson County. Several eastern breeding species, such as Broad-winged Hawk and Yellow-throated Warbler, that are rare elsewhere in the region occur in the Cross Timbers in summer.

In the northernmost portions of the east-central region, the Smoky Hills physiographic region makes up substantial portions of Clay and Washington Counties, and the Glaciated Region makes up nearly all of Marshall, Pottawatomie, and northern Wabaunsee Counties. All the rivers and streams north of Marion and Morris Counties, including much of the northern Flint Hills and all the counties in the Smoky Hill and Glaciated Regions, flow into the Kansas River and its principal tributaries, which are the Big Blue, Republican, and Smoky Hill Rivers.

East-Central Region Hotspots

60. Alcove Spring Park
61. Milford Reservoir
62. Konza Prairie
63. Tuttle Creek Reservoir
64. Echo Cliff Park and Lake Wabaunsee
65. Herington City Lakes
66. Council Grove Reservoir
67. Marion Reservoir
68. Tallgrass Prairie National Preserve
69. Chase State Fishing Lake
70. Lyon State Fishing Lake
71. El Dorado Reservoir
72. Fall River Reservoir
73. Toronto Reservoir
74. Woodson State Fishing Lake
75. Augusta City Lake
76. Chaplin Nature Center
77. Kaw Wildlife Area
78. Winfield City Lake
79. Cross Timbers Scenic Drive

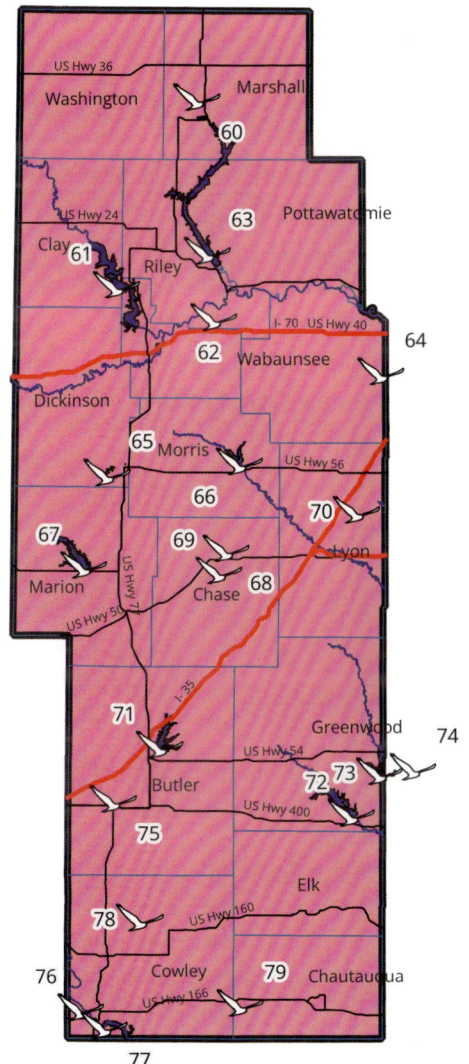

Hotspot 60: In the valley of the Big Blue River, Alcove Spring was a stopover on the Oregon Trail and is on the National Register of Historic Places.

60. Alcove Spring Park

County: Marshall

eBird Hotspot names: Alcove Spring Park, Prospect Hill Cemetery

Hotspot type: Worth a stop

Alcove Spring Park is one of those places that everyone should visit at least once. Located in Marshall County in the valley of the Big Blue River, this 246-acre park is open to the public. It is owned and operated by the Alcove Spring Preservation Association. The spring was a stopover on the Oregon Trail and is on the National Register of Historic Places. This is not really a "destination" birding hotspot, but its combination of beauty, aesthetics, and history makes it an appealing birding stop if you are anywhere in the general vicinity. Marshall County lies entirely within the Glaciated physiographic region.

Just south of Marysville, turn right (east) onto East River Road and go 8 miles south to the parking lot for the park. The East River Road should be driven at a leisurely pace, as the birding from the roadside is usually pretty good. At the park, there is a parking lot with an informational kiosk. A quarter-mile trail leads from the parking lot through a tract of appealing woodland to the spring. The outflow from the spring flows to a 10-foot waterfall into a pool and then downstream through the woods toward the river valley. Although not used often by birders, there are an additional 5 miles of marked and mowed trails in the park that can be explored if you have time.

The 140 species of birds that have been found at Alcove Springs are the same eastern woodland species as those described below for Milford and Tuttle Creek. Blue-gray Gnatcatcher, Louisiana Waterthrush, Northern Parula, Summer Tanager, and Rose-breasted Grosbeak are all found in summer. Purple Finch has been observed in several years during the winter. About a dozen species of warblers have been found during spring and fall migrations.

After birding the park, continue south on the East River Road for another 4 miles to the intersection with 8th Road. Go north a short distance to the small **Prospect Hill Cemetery**, which sometimes has wintering species such as Yellow-bellied Sapsucker and Red-breasted Nuthatch.

The valley of the Big Blue River is flanked on the east by a tall ridge and sometimes acts as a flyway for southbound migrating raptors. In September and October, it is worth keeping an eye on the sky for them.

61. Milford Reservoir

Counties: Clay and Geary

eBird Hotspot names: Kansas Landscape Arboretum, Milford Lake, Milford Lake—Curtis Creek PUA, Milford SP, Milford Lake—Outlet Park and Nature Center, Milford WA—Quimby Creek Wetlands, Milford WA—Steve Lloyd Wetlands, Milford WA—Utah Road, Walla Walla Road Woods, Milford Lake—West Rolling Hills Park, Milford WA—Zac Hudec Wetlands

Hotspot type: Full day

Milford Reservoir is the largest federal reservoir in Kansas. It covers 15,700 surface acres and has a maximum depth of 65 feet. The surrounding wildlife area is huge, covering over 19,000 acres of land. The lake was created by the construction of a dam on the Republican River in the 1960s.

Because of its size and its proximity to populated areas, including Junction City and Manhattan, Milford has been a major birding destination for many years. There are nearly 40 eBird hotspots associated with the lake.

From Junction City, take K-57 northwest several miles to reach the dam. At the south end of the dam, turn right onto Hatchery Drive and follow it along the dam to the **Outlet Park and the Milford Nature Center**, where over 170 species of birds have been seen. The nature center has excellent interpretive displays and a modest raptor display where you can observe hawks and owls at close range. A walking trail at the nature center passes through woodland before reaching the shallow Gathering Pond, which attracts waterfowl, herons, and other aquatic species.

Return to K-57 and cross the dam. Just past the north end of the dam, you can turn into the **Northern Overlook Park**, which offers a good view of the dam and reservoir. Continue east on K-57 for another mile to the entrance of **Milford State Park**. There are several good hiking trails in the state park. Return south across the dam on K-57 to the intersection with K-244 and turn west. Go 2 miles to the intersection with West Rolling Hills Road and turn north to enter the **West Rolling Hills Park**, which has several miles of park roads that offer a variety of habitats, birding opportunities, and lake vantage points. Return to K-244 and go 2 more miles west to Trail Road. Go north on Trail Road for 3 miles, and follow the pavement when it curves east onto 3500 Avenue. After 1 mile, turn south at the sign for the **Curtis Creek Public Use Area**, one of the better birding spots on the west shore of the lake. The lake is shallow in the Curtis Creek arm, and there is often good habitat for shorebirds and waders.

North of Curtis Creek, the **Milford WA** extends for many miles. South of the town of Wakefield are several birding hotspots. Take **Utah Road** south from town for 2 miles to the **Kansas Landscape Arboretum**, where over 100 species of birds have been found. Continue south another mile to 4th Lane and turn east. This road ends after 1 mile after passing the **Quimby Creek Wetlands**. With a diverse mix of brushy areas, wetlands, and woodlands, this 1-mile stretch of road is usually productive.

To reach the 1,100-acre **Steve Lloyd Wetlands** from Wakefield, go west for 1 mile on K-82, then north on Sunflower Road for 2 miles, then east 1 mile on 9th Road, then south on Thunder Road. After a short distance, you can continue south on Thunder Road to the lakeshore or double back on a diagonal road that ends after a mile at an elevated overlook above the wetlands. These wetlands have produced over 170 species of birds. Some of the best birding is in the area along Thunder Road, where you will have a far more favorable light angle later in the day. Return to 9th Road and go west 2 miles to Redwood Road. Turn north and go 6 miles to the bridge over the Republican River. Just before the bridge, turn onto 13th Road. The first mile of 13th west of the bridge is a good place to park and walk. The road passes between a ridge and the river, and there is healthy mature woodland on both sides of the river. Many good birds have been found at this location. Cross the bridge to view the **Zach Hudec Wetlands** east of the road. About 140 species of birds have been seen at Zach Hudec, including 25 species of shorebirds.

The number of bird species reported from Geary County as of 2023 stands at 333, one of the highest numbers of any county in the state. Adjacent Clay County has recorded 287 species. The majority of these have been reported at Milford. Waterfowl are abundant in migration and in winter, and 33 waterfowl species have been recorded, including numerous rarities. Over 30 species of shorebirds have been seen at Milford, especially at Steve Lloyd and Zach Hudec Wetlands but also elsewhere on the lake when drought conditions cause mudflat habitat to appear. Concentrations of migrating Franklin's Gull in October can number hundreds of thousands. Wintering Ring-billed Gull is abundant. Rarer species of gulls are found in most years at Milford and have included Iceland, Lesser Black-backed, Great Black-backed, and Glaucous Gulls. Bald Eagle is common in winter, and the Milford Nature Center usually conducts an all-day Eagle Day event in January. Double-crested Cormorant and American White Pelican are common during migration. Common Loon, Eared Grebe, and Horned Grebe are annual migrants in spring and fall. The western limit of the breeding range for several species of eastern songbirds is at or near Milford. These include Yellow-throated Vireo, Wood Thrush, Kentucky Warbler, and Painted Bunting. Check stands of unburned tallgrass prairie for Henslow's Sparrow in spring and summer. Over 30 species of warblers have been seen in Clay and Geary Counties, mostly during migration.

The **Walla Walla Road Woods** are located about a mile southeast of Milford Dam.

The land adjacent to the road is all private, but local birders consider the woodlands along this road to be one of the best locations for woodland birds in the Milford area. The woodlands are beautiful, making this one of those places that is worth a stop even if the birding is slow.

62. Konza Prairie

County: Riley

eBird Hotspot names: Konza Prairie—Nature Trail

Hotspot type: Worth a stop

The Konza Prairie Biological Research Station is a 9,000-acre tallgrass prairie preserve located just south of Manhattan. It is owned by The Nature Conservancy in Kansas and managed and operated by Kansas State University. Extensive biological field research has been conducted at Konza for over 50 years. Because of its importance as a research facility, much of the preserve is off limits to the public. But on the west side of the property near the headquarters, the **Konza Prairie Nature Trails** allow you to experience the tallgrass prairies of Konza on foot.

At the south edge of Manhattan, just south of the bridge over the Kansas River on K-177, turn west onto McDowell Road and go 6 miles. Turn south at the sign for Konza Prairie and drive to the headquarters, where there is a parking area and a kiosk with trail information. Three trails all begin at the parking lot. You can hike the 2.6-mile Nature Trail Loop or continue on the Kings Creek Loop (4.6 miles total) or the Godwin Hill Loop (6.2 miles total). The Nature Trail Loop goes through lowland gallery forest along Kings Creek before climbing the tallgrass hills above, eventually reaching an expansive view of the Kansas River valley. Be aware that there are a few moderately steep stretches on these trails. An interpretive brochure is available on site or as a download that will enhance your insights into the ecology and wildlife of Konza Prairie.

There are no lakes or wetlands on the trail or elsewhere on Konza, so birds of the gallery forest and the tallgrass prairie dominate the list of 180 species that have been found here. Grassland species, including Greater Prairie-Chicken, Upland Sandpiper, Sedge Wren, Grasshopper Sparrow, Henslow's Sparrow, Eastern Towhee, Blue Grosbeak, and Dickcissel are well represented. The trail is well-known for being one of the few places anywhere in the US where four species of nightjars can be heard from a single spot at dusk in spring and summer. Listen for Common Nighthawk, Common Poorwill, Chuck-will's-widow, and Eastern Whip-poor-will. All the typical breeding species of the gallery forest occur along Kings Creek, including Eastern Wood-Pewee, Red-eyed and Yellow-throated Vireos, Louisiana Waterthrush, Northern Parula, Kentucky Warbler, and Summer Tanager. Twenty species of warbler have been seen during migration. In the winter months, expect fewer birds, but typical eastern woodland species are to be expected, as well as Northern Harrier and other grassland raptors.

63. Tuttle Creek Reservoir

Counties: Marshall, Pottawatomie, and Riley

eBird Hotspot names: Tuttle Creek Lake, Tuttle Creek Lake—Dam, Tuttle Creek Lake—Spillway Cycle Area, Tuttle Creek WA—Black Vermillion Marsh, Tuttle Creek SP—Carnahan Creek Park, Tuttle Creek SP, Tuttle Creek SP—Cedar Ridge Area, Tuttle Creek SP—Fancy Creek Park, Tuttle Creek SP—Randolph Area, Tuttle Creek SP—River Pond East, Tuttle Creek SP—Riverside Trail, Tuttle Creek SP—Stockdale Park, Tuttle Creek WA—Irving Marsh

Hotspot type: Full day

Tuttle Creek Reservoir is the second-largest reservoir in Kansas, with 12,500 surface acres when full and a maximum depth of 50 feet. The lake is long and narrow, extending over 20 miles north from the dam on the Big Blue River just north of the city of Manhattan. The Corps of Engineers operates four park areas at the reservoir, and Tuttle Creek State Park has a total of 1,250 acres divided into four additional areas around the lake. The Tuttle Creek WA covers over 17,000 acres of land, mostly around the northern half of the reservoir, and extends upstream beyond Blue Rapids in Marshall County, over 40 miles from the dam. Tuttle Creek receives a substantial amount of attention from birders, as evidenced by the more than 40 eBird hotspots in and around the lake.

A typical birding visit to Tuttle Creek often begins at the dam. Take US 24 north from Manhattan to the junction with K-13. Just south of the junction, you can turn into the Outlet Park to view the **Tuttle Tubes**. When a large volume of water is being released, the tubes host huge flocks of waterfowl and gulls feeding on stunned fish. Just after turning east onto K-13, turn left onto Observation Point Drive, which ends on top of a hill overlooking the dam and adjacent lake. Cross the dam, turning right onto River Pond Road at the end of the dam. Follow the road downhill to **River Pond State Park**. You can spend a lot of time birding in the River Pond area. The swimming beach near the park entrance is a good place to scan the River Pond for waterfowl and other aquatic bird species. Shorebirds, gulls, and terns often roost on the beach, especially early in the morning when they have not yet been disturbed. The Riverside Trail along the Outlet Channel has good habitat and usually provides quality birding. Take Beach Road around the east end of the pond to reach Willow Lake and the Rocky Ford Campground. This area "behind" the River Pond is much less developed, with more natural habitat. At the Rocky Ford Campground are trailheads for the Eagle Pass Trail and Rocky Ford Trails, both of which provide good birding.

After birding the dam area to your satisfaction, you can explore many good birding spots on both sides of the lake. Between the dam and the town of Blue Rapids almost 40 miles upstream, there is only one bridge over the lake or the river. Consequently, going from one side of the lake to the other represents a significant time investment. On the east shore, turn off K-13 close to the dam onto Spillway Marina Road to reach **Cedar Ridge State Park**, where you have good views of the lake. Return to K-13 and continue east 3 miles to Carnahan Road and turn left (north). Go 5 miles to Park Road and turn left. Follow the road to Carnahan Creek Park, where there is a good variety of habitat and over 200 species of birds have been found. Return to Carnahan Road and continue another 5 miles to the town of Olsburg. At the bridge where Carnahan Road crosses Carnahan Creek, there is usually water on both sides of the road, and good numbers of birds may be present. At Olsburg, turn left (west) onto K-16 and go 6 miles to reach **Randolph State Park**. In addition to land birds in this park, there is a good view of the lake. The adjacent bridge is the only one between the dam and Blue Rapids over 20 miles to the north.

In general, the west shore of Tuttle Creek receives more attention from birders. Starting at K-13 near the west end of the dam, take Tuttle Cove Road north for 3 miles to **Tuttle Creek Cove Park**, which has an exceptional view of the lower end of the reservoir. Return to the dam and take US 24 west for 10 miles to the intersection with US 77. Here turn east onto Stockdale Park Road and drive 6 miles to reach **Stockdale Park**, one of the better hotspots on the west shore of the reservoir, where nearly 200 species of birds have been found. Return to US 24 and drive 10 miles north to the intersection with K-16, then turn east onto the highway. Just past the small town of Randolph, you can turn north into **Fancy Creek State Park** via Gardiner Road or Fancy Creek Road. Other than the River Pond Area, Fancy Creek is probably the best birding location at Tuttle. The area has mature woodland interspersed with open areas. At the north end of the park is the confluence of Fancy Creek and the Big Blue River. Water levels here are highly variable. Sometimes there are mudflats with shorebirds present. However, this area of the lake is shallow enough that conditions can range from nearly dry to completely underwater.

North of the Fancy Creek and Randolph Park areas, the sprawling Tuttle WA begins and extends for over 20 miles to the town of Blue Rapids. The KDWP has developed wetland areas within the wildlife area. A map showing the area in detail is available as a free download at ksoutdoors.com. Two of these wetlands near Blue Rapids, **Black Vermillion Marsh** and **Irving Marsh**, have produced a good variety of shorebirds and other wetland species.

The **Flint Hills Discovery Center** at 315 S 3rd Street in Manhattan is a relatively new interactive museum with exhibits highlighting the Flint Hills ecosystem. Until recently, the center organized and managed Greater Prairie-Chicken viewing opportunities during the spring months, but access to the property it used was lost. As of 2024, the center was attempting to identify new property to allow this activity to continue.

The birdlife at Tuttle is nearly identical to that described in the previous account for Milford. The all-time list of species seen in Riley County stands at 362, exceeded by only four other Kansas counties. An amazing 15 species of gulls have been found at Tuttle, and most of the 31 species of warblers recorded in Riley County have been observed at the reservoir hotspots described in this account.

64. Echo Cliff Park and Lake Wabaunsee

County: Wabaunsee

eBird Hotspot names: Echo Cliff Park, Lake Wabaunsee

Hotspot type: Worth a stop

Echo Cliff Park is a small public park located about 15 miles southwest of Topeka, near Dover in eastern Wabaunsee County. Owned and maintained by the Echo Cliff Park Trust, the park features 50-foot cliffs on the south side of Mission Creek. Echo Cliff lies at almost the precise point where three physiographic regions converge—the Flints Hills, the Glaciated Region, and the Osage Cuestas. The thick layers of rock exposed on the cliff face are layers of ancient river channel deposits that date to the Pennsylvanian subperiod. These strata are 300 million years old. This is a visually appealing location, well worth a stop even if the birding is slow. The park has many mature trees and hardwood forest along the creek. Starting from the Wabaunsee County line on K-4, go 1 mile west to Echo Cliff Road, then south for another mile to the park entrance just north of the bridge. In addition to the park grounds, the old wood bridge next to the new bridge is open to foot traffic. After birding the park to your satisfaction, continue south for another half mile to Goldfinch Road and turn right (west). Go 1 mile until you reach a slight curve in the road. Here there is a substantial high-quality woodland along Massasoit Creek. This is all private land, but walking this stretch of road typically turns up interesting birds. Most birders visiting Echo Cliff also visit this stretch of Goldfinch Road.

The 140 species of birds that have been found at Echo Cliff are woodland birds of the eastern forests. Red-shouldered Hawk, Pileated Woodpecker, and Barred Owl are found year-round. In summer, Eastern Wood-Pewee, Great Crested Flycatcher, Red-eyed Vireo, Yellow-throated Vireo, Blue-gray Gnatcatcher, Louisiana Waterthrush, and Northern Parula are all expected species. Acadian Flycatcher is regular in summer, and Echo Cliff likely marks the western edge of its breeding range in northeastern Kansas. Chuck-will's-widow and Eastern Whip-poor-will have been heard here in early summer. Less common but regular in summer are deep-woods species such as Wood Thrush and Kentucky Warbler, especially on Goldfinch Road.

Another birding hotspot in Wabaunsee County is **Lake Wabaunsee**, located 5 miles west of Eskridge, or 16 miles west of Echo Cliff on K-4. The lake is owned and maintained by the city of Eskridge and covers 216 surface acres. It is among the oldest lakes in Kansas, having been constructed in the early 1930s. Cottages and homes surround much of the lake, but there are public areas as well. Flint Hills Drive encircles the lake. There are parking areas at both ends of the dam that allow good views of most of the lake. The north shore is largely public park land from which you can check the lake and surrounding trees for birds. Near the golf course at the southeastern corner of the lake is an undeveloped woodland area.

This lake has been a popular birding destination for many years, and well over 200 species of birds have been found, including nearly 30 species of waterfowl. There are multiple records of Long-tailed Duck, Surf Scoter, White-winged Scoter, and all three merganser species. When water levels are low, a variety of shorebirds are possible, and 20 species have been recorded. Other waterbirds such as grebes, gulls, and terns are seen annually. In late fall, check the weedy wetland areas at the northeast corner of the lake for Marsh Wren and Swamp Sparrow. Eastern Towhee is often observed in the summer months and is presumably a nesting species. About a dozen warbler species have been found, mostly during May.

65. Herington City Lakes

County: Dickinson

eBird Hotspot names: Herington Reservoir, Lake Herington

Hotspot type: Worth a stop

The city of Herington owns and manages two adjacent reservoirs a few miles southeast of the city that are worthwhile birding stops if you are in the area. The older of the two lakes is usually called the "Old" Herington City Lake but was originally known as Lake Herington. It was created in 1929 by the WPA. It has 367 surface acres and a maximum depth of 18 feet. The "New" Herington City Lake is also referred to as Herington Reservoir. Built in the early 1980s, it has 555 surface acres and a maximum depth of 40 feet.

If only takes two or three hours to fully check both lakes for birds. The **Old City Lake** (Herington Lake) is 2 miles west of Herington on 500 Avenue. There are roads along the entire perimeter of this lake. At the south end of the lake is an extensive area of cattail marsh that sometimes has birds of interest. During drought years, mudflats appear on the southern portions of the old lake and good numbers of shorebirds can be found.

Continue on 500 Avenue for another quarter mile to Sage Avenue, which is the entrance to the **New City Lake** (Herington Reservoir). The New City Lake has roads only on the east shore. The lake is almost 2 miles long, and there are several park roads along the east shore. There are still dead trees standing in the water at the south end, dating to when the lake was created.

These lakes have most of the same waterfowl, shorebird, and gull species that have been detailed for other lakes in the region. Noteworthy birds that have been found here include Sabine's Gull, Neotropic Cormorant, and American Golden-Plover. Listen for booming Greater Prairie-Chickens on spring mornings.

66. Council Grove Reservoir

County: Morris

eBird Hotspot names: Canning Creek RA, Council Grove Dam, Council Grove Lake, Council Gove Lake—Outlet Park, Council Gove Lake—Custer Park, Council Grove Lake—Kansa View, Council Grove Lake—Kit Carson Cove, Council Grove WA—Munkers Creek, Council Grove Lake—Neosho Park RA, Council Grove Lake—Richey Cove Park, Council Grove Lake—Santa Fe RA, Council Grove WA—Slough Creek

Hotspot type: Worth a stop

Council Grove Reservoir was created by the construction of a dam on the Neosho River in 1964. The lake has a surface area of 3,235 acres and a maximum depth of 56 feet. There are a total of eight recreation areas managed by the Corps of Engineers, mostly close to the dam. The KDWP manages an additional 2,638 acres of wildlife area at the upper portions of the reservoir.

K-177 runs north and south along the east side of the reservoir. The entrance to the **Kansa View Park** and the road across the **Council Grove Dam** is 1 mile north of the Council Grove city limits on K-177. Check the lake from Kansa View Park, then drive southwest across the dam to reach the intersection with Lake Road. Here you can turn west to the entrances for the **Canning Creek Cove Recreation Area**, **Neosho Park**, and **Santa Fe Recreation Areas** on the west shore of the reservoir or turn east to reach the entrances for the **Outlet Park** and **Richey Cove Recreation Areas** below the dam. From the Richey Cove area, Lake Road continues southeast to the Council Grove city limits. Just before reaching town, you can cross the river channel on Kahola Road and then turn left (northwest) onto River Road West Road, which follows the river channel back to the dam, ending at the parking area for the Council Grove Off-Road Vehicle (ORV) Area, where there is an area of woodland habitat along the Neosho River channel. On the east shore of the lake, the entrances for **Richey Cove Park** and **Richey Cove North Recreation Area** are north of the Kansa View entrance on K-177. Most of the lake can be viewed by visiting a combination of these park and recreation areas.

In comparison to the wildlife areas associated with most of the other major reservoirs in Kansas, Council Grove WA is smaller, and access is more limited. The heavily wooded **Slough Creek Area** is on the west side of the lake and can be reached via Neosho River Road and Dam Road. On the east side of the lake, the **Munkers Creek Area** has an entrance on K-177 just north of the reservoir. Turn east onto S 900 Road at the KDWP kiosk and follow the road until it ends in a wooded area along the creek. Be sure to check the area east of the bridge on K-177 south of the entrance to Munkers Creek. Depending on water levels in the lake, this can be a good spot for dabbling ducks and shorebirds.

Council Grove City Lake is adjacent to the reservoir and was created in 1942 by a dam constructed on Canning Creek. It has 434 surface acres. Most of the property around the city lake is devoted to private dwellings, but there are several public park areas offering views of the lake. The entrances for Council Grove City Lake are a short distance farther west from the dam beyond Canning Creek Cove on Lake Road.

Council Grove is not birded as frequently as most other Kansas reservoirs and deserves more attention than it receives. The number of eBird checklists submitted from the reservoir is notably low in comparison to nearly any other major reservoir in Kansas. More than 225 species of birds have been found at Council Grove. The total number of waterfowl species is 27. Migrating Common Loon and Horned Grebe are seen in most years. Appropriate habitat is somewhat limited, but the list of shorebirds stands at about 20 species.

Eastern woodland bird species are well represented at Council Grove, especially within the wildlife area. Chuck-will's-widow and Eastern Whip-poor-will are both possible in summer. Other summer species of interest include Fish Crow, Yellow-throated Vireo, Louisiana Waterthrush, Northern Parula, Summer Tanager, and Blue Grosbeak. Twenty-two warbler species have been reported.

67. Marion Reservoir

County: Marion

eBird Hotspot names: Marion WA—Broken Bridge, Marion SP—Cottonwood Point, Marion WA—Durham Cove, Marion WA—French Creek Cove, Marion WA—Kanza Road Causeway, Marion Res., Marion Res.—Dam & Outlet Park

Hotspot type: Half day

Marion Reservoir is a large Corps of Engineers project that was created in 1968 by the construction of a dam on the North Cottonwood River. The reservoir covers 6,100 surface acres at full pool and has a maximum depth of 30 feet. There are four campgrounds located on both sides of the reservoir that are operated by the Corps. Marion WA is managed by the KDWP. It has 3,700 acres of mixed grasslands and woodlands as well as 2,350 acres of cropland. Marion is a popular destination for camping, fishing, and hunting. While it lies on the western edge of the Flint Hills region, actual grassland habitat is sparse, and the private land surrounding the reservoir is almost entirely devoted to cultivation. As a general caution, the composition of the soils at Marion can make unmaintained roads difficult or impossible for driving after rains.

A typical birding trip to Marion begins at the dam. Take US 56 east from Hillsboro and turn north onto Old Mill Road at the sign for **Marion Dam**. There is an overlook at the south end of the dam that offers an unobstructed view along the dam and much of the lake. The Eastshore Area at the north end of the dam is another good place to scan along the dam for waterfowl. Gulls and terns like to perch on the floating boom that borders the swimming beach.

The two Corps park/campsite areas near the dam are always worth checking. The entrance to the **Hillsboro Cove Public Use Area** is 1 mile west of the dam on US 56. The entrance to the extensive **Cottonwood Point** campground is 1 mile north of the dam on Pawnee Road. The stands of trees in these areas usually have multiple Bald Eagles in winter and Osprey during spring and fall migrations. A variety of waterbirds and songbirds can be found, depending on the season. The **Willow Walk Trail** at the north end of Cottonwood Point goes through some good woodland habitat, and over 100 species of birds have been reported from the trail.

Beyond Cottonwood Point on the east shore, there is only limited access to the wildlife area. In general, there are better birding opportunities on the west shore. To reach **French Creek Cove** Public Use Area from Hillsboro, take US 56 east for 1 mile to Kanza Road, then go 1 mile north and 1 mile east to reach the entrance on Limestone Road. French Creek Cove is usually a good stop and has produced over 100 species of birds. Eight hundred acres of the lake adjacent to the cove are designated as a waterfowl refuge during the hunting season. There is also some mature woodland and intermittent wetland habitat. The huge junipers near the entrance provide refuge for sparrows and other birds. When the lake is below full pool, there is mudflat habitat that attracts shorebirds and waders.

Return to Kanza Road and continue north. From Kanza Road, take 220th Street and 230th Street east to reach several shoreline access points. Continue north on Kanza Road for several miles to **Durham Cove** Public Use Area. Durham Cove is not as actively maintained as it was in former years, and native habitats have flourished, making this one of the better birding stops at Marion. In addition to grassland and woodland, Durham offers good views of the upper end of the lake. Just beyond Durham Cove, the road crosses the upper end of the lake on the lengthy Kanza Road Causeway, which can be one of the best birding locations at Marion, depending on the time of year and the water level. Even when the reservoir is at full pool, the water adjacent to the causeway is shallow, and when the lake is low, there are extensive areas of mudflats and shallows. Shorebird and wading bird diversity can be excellent when these conditions exist. One final stop in the wildlife area is the **Broken Bridge** on Jade Road, 1 mile west of Kanza Road. This can be reached from either the north or south, but as the name implies, there is no longer a bridge over the river. Access from the north on Jade Road crosses an area that can be marshy or dry. Broken Bridge is a popular fishing spot and could use more birding attention than it gets.

Over 250 bird species have been found at Marion. Concentrations of waterfowl can be impressive. Mixed-species flocks of geese sometimes number in the tens of thousands from late fall through early spring. Wintering Common Merganser and Common Goldeneye are abundant. All three species of scoter and Long-tailed Duck have been found multiple times. American White Pelican and Double-crested Cormorant flocks can number in the hundreds during migration. Thirty shorebird species have been found, largely at Durham Cove and the Kanza Causeway when conditions are favorable. Gulls are often abundant, including Bonaparte's Gull in late fall. Marion has a history of attracting rare gulls, including multiple records of Sabine's. Overall numbers and diversity of sparrows are high during migration and in winter. Despite the somewhat limited habitat, over 20 warbler species have been found at Marion, including notables such as Hooded, Bay-breasted, and Blackburnian. Prothonotary Warbler has been seen near Kanza Road Causeway during the breeding season.

68. Tallgrass Prairie National Preserve

County: Chase

eBird Hotspot names: Tallgrass Prairie N PRES—Bottomland Trail, Tallgrass Prairie N PRES—Fox Creek Trail, Tallgrass Prairie National Preserve

Hotspot type: Half day

Covering 11,000 acres, the **Tallgrass Prairie National Preserve** was established in 1996 after many years of public and private efforts to set aside this large tract of tallgrass prairie habitat. Today, the preserve is a popular tourist destination. The entrance and visitor center are located on K-177 2 miles north of US 50 near the town of Strong City. The visitor center provides general information and interpretive displays. On the preserve are several developed nature trails and 40 miles of backcountry hik-

© Bruce L. Hogle

Hotspot 68: A limestone rock fence extends across the prairie to the historical Lower Fox Creek Schoolhouse at the Tallgrass Prairie National Preserve.

ing trails. There is a herd of 90 bison on the preserve. About one-third of the prairie is rotationally burned each year.

Several hiking trails on the preserve are popular with birders and other outdoor enthusiasts. Two hiking trails begin at the trailhead near the US 50 and K-177 exit. Turn east onto 227th Road at the entrance to St. Anthony Cemetery. The trailhead parking lot is a short distance to the east. The **Bottomland Trail** is a 1-mile gravel path through an area where there is an ongoing effort to restore bottomland prairie habitat. The 6-mile **Fox Creek Trail** is a mowed trail that connects with the north end of the Bottomland Trail loop and follows the creek for several miles. The north end of this trail connects with the visitor center. The woodlands along Fox Creek are classic Flint Hills gallery forest. The 2-mile **Southwind Nature Trail** and the **Windmill Pasture Scenic Overlook Trail** both begin at the visitor center. The habitat on these trails is largely upland tallgrass prairie, although the Southwind Trail crosses a small creek where there are some cottonwood and hackberry trees.

Despite the limited habitat diversity, 200 species of birds have been recorded at Tallgrass. Obviously, grassland species are of the most interest for visiting birders. Greater Prairie-Chicken leks are reported in most years, especially along the Windmill Pasture and Southwind Trails. Listen for them predawn on spring mornings with light wind. Other summer grassland specialists are Upland Sandpiper, Common Poorwill, Mississippi Kite, Scissor-tailed Flycatcher, Eastern Kingbird, Bell's Vireo, Grasshopper Sparrow, and Lark Sparrow. The sought-after Henslow's Sparrow is a breeding species in most or all years in appropriate prairie habitat. Late-summer nesting Sedge Wren is found in most years. Especially in late fall and early spring, be alert for Smith's

Longspur, often detected by its rattling calls. During migration, flocks of up to 40 of them have been observed, and there are a few winter records as well, indicating that at least a few of them overwinter here. In the winter months, the prairie can seem largely devoid of birds, but grassland raptors such as Rough-legged Hawk, Northern Harrier, and Short-eared Owl are conspicuous.

In the riparian woodland along Fox Creek, summer breeding species include Red-shouldered Hawk, Barred Owl, Pileated Woodpecker, Eastern Wood-Pewee, Louisiana Waterthrush, Northern Parula, and Summer Tanager. There are also summer records for less expected species, such as Yellow-throated Vireo, Wood Thrush, Yellow-breasted Chat, and Kentucky Warbler.

69. Chase State Fishing Lake

County: Chase

eBird Hotspot name: Chase SFL

Hotspot type: Worth a stop

Chase SFL is less than 5 miles from Tallgrass National Preserve. Take K-177 south through Strong City for 2 miles to the town of Cottonwood Falls. At the north edge of town, turn west onto Main Street, which becomes Lake Road at the city limits. After 2 miles, the entrance to the lake is on the south side of the road. The entrance road goes along the entire north shore of the lake. Formed by the construction of a dam on Prather Creek in 1954, the lake has 109 surface acres when full, and the surrounding wildlife area covers 383 acres. This property is managed by the KDWP. Most of the habitat is tallgrass prairie. Where the road ends at the west end of the lake is some woodland and dense shrubbery worth checking for sparrows, warblers, buntings, and others, depending on the time of year. An interesting aspect of this hotspot is the series of terraced waterfalls along the outlet channel below the dam. Dropping 40 feet over a 100-foot stretch of the channel, the falls are in a wooded area and always worth a look.

More than 200 species of birds have been seen at Chase SFL. Many of the species mentioned above for Tallgrass Prairie National Preserve are possible. Additionally, you will see a good variety of waterfowl, gulls, and other aquatic birds. These have included 24 waterfowl, 15 shorebird, and 4 gull species. Common Loon is often found during migration. Greater Prairie-Chicken can be heard booming in the spring. In the summer, drive the park road slowly after dark, looking for the bright red eye shine of Common Poorwills that sometimes perch on the road.

70. Lyon State Fishing Lake

County: Lyon

eBird Hotspot name: Lyon SFL

Hotspot type: Worth a stop

Lyon SFL is in the heart of the Flint Hills north of Emporia. The lake has a surface area of 135 acres and a maximum depth of 34 feet; it is surrounded by a 442-acre public wildlife area. The habitat around the lake is largely tallgrass prairie, but over the years, there has been substantial encroachment of woody vegetation and trees. This is one of the old reservoirs that were built in the 1930s by the WPA and the Civilian Conservation Corps (CCC). Take K-99 north from Emporia for 13 miles, then go east for 2 miles on Road 270 to reach Road T, which is the entrance to the lake. There are park roads on the north and south sides of the lake.

The Smith's Longspur draws birders here each year. Smith's Longspur is relatively common throughout the Flint Hills in migration and winter, but private ownership of land in the region makes it difficult for birders to access these unique birds. Just south of Lyon SFL is a large tract of prairie that is usually mowed in late summer, leaving it short for the winter months. Smith's Longspur is almost always present in this field from late fall through early spring, although it is less numerous in the winter. The pasture is privately owned, making it necessary to remain on the road and wait for the often restless longspur flocks to take flight. Listen for their rattling calls and check flying birds for the small but obvious white wing-patches and buffy underparts. Sprague's Pipit is regular in April and October in these same fields but is almost impossible to find without gaining permission to walk the hay fields.

For years, a large stand of cedars near the lakeshore was reliable for Long-eared Owl in winter, but most of the roosting trees have been removed.

Short-eared Owl is often seen near the lake at dawn and dusk during the winter. In the spring, booming Greater Prairie-Chicken is usually present somewhere in the area surrounding the lake. Drive rural roads slowly while listening for their booming at dawn in late March and April, especially when the winds are light and there is no precipitation. Henslow's Sparrow is another tallgrass prairie specialty that has been observed in summer. There are multiple fall records for LeConte's Sparrow.

The cedars and brush thickets surrounding the lake attract Sharp-shinned and Cooper's Hawks, Golden-crowned Kinglet, Yellow-rumped Warbler, several species of sparrows, and other species during the winter. Townsend's Solitaire has been recorded in multiple winters, well east of its typical winter range. Bell's Vireo and Eastern Towhee nest in these thickets during the summer.

The lake itself usually has a good selection of waterfowl in spring and fall, and 28 species have been reported, including Black Scoter, White-winged Scoter, Long-tailed Duck, and Red-breasted Merganser. There are several spring and fall records for Common Loon.

71. El Dorado Reservoir

County: Butler

eBird Hotspot names: El Dorado Lake, El Dorado Lake—Dam Overlook, El Dorado SP—Bluestem Point Area, El Dorado SP—Boulder Bluff Area, El Dorado SP—Shady Creek Area, El Dorado SP—Teter Nature Trail, El Dorado SP—Walnut River Area

Hotspot type: Half day

Another of the many large Kansas reservoirs created by the Corps of Engineers, **El Dorado Reservoir** is located northeast of the city of El Dorado. Formed by a dam on the Walnut River, the lake has 8,000 surface acres of water at full pool and a maximum depth of 60 feet. Public lands surrounding the lake are divided between the 4,000-acre El Dorado State Park and the 4,000-acre El Dorado WA. This lake was built more recently than most of the other major reservoirs in Kansas and only became fully operational in 1981. Because of its proximity to Wichita, El Dorado State Park is quite busy during the warmer months, especially on holiday weekends, and the state park is extensive, with many loop roads and campgrounds.

To begin a tour of the lake from the town of El Dorado, go east on E 12th Avenue from US 77 for 2 miles to reach the dam. Turn left onto Boulder Bluff Road to reach the entrance to the **Walnut River Area** of the state park and the reservoir outlet channel. There is a lake with a swimming beach in the Walnut River Area and a short nature trail through woodlands that crosses the outlet channel. Over 160 species of birds have been found in the Walnut River Area.

Return to Boulder Bluff Road and follow it west. After passing the end of the dam, turn into the **Boulder Bluff Area** of the state park. This area includes several miles of shoreline with numerous loop roads and access points where you can scan the lake. Returning to the intersection with 12th Street at the dam, proceed east onto NE Shady Creek Access Road. You will pass the parking area for the **Teter Nature Trail**, which traverses woodland habitat on the slopes below the dam. Continue a short distance to the **Dam Overlook** at the east end of the dam, where you can scan a large part of the lake. Continue east on NE Shady Creek Access Road for another half mile to the entrance for the **Shady Creek Area** of the state park. Like the Boulder Bluff Area, the Shady Creek Area extends along the lake, with multiple loop roads and pullouts. Return to Shady Creek Road and continue east another 2 miles to K-177. Turn left (north) onto the highway. The highway crosses a portion of the lake on a long causeway. There are often a variety of birds in the shallow area of the lake east of the highway. Park in the small pullout at the south end of the causeway to safely scan this part of the lake. Continue another mile north on the highway to the entrance of the **Bluestem Point Area** of the state park. There are numerous park roads leading to the shoreline on both sides of Bluestem Point.

The woodlands you see east of the causeway on K-177 are within the **El Dorado WA**. This stand of mature woodland has a lot of good birds, but all access is by foot. If you want to explore this interesting area, turn east just south of the causeway onto NE 10th Street and go 1 mile to the intersection with NE Ellis Road. Go north on Ellis Road until it ends at Shady Creek. Park here and hike north either along the creek or along the abandoned stretch of Ellis Road. Birding this area thoroughly on foot can take several hours. This area is seldom visited and has few trails or any other amenities but is a noteworthy area of woodland and edge habitat. The entire wildlife area is open to hunting, so take appropriate precautions during hunting seasons.

More than 250 bird species have been reported at El Dorado. Twenty-nine water-fowl species have been reported. The lake is apparently attractive to the "sea ducks." Long-tailed Duck and Black, Surf, and White-winged Scoters have been seen in multiple years, usually in the deeper water near the dam. Shorebird habitat is limited, but more than 20 species have been reported. Gulls are numerous in migration and in winter. Migrating Bonaparte's Gull can number in the hundreds on occasion. Glaucous Gull has been seen in multiple years. In the winter months, many gulls feed at the Butler County landfill southwest of El Dorado and use the lake as an overnight roosting area. Flocks of Double-crested Cormorant and American White Pelican often number in the hundreds during migration.

Tallgrass prairie is the dominant habitat in the undeveloped areas at El Dorado. Greater Prairie-Chicken is now largely absent from the immediate area, although it still occurs in Butler County north and east of the reservoir. Northern Harrier, Rough-legged Hawk, Short-eared Owl, and Merlin are present in winter. Smith's Longspur is possible in mowed prairie, especially in early spring. Both kingbirds and Scissor-tailed Flycatcher are common in summer.

For woodland species, concentrate on Walnut Grove, Teter Nature Trail, and El Dorado WA. Typical eastern woodland species can be expected according to season. Red-shouldered Hawk, Barred Owl, and Pileated Woodpecker are all well established. Seven vireo, 19 sparrow, and 20 warbler species have been seen, mostly during spring and fall migrations. Louisiana Waterthrush, Northern Parula, Prothonotary Warbler, Summer Tanager, and Painted Bunting are among the many summer resident species

72. Fall River Reservoir

County: Greenwood

eBird Hotspot names: Fall River Dam, Fall River Lake, Fall River WA, Fall River SP—Fredonia Bay Campground, Fall River WA—Ladd Bridge, Fall River Lake—Outlet Park, Fall River SP—Quarry Bay Area

Hotspot type: Half day

Located in Greenwood County, Fall River Reservoir is a Corps of Engineers reservoir that covers 2,450 surface acres when full and has a maximum depth of 25 feet. This is one of the oldest of the large federal reservoirs in Kansas, having been completed in the late 1940s. Surrounding public lands consist of the 8,400-acre Fall River WA and the 1,100-acre Fall River State Park. The wildlife area is substantial in size and extends along 14 miles of the Fall River to within a few miles of the town of Eureka as well as along 6 miles of Otter Creek. The wildlife area has 2,500 acres of prairie, 2,300 acres of riparian woodland, and 3,000 acres of cropland. Fall River State Park consists of the Fredonia Bay area on the west shore and the Quarry Bay area on the east shore.

From the town of Severy, go 8 miles east on US 400 and turn north at the sign for the state park. Follow Lake Road north for 2 miles to the **Fredonia Bay** area of the state park, where there are multiple campgrounds, good views of the upper half of the reservoir, and two hiking trails. After exploring this area, return to the park entrance and go south for 1 mile, then east for 1 mile to reach the dam overlook. With a spotting scope, you can see birds on most of the lake from the overlook. From here you can take a road down to the **Outlet Park**, which is usually a good birding stop. Check the spillway and channel for herons, gulls, and shorebirds, and check out the wooded areas along the river channel and the adjacent slope. Return to the dam and drive across it to reach the entrance to the **Quarry Bay** area of the state park. Follow the park road along the shore past the Bluestem Campground. Be sure to check the long wetland area to the left of the road, where you might see dabbling ducks, waders, and shorebirds. This road ends at the trailhead for the Bluestem Trail, which primarily crosses tallgrass prairie. On the park road just past the Quarry Bay entrance is the trailhead for the Post Oak Trail, which is usually the best of the foot trails for birding. This trail goes through classic Cross Timber oak woodlands with prominent sandstone outcrops.

The Fall River WA stretches for many miles, and there are numerous access roads into good habitat. A popular birding road is County Road 616. One mile south of the town of Climax, turn east from K-99 onto Road 616 and go about 3 miles to where the road merges with Road 18. Continue another 2 miles to the **Ladd Bridge**. The area near the bridge has a variety of habitats and is a good place to stop and spend some time birding on foot. Eastward from the bridge, Road 18 zig-zags along the north side of the lake for several miles through croplands, prairies, and woodlands. Return to Climax and go north 2 miles on K-99 to Road 610. Turn east and go 2 miles to the **Rice Bridge**, which is another good birding stop within the wildlife area.

Fall River attracts abundant waterfowl in migration and winter. Flocks of Common Merganser in the winter sometimes number in the thousands and Common Goldeneye in the hundreds. This is one of the reservoirs where flocks of Franklin's Gull in late October can number in the hundreds of thousands. In spring, if you encounter any recently burned prairies, check them carefully for flocks of American Golden-Plover, Buff-breasted Sandpiper, and other shorebirds. In addition to typical eastern woodland birds, and especially in the summer months, southeastern species such as Black Vulture, Acadian Flycatcher, Yellow-throated Vireo, and Fish Crow are possible. Check all areas of prairie that do not appear to have been burned for a few years, as there are multiple records of Henslow's Sparrow from various locations in the Fall River area. Other tallgrass specialists that you may encounter are Greater Prairie-Chicken and Upland Sandpiper. This is in the contact zone between Black-capped and Carolina Chickadees, and both can be seen during a single day. Numbers and diversity of warblers begin to increase in the Cross Timbers, and more than 20 species have been found at Fall River, especially in early May.

73. Toronto Reservoir

Counties: Greenwood and Woodson

eBird Hotspot names: Holiday Hill Area, Cross Timbers SP—Toronto Point Area, Cross Timbers SP—Woodson Cove Area, Toronto Lake, Toronto Lake—Outlet Park, Toronto Lake—Dam, Toronto WA

Other hotspots: Toronto WA—Cedar Creek Area, Toronto Lake—East Spillway Road, Toronto WA—US 54 Wetland, Toronto WA—Walnut Creek Bridge

Hotspot type: Half day

Toronto Reservoir in Woodson County was created by the construction of a dam on the Verdigris River and Walnut Creek that was completed in 1960. The lake has 2,800 surface acres and a maximum depth of 18 feet. Surrounding the lake are the

4,700-acre Toronto WA and the 1,075-acre Cross Timbers State Park, which is composed of five areas. Toronto and Fall River are only 15 miles apart, and a state park vehicle day permit purchased at one is valid for state parks at both reservoirs.

The state park has several components. The **Toronto Point Area** is 2 miles south of the town of Toronto on South Point Road and offers an unobstructed view of the lake. In summer, there are usually Chipping Sparrows nesting in the small cemetery along South Point Road. Be sure to hike the Ancient Oaks Trail, which begins and ends at the small parking area located just outside the entrance to Cross Timbers State Park, about 2 miles south of the town of Toronto. Because of the rocky soils, the post and blackjack oaks of the Cross Timbers grow very slowly, and some of the trees on this trail are over 400 years old. Interpretive signs on the trail provide information about this unique habitat. Look for oak specialists such as Red-headed Woodpecker, Summer Tanager, and others. There are multiple records of Scarlet Tanager on this trail, mostly from early May. Spend some time birding the campground area in addition to scanning the lake. In drought years, there are major mudflats south and west of the point that attract shorebirds, gulls, terns, and other waterbirds.

K-105 loops around the east side of the lake from the town of Toronto to the **Woodson Cove Area** at the east end of the dam, crosses the dam, and ends at the entrances to the **Dam Outlet** and **Holiday Hill** areas of the state park. These all provide vantage points for scanning the lake and opportunities for woodland birds.

While the state park areas provide quality birding, multiple locations within the **Toronto WA** offer equal or better opportunities. The **Cedar Creek** area is reached by taking K-105 north from Toronto (town) for 1 mile. Immediately after crossing the bridge over Cedar Creek, turn west onto a gravel road that reaches a dead end after 1 mile. There is a large, privately owned pecan grove on the north side of the road that attracts many species of birds throughout the year. Cedar Creek is parallel to the road on the south side of the road for this entire mile, and all land south of the road is within the wildlife area. The combination of the pecan orchard and the riparian habitat makes this an excellent stop at any season. If you have time, walking this road can be very productive. This has been a reliable location for Yellow-throated Vireo during the breeding season in most years. Many interesting birds have been reported here.

The **US 54 Wetland** can be a surprisingly good stop. From the town of Toronto, go north on K-105 to US 54 and turn left (west). Go 1 mile (crossing the bridge over the river) to a small parking area at the intersection with GG Road (marked as Road 33 on some maps). The developed wetland south and east of the parking area attracts waterfowl, shorebirds, and other wetland species when conditions are good, although it can be bone-dry in drought years.

To reach the **Walnut Creek Bridge**, continue south on Road 33 for 3 miles to the T intersection with 130th Street, then go west 1 mile on 130th to FF Road and turn south again. For the next mile, the road is elevated above the mature floodplain forest, allowing views of birds in the canopy. At the bridge over Walnut Creek, there is a good view of the creek and woodlands in both directions. This is another good area to do some birding on foot.

Another prime location in Toronto WA is just below the reservoir dam. Near the east end of the dam, where K-105 turns west, continue straight south on Deer Road for a short distance and turn west onto 10th Road. After a quarter mile, the road curves north and becomes **East Spillway Road** and the land on both sides of the road is public. Starting at this curve, the road goes through exceptional hardwood forest that is on the slopes and also the bottomland along the Verdigris River. Eastern shrubs and plants such as buckeye and wahoo thrive in this woodland, where deep-woods species such as Acadian Flycatcher, Wood Thrush, and Kentucky Warbler are present during the breeding season. If you have time, walk this short stretch through the big timber, if only to fully appreciate the habitat. Spillway Road continues west along the length of the dam to the spillway, then curves back eastward along the river channel through brushy areas and woodlands before reaching a dead end.

Over 200 species of birds have been found at Toronto Reservoir. In general, the expected birds are as you would expect at Fall River, but because of the more extensive Cross Timber oak forest, woodland species are more numerous and grassland species less so at Toronto. Additionally, waterfowl diversity and overall numbers tend to be lower in general than at Fall River. Most of the 20 or more shorebird species reported from Toronto have been seen either at Toronto Point or at the US 54 Wetland. These have included American Golden-Plover, Black-bellied Plover, Piping Plover, Hudsonian Godwit, and Buff-breasted Sandpiper. Black Vulture has started to appear in recent years at Toronto and is likely to become more common in summer. Chuck-will's-widows and Eastern Whip-poor-will are both breeding species and can be heard at dusk and dawn. Along with Fall River, this area is in the contact zone for the two chickadee species, and there are multiple records for both. Over 20 species of warblers have been reported during migration. Summer populations of Northern Parula and Prothonotary Warbler are robust.

74. Woodson State Fishing Lake

County: Woodson

eBird Hotspot name: Woodson SFL

Hotspot type: Worth a stop

Woodson SFL is small in comparison to Fall River and Toronto but is one of the best tracts of public land for birding in the Cross Timbers region. The area is 2,885 acres in size. The lake covers 180 surface acres and has a maximum depth of 45 feet. This is one of the old WPA lakes that were constructed by the CCC in the 1930s and was originally known as Lake Fegan.

The lake is less than 5 miles from Toronto. Just east of Toronto (town) on K-105, turn east onto 50th Road. Go east for 4 miles to reach the entrance on the west side of the lake. The lake can also be reached from Yates Center by taking US 54 west 8 miles to the intersection with Fox Road, then driving south for 3 miles to 50th Road, and then driving 2 miles east.

Fegan Road encircles the lake, allowing easy access. The most extensive woodlands are on the east side of the lake and include upland Cross Timber oak forest drained by rocky draws filled with large boulders and water-smoothed rock surfaces. There is a

Hotspot 74: A good place to explore the Cross Timbers oak forest is along the east side of Woodson State Fishing Lake. Here mosses, ferns, and lichens create a spongy ground cover through rocky draws filled with large boulders and water-smoothed rock surfaces.

surprising amount of spongy ground cover made up of ferns, mosses, and lichens. It's a great place to hike and explore, but be aware that this is Copperhead habitat. There are also extensive quality woodlands below the dam to explore on foot. This riparian hardwood forest can attract different birds than those found in the upland oak habitat on the east side of the lake.

At least 212 species of birds have been observed at Woodson SFL, which puts it on an equal footing with the much larger Toronto Reservoir. The species described for Toronto are largely the same at Woodson SFL. Here again there are multiple records for both chickadee species. Chuck-will's-widow, Eastern Whip-poor-will, Barred Owl, and Eastern Screech-Owl can be heard at dawn and dusk. Deep-woods species such as Acadian Flycatcher, Wood Thrush, and Kentucky Warbler are breeding species in the summer months. Other breeding warbler species include Louisiana Waterthrush, Northern Parula, Black-and-white Warbler, Prothonotary Warbler, and Common Yellowthroat. More migratory species of warblers have been observed here than at either Fall River or Toronto. Yellow-breasted Chat is often present in brushy habitats in summer. Summer Tanager, Indigo Bunting, and Blue Grosbeak are all common from spring through fall.

About half of the area is tallgrass prairie, with typical grassland birds such as Upland Sandpiper, Common Nighthawk, Scissor-tailed Flycatcher, Grasshopper Sparrow, and Dickcissel present from spring through fall. Greater Prairie-Chicken was a former resident, but there are no recent records. Henslow's Sparrow can be found in most years in unburned prairie. This is one of the most reliable locations for it in Kansas.

75. Augusta City Lake

County: Butler

eBird Hotspot name: Augusta Lake

Hotspot type: Worth a stop

Despite its urban location, Augusta City Lake is a surprisingly good birding hotspot that is usually worth a stop if you are in the area. Over 200 species of birds have been seen at the lake, which is located at the northwest edge of Augusta in Butler County. At full pool, it covers 190 surface acres and has a maximum depth of 41 feet. There are 75 acres of park land surrounding the perimeter of the lake, including a small area of woodland along Elm Creek at the upper end of the lake.

To reach the lake from US 54, go north on Ohio Street for 1.5 miles and turn left (west) at the entrance to the lake. There is a perimeter road around the entire lake. The east shore has playgrounds and picnic areas. The west shore is kept mowed but is mostly undeveloped.

The upper end of the lake is shallow. Teal, other dabbling ducks, and shorebirds can frequently be found here during migration. Most of the 24 shorebird species that have been seen at the lake have been found in this area of shallow water. These have included Black-bellied Plover, Piping Plover, Hudsonian Godwit, Dunlin, and Buff-breasted Sandpiper. In years of severe drought, additional mudflats are created by receding surface water.

Stop at the bridge located at the north end of the lake, where woodland species such as Red-shouldered Hawk, Barred Owl, Pileated Woodpecker, Red-eyed Vireo, Hermit Thrush, Fox Sparrow, and Summer Tanager are possible, depending on the season. This is not a large woodland, but it has attracted 10 warbler species, including Ovenbird and Canada Warbler.

The lake itself is of the most interest for birders. Twenty-seven waterfowl species have been seen, including Trumpeter Swan, Cinnamon Teal, Black Scoter, and all three merganser species. Significant numbers of diving ducks occur in early spring and late fall. Eared and Horned Grebes are seen most years in spring and fall, and Western Grebe is possible. Large numbers of Ring-billed Gull are seen all winter, and Franklin's Gull is abundant in spring and fall. Bonaparte's Gull, Black Tern, and Forster's Tern are reported in most years. There are several homes with Purple Martin houses near the lake, and martins are abundant in summer.

76. Chaplin Nature Center

County: Cowley

eBird Hotspot names: Chaplin Nature Center

Hotspot type: Half day

Located on the west bank of the Arkansas River, **Chaplin Nature Center** is owned and operated by the Wichita Audubon Society. It preserves 200 acres of natural habitats that include bottomland forest, broad sandbars, and native prairies. A visitor center has a small museum and provides information about the nature center and

Hotspot 76: Located on the west bank of the Arkansas River, Chaplin Nature Center preserves 200 acres of natural landscape. Five miles of hiking trails provide access to explore bottomland forest, broad sandbars, and native prairie.

its mission. A staff naturalist lives on the grounds, and a variety of free programs are offered throughout the year. Five miles of hiking trails are open year-round.

To reach the center from Arkansas City, go west for 4 miles on US 166 to 31st Road, turn north, and go 3 miles. The road curves west into 272nd Road. The Chaplin entrance is just past the curve. Park at the visitor center.

Over 225 species of birds have been observed on the property. Expected species in summer include Red-shouldered Hawk, Barred Owl, Pileated Woodpecker, Fish Crow, Chuck-will's-widow, Eastern Bluebird, Wood Thrush, Summer Tanager, and Painted Bunting. A wide variety of migrants occur in spring and fall. Nearly 20 shorebird species have been seen on the sandbars in the river. At least 5 vireo, 17 warbler, and 18 sparrow species have been identified at Chaplin. During the winter, look for typical winter woodland species, including Brown Creeper; both kinglets; Hermit Thrush; Fox, Harris's, and White-throated Sparrows; Spotted Towhee; and in some years Purple Finch. Bald Eagle can usually be observed foraging along the Arkansas River. Displaying American Woodcock have been observed in early March.

77. Kaw Wildlife Area

County: Cowley

eBird Hotspot names: Kaw Wildlife Area, Kaw Wildlife Area—Lower Grouse Creek

Hotspot type: Half day

East of Arkansas City is the **Kaw WA**, which encompasses almost all the land on both sides of the Arkansas River between Arkansas City and the state line and additional substantial acreage along Grouse Creek. The public land covers 4,341 acres, extending for 9 miles on both sides of the river and an additional 8 miles of Grouse Creek.

Both sides of the river offer excellent birding habitat, but birders tend to visit the north bank areas most often. To explore the north bank, take US 166 east from Arkansas City. There are access roads into the wildlife area at the 4.5-mile mark (at the second curve) and the 5.5-mile mark. At the 6.5-mile mark, turn south onto 304th Road at the sign for Camp Horizon. Follow 304th for 3 miles as it zigzags south and east to the intersection with 141st Road. Look for Painted Bunting, Summer Tanager, and other colorful birds along this road in summer. Take 141st south for 2 miles to the T intersection at the state line. Follow the state line road 1 mile west to reach the **Lower Grouse Creek** area, which is one of the best birding spots within Kaw WA. Here the river is broad and deep, and there are steep slopes covered with mature hardwood forest that extends for miles in both directions from the parking lot. In addition to the species listed above for Chaplin, Acadian Flycatcher, Wood Thrush, Prothonotary Warbler, and Kentucky Warbler all breed at this location in summer. Black-and-White Warbler is probably a breeding species as well. In addition to Red-shouldered Hawk, look for Black Vulture soaring overhead. They have recently colonized this area and are becoming more common each year.

78. Winfield City Lake

County: Cowley

eBird Hotspot name: Winfield City Lake

Hotspot type: Worth a stop

Winfield City Lake is formed by a dam on Timber Creek and was originally referred to as Timber Creek Lake. The lake covers 1,200 surface acres and has a maximum depth of 41 feet. The public lands around the lake cover an additional 1,150 acres. There is a considerable amount of developed park land, especially close to the dam. Most of the remaining property is tallgrass prairie. The coves, draws, and shorelines surrounding the lake have trees and brush, and at the east end of the lake is a substantial woodland along Timber Creek where it enters the lake. The lake is owned and operated by the city of Winfield.

To reach the lake, take US 77 8 miles north from Winfield to 82nd Road, then go east for 6 miles. When you reach the intersection with 141st Road, you will be at the north end of the dam with much of the lake visible. You can continue straight on park roads along the north shore for 3 miles, with several loop roads and access points where you can view the lake. Return to the dam and drive south on 141st Road across the dam. The entrance to the south shore area is a quarter mile south of the dam. South Lake Road winds along the full length of the south shore, with many turnouts for viewing the lake. Toward the east end of the lake, the bluffs are tall, allowing a panoramic view of most of the lake. This road ends at 171st Road. Go south on 171st for a quarter mile until it dead-ends at Timber Creek, where there is a small parking area and the trailhead for the Timber Creek Lake Nature Trail. The lake property extends for over a half mile to the east of 171st Road, so you can explore the woodlands along the creek if you wish to. Migrating shorebirds have been seen in the shallow areas at this part of the lake, especially on the exposed mudflats when the lake level is lower than normal.

Because of the southern location, Winfield Lake often remains open in winter, when many lakes and reservoirs farther north have frozen over. Large numbers of waterfowl and gulls can be present when these conditions exist.

The lake has been a birding destination for more than 60 years. At least 225 species of birds have been reported from this hotspot to eBird. Including the substantial volume of records that predate eBird, the all-time list likely exceeds 250 species. Large flocks of waterfowl occur in early spring and late fall. At least 30 waterfowl species have been seen at the lake, including multiple records of Tundra Swan, all three scoter species, and Long-tailed Duck. Despite the relative scarcity of suitable habitat, at least 18 shorebird species have been found, especially in drought years when there are open mudflats at the upper end of the lake. In addition to common shorebird species, these have included Black-bellied Plover, American Golden-Plover, Snowy Plover, Ruddy Turnstone, and Dunlin. From fall through spring, the lake attracts good numbers of gulls, which remain all winter if there is open water. These have included rarities such as Iceland, Lesser Black-backed, and Great Black-backed Gulls. Other noteworthy finds have included Western Grebe, Neotropic Cormorant, and Caspian Tern. Twenty sparrow species have been seen, including LeConte's. Despite somewhat limited woodland habitat, at least 12 warbler species have been observed at the lake.

79. Cross Timbers Scenic Drive

Counties: Chautauqua and Montgomery

eBird Hotspot names: Cedar Creek—Bronco Rd., Caney River—Elgin Bridge, Caney River—Hart's Mill, Caney River—Wilson Cemetery, Hewins Park & Cedarvale WTP, Copan WA

Hotspot type: Half-day auto tour

In Chautauqua County, there are few public lands, but exploring its winding roads provides an excellent opportunity to experience this beautiful country. You can't really go wrong just exploring on your own, but the tour described here between the towns of Cedar Vale and Caney has been popular with birders for years, especially in early April when the first warblers have arrived and the redbud trees are in bloom. It crosses the Caney River multiple times and passes through prime Cross Timber oak habitat.

Start your tour at Cedar Vale on US 166. At the southeastern corner of town is **Hewins Park** and the adjacent **Cedar Vale WTP**. The ponds are easily viewed from the adjacent ball fields and attract a variety of waterfowl in spring and fall. Check the huge cedar trees in the park for interesting birds. From the park, turn east onto Sale Barn Road, which curves into County Road 3. Take Road 3 south for 4 miles to Dalton Road and turn east. Continue east on Dalton Road for 8 miles, passing through the near-ghost town of Hewins. In Hewins, Dalton Road jogs north one block. Continue east on Dalton Road to **Hart's Mill**, where there is a bridge over the Caney River. This is a good place to stop and do some birding on foot. Black Vulture has been seen at this bridge several times. Continue on Dalton Road for another 3 miles to the old **Wilson Cemetery**. Park at the cemetery. From here, you can hike along the seldom-used Wilson Lane, which is between the river and some steep slopes. This is an excellent birding stop. Remain on Dalton Road for 2 more miles to the **Elgin Bridge**, where most birders stop and spend some time on foot. After another 2 miles, Dalton Road reaches the old town of Elgin. Just before you reach the town, the road passes through dense forest and is elevated, allowing easy viewing of birds in the canopy. From Elgin, continue east on Bronco Road for 6 miles to the town of Chautauqua, stopping at the **Cedar Creek Bridge** on the way. From Chautauqua, remain on Bronco Road for another 15 miles to the town of Caney. This stretch of the tour passes through a lot of the beautiful upland oak forest that typifies the Cross Timbers physiographic region.

The 2,300-acre **Copan WA** is just west of the Caney city limits and has some high-quality hardwood timber along the Little Caney River. There are several access points to Copan from the west side, but birders typically use the access on the east side of the wildlife area. Bronco Road becomes 6th Avenue in Caney. Go three blocks and turn north on Wood Street. Drive to the north edge of town, turning left onto 1600 Road. Follow the road for about a quarter mile as it crosses a creek and curves north. Turn left (west) at the Copan WA sign and follow the entrance road to the parking area adjacent to the river. Park here and walk into the woodlands to the north and west of the parking lot. Yellow-throated Warbler is possible at all the Caney River stops mentioned above, but the mature woodland along the river at Copan WA is the most reliable location for them. They typically arrive in early April. Listen for their distinctive song.

Except for the Cedar Vale WTP, opportunities for aquatic birds are limited on this tour, and the focus is largely on woodland birds. Be alert for Black Vulture, which is gradually becoming more common in the area. Red-shouldered Hawk and Pileated Woodpecker are common all year. Report all summer records of Broad-winged Hawk to eBird, as there is apparently a poorly understood small breeding population of them in the Cross Timbers. Several breeding species of eastern Kansas are near the western edge of their range in this part of the state. These include Acadian Flycatcher, White-eyed Vireo, Yellow-throated Vireo, Wood Thrush, and Kentucky Warbler. Early-April trips on this tour route nearly always produce good numbers of the early-arriving breeding-season warblers: Louisiana Waterthrush, Black-and White Warbler, and Northern Parula. Bell's and Red-eyed Vireos, Summer Tanager, Blue-gray Gnatcatcher, Carolina Chickadee, Prothonotary Warbler, Blue Grosbeak, and Painted Bunting are common from spring through fall. Bewick's Wren and Eastern Towhee are low-density permanent residents. During spring and fall migrations, a variety of migratory sparrows and warblers have been reported. Wintering species are in general typical of eastern woodlands. If it has been a good mast year, Red-headed Woodpecker can be abundant in winter.

Eastern Region

The east region consists of the eastern three tiers of counties in Kansas. It includes all or most of several major physiographic regions in Kansas. The majority of the Glaciated Region within Kansas is in this part of the state, which includes the entire area north of the Kansas River. The Osage Cuestas make up much of the rest of the region from the Kansas River south to the Oklahoma border. In the southeastern corner of the state are the Cherokee Lowlands, which cover about 1,000 square miles, mostly in Bourbon, Cherokee, and Crawford Counties. In a small area of southeastern Cherokee County, the Ozark Plateau barely intrudes into Kansas. About half of the narrow Cross Timbers region in Kansas lies in the eastern region, primarily in eastern Montgomery and Wilson Counties. Each of these regions has unique plant and animal communities.

The broad, deep Missouri River delineates the northeastern border of Kansas from the Nebraska state line to Kansas City, where it has its confluence with the Kansas River. Farther south is the Marais des Cygnes River, which drains a substantial area of the state and flows east into Missouri. The Neosho and Verdigris Rivers and their tributaries are the principal streams in the southern part of the region and flow south to the Arkansas River in Oklahoma. There are numerous major reservoirs and smaller man-made lakes on all of these rivers and their tributaries. Scattered throughout the region are a variety of wetlands, of which the Baker Wetlands and Neosho WA are the largest and best known. There are wetland habitats at the upper ends of many of the federal reservoirs, offering additional habitat for shorebirds, herons, rails, and many other wetland species. Many of these reservoir wetlands have been deliberately created by the KDWP and the Corps of Engineers. Others have naturally developed because of ongoing siltation and the subsequent encroachment of wetland plant life.

Woodland habitats are more prevalent in the east region than anywhere else in the state, and so are the woodland species that inhabit them. Of greatest interest to many birders, migratory warblers are at their greatest abundance in the eastern counties, far more than in any other part of state. This is also true for many other neotropical migratory songbirds. In this region, breeding species of the eastern forests are at their greatest density in Kansas, including species such as Eastern Whip-poor-will, Chuck-will's-widow, Broad-winged Hawk, Acadian Flycatcher, Wood Thrush, Kentucky Warbler, and Yellow-throated Warbler. Rare but possible summer species include American Redstart and Scarlet Tanager, especially in the Missouri border counties. Black Vulture was formerly rare in Kansas but is now an expected species in the southernmost part of the region, especially in the warmer months. Neotropic Cormorant and Fish Crow are other species of the southeastern US that were formerly considered rare in Kansas but now occur much more regularly.

Eastern Region Hotspots

80. Nemaha Wildlife Area
81. Pony Creek Lake
82. Wathena Area
83. Banner Creek Lake
84. Atchison State Fishing Lake
85. Perry Reservoir
86. Kansas City Area
87. Topeka Area
88. Lawrence Area
89. Flint Hills Trail State Park
90. Hillsdale Reservoir
91. Melvern Reservoir
92. John Redmond Reservoir and Flint Hills National Wildlife Refuge
93. Marais des Cygnes Wildlife Area and Marais des Cygnes National Wildlife Refuge
94. Lehigh Portland State Park
95. Elm Creek Lake
96. Neosho Wildlife Area
97. Bone Creek Lake
98. Crawford State Park
99. Elk City Lake
100. Cherokee County Area

80. Nemaha Wildlife Area

County: Nemaha

eBird Hotspot names: Nemaha WA

Hotspot type: Worth a stop

Nemaha WA covers 700 acres and is located 4 miles south of Seneca on K-63. Originally known as Lake Nemaha, it is a WPA project constructed in the 1930s. It was created by the construction of a dam on the South Fork of the Nemaha River. The 90-year-old masonry guardrails on the dam date to its construction and are listed on the National Register of Historic Places. In 1986, the spillway collapsed and the lake was completely drained. In 2002, a restoration project funded by Ducks Unlimited and multiple local, state, and federal agencies created a 125-acre wetland area adjacent to the dam that is separated by a berm from an 18-acre fishing pond at the east end of the original lake bed. Because the lake was created over 90 years ago, considerable mature woodland surrounds it, creating an island of habitat in a sea of cultivated farmland. Significant siltation from the surrounding farmland is ongoing.

The main entrance to the wildlife area is at the north end of the dam, where you turn east onto 116th Road. After a half mile, there is a substantial wet meadow behind trees on the south side of the road. This is a good spot for displaying American Woodcock in early spring and for Swamp Sparrow, Marsh Wren, and Common Yellowthroat during spring and fall migrations. 116th Road turns south, and after another half mile, there is an old stone picnic shelter with parking. From the parking area, walk to the fishing jetty, which is an excellent vantage point for scanning the fishing pond and adjacent wetland areas. Continue on 116th for another half mile to the T with L4 Road. Turn south and follow L4 to the T intersection with 104th Road. There are private homes and cottages on the east side of the road, but all land on the west side is public. There is some good woodland along this stretch. If there has not been recent rain and the roads are dry, you can turn west onto 104th and go 1 mile to return to K-63. This stretch of 104th has the best and least disturbed hardwood timber in the wildlife area, but it can be hair-raisingly treacherous when muddy. If it is wet, just don't do it. Return north on K-63 for 1 mile to the south end of the dam. Here there is a parking area from which you can scan the wetland area with a spotting scope.

More than 200 species of birds have been found at Nemaha WA. These include 24 waterfowl, 15 shorebird, 4 tern, 6 heron, 7 woodpecker, 18 sparrow, and 16 warbler species. Waterfowl numbers are substantial, especially in early spring and late fall, and often include large numbers of diving ducks. Shorebirds are common in spring in fall, but typically in lower numbers than at some of the other Kansas wetlands. Noteworthy finds at Nemaha have included Trumpeter Swan, Dunlin, Least Tern, Common Tern, Veery, Purple Finch, and Lazuli Bunting. In addition to the expected summer species, possible breeding species include Blue-winged Teal, Northern Shoveler, Least Bittern, Yellow-throated Vireo, Eastern Towhee, Yellow-breasted Chat, Summer Tanager, and Rose-breasted Grosbeak. Winter species are typical of eastern Kansas and include Bald Eagle, Winter Wren, Hermit Thrush, Swamp Sparrow, and Rusty Blackbird. Despite its habitat diversity and birding potential, Nemaha WA is not often visited by birders, and the potential for new finds is good.

81. Pony Creek Lake

County: Brown

eBird Hotspot name: Pony Creek Lake

Hotspot type: Worth a stop

Pony Creek Lake is located 2 miles north of Sabetha in Brown County, just a few miles south of the Nebraska state line. The lake covers 170 surface acres and has a maximum depth of 14 feet. It is among the most recently built reservoirs in Kansas, having been completed only in 1992 to provide a stable source of drinking water for the city of Sabetha. This is a quality birding hotspot that usually produces a good trip list.

All public access to the lake is on the west shore. From Sabetha, go 3 miles north on US 75 to 305th Street and turn right (east). Go a half mile to Angler Road, which is the primary entrance. There are several parking areas and fishing jetties where you can view the lake. Closest to the dam is an area of tallgrass that is sometimes left unmowed and attracts grassland sparrows. A small colony of Bobolink attempted to nest in this area during several years in the 1990s, but mowing destroyed the colonies each time. Bobolink has been reported in this area in several recent years as well.

If you want to add some woodland birds to your trip list, return to 305th Street and go east 1 mile to the intersection with Antelope Road, where there is a bridge over Pony Creek. All the adjacent land is private, but you can park near the bridge and walk along Antelope Road for a quarter mile in either direction. The riparian woodland along Pony Creek is generally undisturbed and attracts a variety of bird species.

To access the upper (southern) end of the lake, return to US 75 and go south for a half mile. After passing the large church, turn on the south entrance road, which has several branch roads leading to the lake. The southernmost road ends at the southern tip of the lake, where there is a small peninsula with woodlands and an adjacent area of wetland habitat. Shorebirds, herons, and other wetland species are sometimes found in this area of the lake.

Pony Creek has produced at least 220 species of birds. The diversity of waterfowl is above average, with 29 species having been seen, including Trumpeter Swan, Surf and White-winged Scoters, and all three merganser species. Long-tailed Duck has been found in multiple years in late fall and early winter. Among the more than 20 shorebird species reported have been American Avocet, American Golden-Plover, Black-bellied Plover, Hudsonian Godwit, Dunlin, and Buff-breasted Sandpiper. Bald Eagle is possible year-round and has nested in recent years. The woodlands along Antelope Road can produce Red-shouldered Hawk, Barred Owl, and Pileated Woodpecker year-round. Chuck-will's-widow and Eastern Whip-poor-will are both possible in summer. Summer species include typical eastern woodland birds such as Yellow-throated Vireo, Northern Parula, Eastern Towhee, and Summer Tanager. Migrating Osprey, Horned Grebe, Common Loon, and American White Pelican are reported annually. Uncommon species have included Western Grebe, Caspian Tern, Least Tern, and Merlin. During spring and fall migrations, 17 sparrow and 15 warbler species have been observed. Winter species include Winter Wren, Fox Sparrow, Rusty Blackbird, and Purple Finch.

82. Wathena Area

County: Doniphan

eBird Hotspot names: Browning Lake—165th Road Access, Browning Lake, Elwood WA, Browning Lake—Treece Road Access

Hotspot type: Worth a stop

Doniphan County has a look and feel quite unlike any other part of the state. It is bordered on three sides by the Missouri River, which is by far the largest of all Kansas rivers. Steep, rolling hills with old hardwood forest give way to extensive floodplains along the course of the river. Several locations near the town of Wathena allow you to sample this unique area.

Take **Monument Road** southwest out of Wathena for 2 miles as it follows the Missouri River below a steep wooded slope. There are several parking areas on the riverbank where you can scan for Bald Eagle, waterfowl, and other birds. Don't be surprised if you see a tugboat towing a cargo barge up or down the river! Continue on Monument Road past the old community of Palermo. Over the next several miles, you can turn left (south) onto Sheridan, Saratoga, or Randolph Roads, all of which go through the bluffs and down to the river through good habitat.

Return to Wathena. At the east edge of town on US 36, turn north onto Treece Road. Go 2 miles to the historical marker for Bellemont Crossing and park. This spot is marked as the **Treece Road Access** in eBird. Here you have an excellent view of the long oxbow lake known as **Browning Lake**. The lake is several miles long, and much of it can be seen from this location, especially if you walk back south from the parking area. The lake gradually transitions to an area of shallow water and mudflats. Sometimes there are shorebirds here. At the historical marker, the lake becomes an extensive area of cattails and other wetland plants. Interesting birds have been found here that are scarce elsewhere in Doniphan County, including Sora, Least Bittern, Marsh Wren, Swamp Sparrow, and other wetland birds. Birds seen near the east side of the lake are in Missouri.

Return to the Treece Road/US 36 intersection, go east for 2 miles on 170th Road to the intersection with Voltaire Street, and turn south. The southern arm of **Browning Lake** is on the west side of the road for the next half mile until you reach 165th Road. Depending on water levels, the portion of the lake closest to 170th Road is frequently shallow enough to attract shorebirds. There is a parking area south of the corner for viewing this part of the lake. Continue south on Voltaire to 165th Road and turn west for a quarter mile to reach the **165th Road Access**, which crosses the south end of Browning Lake. On the south side of the road, the lake is shallow and there is usually some mudflat habitat. Reverse directions and drive east for 1.5 miles to K-238, then north for 1.5 miles to the bridge over the main body of Browning Lake. There is a parking area at the south end of the bridge where you can check the lake for birds. Browning Lake attracts diving ducks, grebes, Bald Eagle, shorebirds, gulls, terns, cormorants, and pelicans. Less common species have included Black-necked Stilt, Hudsonian Godwit, Sanderling, Dunlin, and Neotropic Cormorant.

East from Wathena to the town of Elwood, nearly all the land south of US 36 is within the **Elwood WA** and can be explored as you wish. This is all floodplain that has experienced multiple heavy floods in past decades. There is a frontage road south of the highway that allows access to the entrance roads that go south into the wildlife area toward the major levees along the river. The second-growth cottonwood forest has produced warblers and other bird species. This area is closed in the winter and open to hunting in season.

If you are seeking woodland birds and you are in an exploratory mood, you can continue north and west from Wathena for 10 miles on Treece Road to Columbus Road, north 1 mile to 255th Road, then west for 2 to 3 miles to connect with several Doniphan County roads near the Missouri River. Several of these roads pass through a sparsely populated and largely undisturbed area of mature woodland on the hills south of the river. Space does not permit more detailed directions, so you will need an accurate map or GPS app to navigate this area. Internet access is poor or absent in places, so a printed map is recommended. The best roads for birding this wooded habitat are Ottumwa, Peck, Randolph, and Runnymede Roads. Some of these roads are exceptionally treacherous after rain events. This woodland area provides excellent birding. In summer, Broad-winged Hawk, Acadian and Willow Flycatchers, Wood Thrush, Ovenbird, Louisiana Waterthrush, Kentucky Warbler, Yellow-throated Warbler, and Scarlet Tanager have been found, and rarer warblers such as Black-throated Blue Warbler have been seen during spring migration.

Over 260 species of birds have been observed in Doniphan County. Many of them are possible if you visit the hotspots described here at the right time of the year. These have included 25 waterfowl, 21 shorebird, 7 heron, 10 hawk, 7 vireo, 5 thrush, and an impressive 27 warbler species.

83. Banner Creek Lake

County: Jackson

eBird Hotspot names: Banner Creek Lake

Hotspot type: Worth a stop

Owned and managed by the city of Holton, Banner Creek Lake is located 2 miles west of Holton just south of K-16. The reservoir covers 535 surface acres when full and has a maximum depth of 35 feet. There are several hundred acres of public land surrounding the lake. This is one of the newest reservoirs in Kansas, as the dam on Banner Creek was completed in 1997.

There are two developed recreation areas near the dam. The North Entrance is 1 mile west of Holton on K-16. The South Entrance is a half mile west of Holton on 222 Road. These two recreation areas offer the best views of the main body of the lake. Unlike many Kansas reservoirs, there is not a road across the dam, although there is a hiking trail. From the South Entrance, continue south and then west for 2 miles on 218 Road to the intersection with N Road. At the corner is a small parking lot adjacent to one of the access points for the 14-mile Banner Creek Hike and Bike Trail, which encircles most of the lake. This "Turkey Hollow" segment of the trail goes south for a half mile and then north again for another half mile through an area of mature hardwood forest. This is an excellent location for warblers and other migratory songbirds, especially in May and September. To reach another productive stretch of the trail, continue south on N Road to 214 Road and go west for 1 mile to M Road, then north for 1 mile to the parking area at Grange Bridge Crossing over Banner Creek, where there is another parking area with access to the Main Trail. The leg of the trail on the south side of the creek is another stretch of quality woodland habitat and is brushier than Turkey Hollow. The portion of trail on the north side of the creek follows the shoreline for over 2 miles to the Banner Creek North Entrance area. From Grange Bridge, you can drive north a short distance to 222 Road, then east for 1 mile to N Road, then north for another mile to return to K-16.

More than 225 bird species have been reported at Banner Creek. The list of waterfowl, shorebirds, gulls, terns, herons, and other aquatic bird species is in general like that of other eastern Kansas reservoirs. Waterfowl are abundant in early spring and late fall, often including large numbers of diving ducks. Eared and Horned Grebes are seen in most years, and there are several records for Western Grebe. Shorebird habitat is usually limited but is more widespread in drought years. Notable shorebird sightings have included American Avocet, Black-bellied Plover, Marbled Godwit, Ruddy Turnstone, and Buff-breasted Sandpiper. Migratory songbirds are well represented. Seven vireo, 18 sparrow, and 20 warbler species have been observed. Rarer warblers have included Golden-winged, Blue-winged, and Chestnut-sided. Several summer records of Wood Thrush suggest that the woodland is mature enough to support a small breeding population. Be alert for Henslow's Sparrow in summer in areas with appropriate habitat.

84. Atchison State Fishing Lake

County: Atchison

eBird Hotspot names: Atchison SFL, Jackson Park, Warnock Lake

Hotspot type: Worth a stop

Atchison SFL is located near the city of Atchison. The lake dam, on a tributary of Independence Creek, was constructed in 1957. The lake covers 66 surface acres and has a maximum depth of 30 feet. Adjacent public land includes a 130-acre wildlife area and an additional 50 acres set aside for camping and picnicking. The habitat is predominantly oak-hickory forest, but there is grassland habitat on the northwest side of the lake. In autumn, the surrounding woodlands provide a spectacular display of color.

To reach the lake from Atchison, go north for 3 miles on K-7 to 318th Road, then west for 2 miles to Pawnee Road, then north for a half mile on Pawnee Road to the south entrance to the lake. Just north of the entrance, turn left (west) onto 322nd Road, which follows the south shore of the lake for a half mile before ending at a turnaround loop. You are free to hike in from the loop for up to a quarter mile to the west through the woodlands, but there are no developed trails. The adjacent portion of the lake is quite shallow, and shorebirds, herons, and other wetland species are sometimes found. Return to Pawnee Road and follow it north along the shore of the lake. The entire lake can be viewed from the campgrounds and the several fishing jetties that are located along this stretch. The steep hillsides to the east are covered with mature oak-hickory forest. When you reach the dam, you can cross it on foot to reach the trailhead for a hiking trail that follows the north shore of the lake. Continue north on Pawnee Road as it drops below the dam through brushy woodland. You are on public land until the road turns straight north.

More than 220 species of birds have been identified at Atchison SFL. These include 23 waterfowl species, but the lake is not known as a major waterfowl location. Twenty shorebird species have been found, mostly in drier years when water levels are below normal. Listen for Eastern Whip-poor-will calling at dusk in spring and summer. Other summer residents include deep-woods specialists such as Acadian Flycatcher and Wood Thrush. At least 22 migratory warbler species have been found, including Magnolia, Blackburnian, and Chestnut-sided. Winter species are in general typical for eastern Kansas and can include Yellow-bellied Sapsucker, Hermit Thrush, Winter Wren, and White-throated Sparrow.

The town of Atchison is located on the bank of the Missouri River and is among the oldest towns in Kansas. The appealing riverfront area downtown is a good place to watch the river for waterfowl, gulls, Bald Eagle, and other waterbirds. At the southwest edge of town, **Warnock Lake** takes only a few minutes to check for birds but can be a surprisingly good stop and has produced 208 species of birds. It is often a better location for waterfowl than the state lake, having produced almost 30 species. **Jackson Park** on the south edge of Atchison is on top of the bluffs looking out over the river and has numerous huge oak trees. From the park entrance at the corner of S 6th and College Streets, you can follow Jackson Park Loop as it wanders through the park and down to the bottomland along the river. This park has produced at least 22 species of warblers, mostly during spring migration in May.

85. Perry Reservoir

County: Jefferson

eBird Hotspot names: Perry Lake, Perry Lake—Dam, Perry Lake—Old Military Trail & Campground, Perry Lake—Outlet PUA & Delaware Marsh, Perry Lake—Rock Creek PUA, Perry SP—Jefferson Point SP, Perry WA—Kyle Marsh, Perry WA—Lassiter Marsh, Perry PUA, Topeka Audubon Society Bird Sanctuary

Hotspot type: Full day

Located conveniently close to both Lawrence and Topeka, **Perry Reservoir** is among the most popular birding destinations in eastern Kansas. It is one of the largest reservoirs in the state, with a surface area of 11,146 acres and a maximum depth of 45 feet. The total length of the public land from the dam to the north end of the wildlife area is over 25 miles. There are two state parks and six Corps of Engineers parks. The 11,000-acre Perry WA extends for 10 miles beyond the north end of the reservoir. Within the wildlife area are 13 wetland areas managed by the KDWP.

A typical visit begins at the town of Perry. From the north edge of town, take

Ferguson Road north for 3 miles to 39th Street, then west for a mile to Perry Park Drive. Turn right onto Perry Park Drive to access the road across **Perry Dam** or continue on 39th below the dam to reach the spillway area. Before you reach the spillway, there is a short road that dead-ends near the edge of **Delaware Marsh**, which has a 2-mile hiking trail that is good for wetland birds when conditions are favorable.

The parking lot at the west end of the dam is a good vantage point for scanning the lake. The deep water offers the best opportunity for loons, grebes, diving ducks, and gulls. After scanning the lake, turn north onto Rock Creek Park Road. For the next several miles, this road follows the lakeshore through the mature woodlands of the **Rock Creek Public Use Area**. These woods are one of the best locations for woodland species at Perry. When you reach the intersection with West Lake Road/K-237, turn right and go north for 2 miles to reach **Jefferson Point State Park**. Cross the bridge over the Rock Creek arm of the lake to reach the **Rock Creek Marina**, where you can park and scope the lake and shallows north and west of the marina.

To explore the east side of the reservoir, return across the dam to Perry Park Road and turn left. In a short distance, you reach the entrance to the **Perry Marina**, where you can scan the water along the dam. Perry Park Road winds through the wooded **Perry Public Use Area**, eventually merging into 43rd Street. Go east on 43rd for 1 mile to Ferguson Road, then north for several miles to Lakeside Village. At the north edge of Lakeside Village, turn east onto 74th Street and follow it east and then south until it ends at the entrance for the 83-acre **Topeka Audubon Bird Sanctuary**. The sanctuary is leased and maintained by the Topeka Audubon Society. A hike on the maintained trails at the sanctuary offers an appealing mix of grasslands, oak-hickory forest, and shoreline. Return to Ferguson Road and go north for another 4 miles to the parking area at the trailhead for the 7-mile (out and back) **Old Military Trail**. The Old Military Trail has the most extensive and least disturbed area of mature oak-hickory woodland at Perry. During migration, it is often the best location at Perry for migratory vireos, warblers, and other neotropical migrants.

To explore some of the Perry WA wetlands, continue north on Ferguson Road for another 6 miles to the fork with Jackson Road. Take the left fork onto Jackson and continue for a quarter mile to reach the **Upper Ferguson Marsh**, which can be worth a brief check if there have been recent heavy rains. From here, continue north for 3 more miles to the intersection with K-16. Turn left (west) and go 4 miles (passing through the town of Valley Falls) to the intersection with Finney Road and turn north. Finney Road ends after 2 miles. Park here and walk west for a quarter mile to reach a good vantage point for the **Northwest Marsh**, which in recent years has been one of the more productive wetland impoundments at Perry for birding. After birding the marsh, you can continue west on a walking path that goes between the marsh and riparian woodland along the Delaware River.

Return to Valley Falls and at the intersection of K-16 and K-4, turn right (south) and go 3 miles to 142nd Street. Turn left onto 142nd and go east 1.5 miles to the intersection with Geary Road, then south a mile to the parking area for the **Kyle Marsh**, which has produced at least 210 species of birds. If you have time for only one of the Perry wetlands, Kyle should be the one you choose. It is the highest-quality wetland within the Perry WA. It has produced nearly 30 shorebird, 10 heron, and many other wetland bird species, including rarities such as Glossy Ibis, Sabine's Gull, and Nelson's Sparrow.

The all-time list for Jefferson County is one of the highest of all Kansas counties, at 342 species of birds. More than 300 of them have been recorded at Perry Reservoir and the adjacent public lands. These include 32 waterfowl, six grebe, 32 shorebird, 14 gull, nine heron, 13 hawk, 22 sparrow and 32 warbler species. The shorebird diversity at Perry is one of the highest of any eastern Kansas hotspot. Most of the best shorebirds have come from the Kyle, Northwest, and Upper Ferguson Marshes. Other wetland specialists to look for at the marsh hotspots are American and Least Bitterns, Little Blue Heron, Sedge and Marsh Wrens, LeConte's and Nelson's Sparrows, and Yellow-breasted Chat. While gulls are possible anywhere on the lake, the greatest numbers are typically observed from the Rock Creek Marina and southward to the dam. Nearly all records of the rarer gulls, such as Glaucous, Iceland, Lesser Black-backed, and Great Black-backed, come from this deepwater area of the lake. For woodland species, and especially migrant flycatchers, vireos, and warblers, concentrate on the Rock Creek Public Use Area and the Old Military Trail.

86. Kansas City Area

Counties: Johnson and Wyandotte

eBird Hotspot names: Ernie Miller Park & Nature Center, KCP&L Wetlands, Mill Creek Streamway Trail—Midland Access, Mill Creek Streamway Trail—Wilder Rd. Access, Nelson Island, Overland Park Arboretum, Shawnee Mission Park, Wyandotte County Lake

Hotspot type: Various

Kansas City and its suburbs encompass several counties in Kansas and Missouri. Despite extensive urbanization, there are outstanding birding hotspots within the metro area. Johnson County has done an exceptional job of preserving and providing access to natural habitats with its extensive and visionary park system. There are 223 official eBird hotspots in Johnson County alone, probably the most of any Kansas county. The Burroughs Audubon Society is an active organization and conducts many birding field trips year-round.

One of the most birder-friendly area hotspots is the 300-acre **Overland Park Arboretum**. The riparian habitat along Wolf Creek is some of the best in the Kansas City metro area. There are beautiful botanical gardens, but most of the property is devoted to maintaining natural habitats, which include steep limestone bluffs, native prairies, and hardwood forests. There are over 5 miles of wood-chipped trails, allowing easy access to much of the arboretum property. This is an excellent site for observing migratory songbirds in both spring and fall. Among the more than 200 species of birds that have been observed at the arboretum are 31 species of warblers. Nesting species include Pileated Woodpecker, Yellow-throated Vireo, Kentucky Warbler. Northern Parula, and Summer Tanager. During the winter months, bird feeders near the entrance building are always filled and attract many birds, often including Pine

Siskin and Purple Finch. To reach the arboretum, go south on US 69 from Overland Park. Exit at 179th Street and proceed west approximately 1 mile to the arboretum entrance on the left (south). The arboretum opens at different times based on season and day of the week, so plan ahead, as early-morning birders often start their day elsewhere.

Shawnee Mission Park is administered by Johnson County Parks and Recreation. Together with access to the adjoining **Mill Creek Streamway**, this 1,250-acre site offers the greatest habitat diversity of any single hotspot in the Kansas City area, including a 120-acre lake, extensive upland and riparian woodlands, prairie grassland, and wet meadows. To reach the park, take the 87th Street exit from I-435 and go west for a short distance to Renner Road, then north on Renner to the main entrance at 79th Street. A perimeter road encircles the lake, as do the hiking trails. In the upland forest north of the lake are over 10 miles of hiking trails. The lake can be easily viewed from the dam and from numerous locations on the perimeter road.

Nearly 270 species of birds have been found at this park alone, and one ardent local birder has recorded 252 of them! The lake attracts loons, grebes, waterfowl, Osprey, Bald Eagle, shorebirds, and a variety of gulls and terns. Thirty-three waterfowl species have been observed, including Cinnamon Teal, all three scoter species, and Long-tailed Duck. The woodlands and prairies attract migratory thrushes, vireos, warblers, and others in spring and fall. Over 30 warbler species have been found in the park, including rarities such as Blue-winged, Cerulean, Golden-winged, and Prairie Warblers. In winter and summer, a trip to this park will produce most of the expected bird seasonal species of eastern Kansas.

The general area below the dam is one of the most productive birding locations in the park and is a reliable location to observe displaying American Woodcocks on calm evenings in March. In spring and summer, be alert for Yellow-throated Warbler in the adjoining woodlands along the creek. When the habitat conditions are favorable in late fall, wet meadows here attract Marsh and Sedge Wren, LeConte's Sparrow, and occasionally Nelson's Sparrow. A colony of Henslow's Sparrow nested here at least once.

Adjoining Shawnee Mission Park below the dam is one of many access points for the **Mill Creek Streamway Park**, which is a narrow band of public land extending for 17 miles along Mill Creek between the city of Olathe on the south and the Kansas River at Nelson Island on the north. Detailed information on the trails is available at www.jcprd.com. There are eight primary access points on the trail. The streamway habitat is mostly riparian woodland for most of its length. Several segments of the streamway trail system are popular with birders. The **Midland Drive** access point is in Shawnee at 19405 Midland Drive, near the intersection of Shawnee Mission Parkway and Midland Drive. From the parking lot, walk the trail to the west along Little Mill Creek, then cross the footbridge and follow the trail south into the upland forest. Follow the trail as your time allows, then either retrace your steps back to your car or continue on the 2.2-mile paved loop trail, which will cross paths with those at Shawnee Mission Park where there is a portable toilet near a small pond.

The riparian areas along the creek and the trail into the woods can be excellent in spring, fall, and summer. Reach the **Wilder Road** access point at 19425 Wilder Drive in Shawnee by taking Holliday Drive west from I-435 and following the signs. From the parking lot, follow the trail north to the north terminus of the Streamway Trail, where you can view **Nelson Island** in the center of the river channel. The island serves as a staging and roosting location for hundreds or thousands of gulls that forage at the nearby Johnson County Landfill. While most of them are Ring-billed and Herring Gulls, at least eight gull species have occurred. Glaucous and Lesser Black-backed Gulls are seen annually. Shorebirds, terns, and other waterbirds are also likely, depending on the season. The adjacent woodlands on this segment of the trail typically produce a good list of birds, and this is another Kansas City area location where 30 or more species of warblers have been observed during spring and fall migrations.

The **Ernie Miller Park and Nature Center** is small at 114 acres, but it is a popular birding destination. The park consists mostly of upland and riparian woodland but also contains some grassland and transitional habitat. A nature center on the grounds includes bird feeders and a water feature that can be viewed from indoors. There are 3 miles of nature trails. This park is most noteworthy for producing an excellent variety of warblers during spring migration but offers good birding throughout the year. The entrance to the nature center is at 909 N K-7, 1 mile north of the intersection of K-7 and Santa Fe Street in Olathe.

Wetland habitat is scarce in the Kansas City area. The 55-acre **KCP&L Wetland Park** provides Kansas City birders with an accessible location for shorebirds, rails, herons, and other wetland species. This is a mitigation wetland that was developed by the Kansas City Power and Light Company. It is a good birding hotspot, but be aware that it is located within an industrial/heavy commercial area that experiences substantial railroad and truck traffic, making most visits a bit less than pastoral. To reach the wetlands, go west from Gardner on US 56 and turn south onto Waverly Road. The wetlands will be on your left (east) after you cross the first set of railroad tracks. There is a viewing blind along the perimeter trail. The best view of this area is from the east-west levee between the two main pools.

At this wetland, 240 species of birds have been seen, an impressive number given its modest size and lack of any significant woodland habitat. The 31 shorebird species include uncommon species such as Black-necked Stilt, Black-bellied and Piping Plovers, American Golden-Plover, Hudsonian and Marbled Godwits, Ruddy Turnstone, Dunlin, Short-billed Dowitcher, and Red-necked Phalarope. Other wetland specialties recorded here have included Neotropic Cormorant, American and Least Bitterns, Virginia Rail, Sora, and Least Tern.

Wyandotte County Lake and Park, located at 9100 Leavenworth Road, is in some respects the best birding hotspot in the Kansas City region. The lake covers 407 surface acres and has a maximum depth of 50 feet. There are 1,500 acres of public land surrounding the lake, consisting mostly of mature oak/hickory forest. There are 325 acres of developed park, picnic, and activity areas along much of the shoreline. This is another of the many old WPA dam projects in Kansas. The construction of the dam began in the late 1930s and after major setbacks (including the complete collapse of the nearly completed dam in 1937) was completed in 1942. The lake and park both receive heavy recreational use throughout the year.

Most visitors to the lake travel north past the Kansas Speedway on I-435 to the K-5/93rd Street exit and go southeast on N 93rd Street for about 1 mile. At the fork with Nelson Road, bear left to remain on 93rd, and after a short distance, you will reach the lake. The 6.5-mile primary perimeter roads encircling the lake are West Drive and East Drive, which have numerous branch roads allowing access to the lakeshore, park shelters, and woodlands. With over 30 miles of miles of hiking and equestrian trails throughout the park, you have access to most of the public land. The parking areas at both ends of the dam are the best places to scan the lake for waterfowl, loons, grebes, and other swimming birds, and they also overlook the broad floodplain of the Missouri River below. At the south end of the lake, the water is shallow and attracts teal, Northern Shoveler, other dabbling ducks, and sometimes a few shorebirds.

On the west side of the lake is the **Mr. & Mrs. F. L. Schlagle Library and Environmental Learning Center** (www.kckpl.org/libraries.flschlagel.html). This is a refreshingly unique institution offering a variety of resources and programs for students and the public. The bird feeders near the observation deck are kept stocked year-round. In addition to the library itself, there are excellent interpretive displays and a modest bookstore. Special events typically include Eagle Days in January and a Butterfly Festival in September.

Nearly 250 species of birds have been reported at Wyandotte County Park. The lake attracts good numbers of waterfowl from October through April. Thirty waterfowl species have been seen here, including both Tundra and Trumpeter Swans in most years. Wintering Trumpeter Swan has increased in numbers in recent years, and on occasion over 100 can be present along with thousands of geese and ducks. Osprey is common in spring and fall. Bald Eagle is numerous in winter, and a pair often remains to nest in the vicinity. This is one of the most reliable locations in Kansas for Broad-winged Hawk during the summer months, and confirmed breeding has been documented in multiple years. Red-headed Woodpecker is common in winter unless the mast crop has failed, and this is the one of the most reliable places for Pileated Woodpecker in the Kansas City metro area. The maturity and quality of the woodland is illustrated by the presence in summer of species such as Acadian Flycatcher, Yellow-throated Vireo, Wood Thrush, Kentucky Warbler, Yellow-throated Warbler, and Scarlet Tanager. As with most of the other Kansas City area hotspots, the species diversity of warblers possible in May is high, with at least 30 species having been reported.

87. Topeka Area

County: Shawnee

eBird Hotspot names: Felker Park, Governor's Mansion Trails & MacLennan Park, Kaw River SP, Lake Shawnee, Shawnee SFL

Hotspot type: Full day

The Topeka area offers numerous birding opportunities. Within the city are over 100 parks and green spaces and many miles of well-planned walking trails. The Topeka Audubon Society offers numerous programs and field trips throughout the year. A local specialty is the pair of Peregrine Falcons that nest on high-rise buildings in the downtown area most years. The nesting birds are sometimes viewable online on a 24-hour video camera.

Felker Park, located in central Topeka at SW 25th Street and SW Gage Boulevard, is one of the most popular birding destinations. The park entrance and parking lot are on the east side of Gage Boulevard. This park has a substantial area of riparian woodland as well as some grassland habitat and a wetland in the adjoining Warren Natural Area. The 8-mile paved Shunga Trail spans most of the city and passes through the full length of the park. The Topeka Audubon Society monthly bird walks are conducted on the segment of the Shunga Trail within Felker Park and the Warren Natural Area. The best woodland birding is along the unpaved Orville Rice Nature Trail, which winds along Shunga Creek on the south edge of the park.

Over 200 species of birds have been recorded at Felker Park. It is especially popular with birders at the peak of spring warbler migration during the first half of May and during the southward migration in late August and early September. Summer and winter birds are typical of eastern Kansas. The wetland area does not attract many waterfowl or shorebirds, but depending on the season, look for other wetland specialists, such as Wood Duck, American and Least Bitterns, Marsh and Sedge Wrens, and LeConte's Sparrow. More than 20 sparrow and 30 warbler species have been recorded at Felker Park, including many sought-after species. Deep-woods breeding species possible in summer include Acadian Flycatcher, Yellow-throated Vireo, Wood Thrush, and Black-and-white Warbler.

Located at the southeastern edge of Topeka at 3131 SE 29th Street, **Lake Shawnee** has produced over 250 species of birds, including 33 waterfowl, 5 grebe, 8 gull, and other swimming bird species. The lake covers 410 acres and is surrounded by 1,100 acres of public recreation lands administered by Shawnee County. Constructed as a WPA project in 1935, it is the best location in the immediate Topeka area to observe waterfowl, loons, grebes, and shorebirds. Migratory songbirds are often seen on the grounds as well. There are access roads along the eastern and western shores, where you can scan the lake. At the north end of the lake on the west shore, you can park at the dam overlook and walk down to the overflow to search for shorebirds. Bald Eagle has nested at the lake in recent years. Be sure to watch the nest from a safe distance to avoid disturbance. The gazebo "point" on the west shore of the lake doesn't look special but has a history of attracting migrating warblers and other passerines in May. The point offers a good vantage point from which to scan the lake for waterbirds. Be sure to check the big oak trees in the center of the point. Despite the general scarcity of woodlands, at least 24 species of warblers have been found around the lake, including zingers such as Bay-breasted, Cape May, and Pine Warblers.

Near the northwestern corner of Topeka is the 76-acre **Kaw River State Park** and the directly adjacent 244-acre **MacLennan Park** (operated by the Topeka Parks Department), which wraps around the Kansas governor's Cedar Crest residence. Together these two parks offer public access to more than a mile of the prominent hills that slope down to the south bank of the Kansas River. There are miles of trails in these two parks, and the habitat is largely hardwood forest. Kaw River is one of the newest state parks in Kansas and was opened to the public in 2010. There are a parking lot and trailhead on the east side of MacLennan Park on SW Fairlawn Road, but most birders enter via Kaw River State Park Road from SW 6th Avenue and drive north to the parking area near the river to access the trail system.

Nearly 200 species of birds have been seen in these two parks. While the emphasis is on woodland species, the river access adds some variety. While generally scarce, 15 waterfowl, 15 shorebird, and other wading and swimming bird species have been observed on the river, sandbars, or flying overhead. Because it is situated on the hills overlooking the river, the area is conducive to raptor watching. In late summer, many Mississippi Kites can be seen soaring overhead, foraging for cicadas and dragonflies. Warblers are well represented, with over 25 species having been found, and other migratory songbirds such as vireos, thrushes, buntings, and grosbeaks are also common during migration.

Shawnee SFL, located northwest of Topeka, is visually appealing and is the most productive birding hotspot in Shawnee County, having produced more than 250 bird species. To reach the lake from Topeka, take US 75 north to 62nd Street, then go west 3 miles to Landon Road, then north 2 miles to 86th Street, then west for another mile to the lake entrance. Park roads extend for the full length of the lake on both the east and west sides. The lake itself has 135 surface acres and a maximum depth of 25 feet. The surrounding public land covers 640 acres, which includes a large area of tallgrass prairie, primarily on the west side of the lake. Scattered stands of trees and brushy habitat can be found all around the lake. The north end of the lake has the largest area of riparian woodland.

Check the marshy area below the dam in October and November for species such as Marsh Wren, LeConte's Sparrow, and Swamp Sparrow. In spring and fall, the lake attracts migrating waterfowl, loons, grebes, gulls, and shorebirds. During the summer months, there are numerous Dickcissel, Grasshopper Sparrow, and other grassland species. Listen for singing Sedge Wren, which is sometimes found in good numbers, especially in late summer. In years when the tallgrass habitat is favorable, Henslow's Sparrow nests on the west side of the lake and has also nested several times on private land just south of the public land. If you are present at dawn or dusk from late fall through early spring, look for Short-eared Owl foraging over the tallgrass prairie on the west side of the lake. Greater Prairie-Chicken is scarce but possible in this general part of Shawnee County, but its lek locations change from year to year.

88. Lawrence Area

County: Douglas

eBird Hotspot names: Baker Wetlands, Burcham Park (Lawrence), Clinton Lake, Clinton Lake—Bloomington East, Clinton Lake—Dam, Clinton Lake—Model Airport & Marsh, Clinton Lake—Overlook Park, Clinton SP, Clinton SP—Marina, Clinton WA-Wakarusa Causeway, Fitch NHR, Rockefeller Native Prairie

Hotspot type: Various

Home to the University of Kansas (KU), the city of Lawrence has a rich birding history in Kansas dating back to the late nineteenth century. Many of the most respected ornithologists in the state and even in the nation have studied or taught at KU as well as at Baker University in the nearby city of Baldwin. The curated collection of bird specimens at KU is among the most extensive in the nation. The combination of a major reservoir, a significant wetland, grasslands, and extensive woodlands has given Douglas County a cumulative checklist of 369 species of birds, a total exceeded by only two other Kansas counties.

Clinton Reservoir, located just west of Lawrence, is one of the most frequently visited local birding hotspots. Over 275 bird species have been found there, including 32 waterfowl, 30 shorebird, 12 gull, and 20 sparrow species as well as many other birds of interest. The lake covers 7,000 surface acres and has a maximum depth of 55 feet. There is a 1,500-acre state park, and the public wildlife area covers over 9,000 acres. Enter the **Overlook Park** at the north end of the dam from E 900th Road, where you can scan the lake near the dam. The Free Trailhead at the parking area has access to several miles of hiking trails through the woods and along the shore. N 1415 Road follows the north shore of the lake, with access to the Clinton Marina and Clinton State Park. Painted Bunting has become much more common around Clinton in recent years, and many of the records come from the general area near the marina. The area of the lake near the marina often attracts rare gulls and waterfowl. Below the south end of the dam is the parking area for the **Model Airport**. In spring and fall, the wet meadows in the vicinity of this parking area often produce rails, Sedge and Marsh Wrens, LeConte's and Nelson's Sparrows, and other sought-after wetland species. Reached from the west end of the reservoir, the small town of Clinton is located on a peninsula of the lake. The **Bloomington Beach** area and nearby sandbars at the east end of the point often attract shorebirds, gulls, and terns, which have included numerous rarities over the years. Bloomington Beach is the best shorebird spot at Clinton. The **Wakarusa Causeway** on the west end of the reservoir is a good place to see a diverse list of birds, including nesting Bald Eagle and Painted Bunting in the summer and shorebirds in drier years when there are extensive mudflats. Bald Eagles are often abundant in winter, and their first nest anywhere in Kansas during the modern era was at Clinton.

The 590-acre **Fitch Natural History Reservation** is a field station of KU located about 5 miles north of Lawrence on E 1600 Road. It lies within one of the largest stands of intact woodland in Douglas County. There are 5 miles of hiking trails. Fitch is renowned as one of the best locations for migratory warblers anywhere in Kansas.

Hotspot 88: The Wakarusa River flows into the west side of Clinton Reservoir. The causeway is a good place to view eagles, shorebirds, waders, pelicans, and more.

At least 30 species have been recorded, and overall warbler numbers are often impressive. A short distance to the north on 1600 Road is the trailhead for the **Rockefeller Native Prairie**, where a tract of preserved tallgrass prairie can be explored that often produces a good set of grassland bird species.

The **Baker Wetlands** on the southern edge of Lawrence is the largest and best-known publicly accessible wetland in northeastern Kansas. The 927 acres are owned and managed by Baker University. The wetlands extend for about 2 miles on the south edge of Lawrence between K-10 and the Wakarusa River. Starting in 2014, the wetlands were substantially altered by the expansion of K-10, which eliminated some habitat but also created substantial new mitigation wetlands. The construction of the Baker University Wetlands Discovery Center was a component of the larger project. The Discovery Center at 1365 N 1250 Road has interpretive displays and offers a variety of programs and nature walks throughout the year. Several of the more than 11 miles of hiking trails begin at the Discovery Center. Additional parking for the trails can be found on the east side of the wetlands south of K-10 on Haskell/E. 1500 Road as well as on the west side of the wetlands on E 1400 Road just north of the Wakarusa River. There is a detailed trail map on the Baker University website. Adjoining the marsh on the south is riparian woodland bordering the Wakarusa River, where many woodland species have been found. Nearly 280 species of birds have been observed at Baker, which is noteworthy for a hotspot of modest size. Waterfowl, herons, and shorebirds are seen from early spring through late fall. Thirty-three species of shorebirds have been seen, including some exceptional rarities. Specialties of the area include American Bittern, Least Bittern (rare), Yellow-crowned Night Heron, Sora, Virginia Rail,

Sedge Wren, Marsh Wren, Swamp Sparrow, LeConte's Sparrow, and Nelson's Sparrow (rare). Fifteen or more sparrow species can be seen on favorable days in spring and fall. Rarities at Baker have included Cinnamon Teal, King Rail, Common Gallinule, Whimbrel, and Red-necked Phalarope.

If you are in Lawrence with limited time for birding, **Burcham Park** on the west bank of the Kansas River near downtown is a good spot to spend an hour or two, as over 200 species have been seen there. Walk the path that begins near the south end of Bowersock Dam and goes north for about a half mile along the river to the Rowing Boathouse. Although "improvements" have eliminated a lot of the best brushy habitat along the trail, this is still a worthwhile birding hotspot, especially during migration, when 18 sparrow and 28 warbler species have been seen. Over 20 species of waterfowl have been seen on the river, which is deep behind the dam. In the winter months, there are sometimes many gulls below the dam, at times including rarer species.

89. Flint Hills Trail State Park

Counties: Andersen, Franklin, and Miami

eBird Hotspot names: Flint Hills Nature Trail—Rantoul, Mount Hope Cemetery, Prairie Spirit Trail—NW Mitchell Road

Hotspot type: Various

An encouraging development for Kansas outdoor enthusiasts in recent years has been the conversion of several abandoned railroad rights-of-way in eastern Kansas to hiking and biking trails. These trails are now known collectively as the **Flint Hills Trail State Park**. The trails are typically paved within city limits and surfaced with hard-packed crushed limestone in rural areas. Stretching through the Osage Cuestas and Flint Hills for 118 miles from Osawatomie west to Herington, the **Flint Hills Trail** is the longest public hiking trail in Kansas and the eighth-longest rail trail in the US. Development of the trail is ongoing, and as of 2023, at least 100 miles had been fully completed. This trail connects in Ottawa with the 52-mile **Prairie Spirit Trail**, which runs south to Humboldt and connects there with the **Southwind Rail Trail** between Humboldt and Iola. These trails were conceived and are maintained by a variety of state, county, and municipal entities, nonprofit organizations, and local volunteers, spearheaded by the Sunflower Rail-Trails Conservancy. Several segments of these trails stand out as birding hotspots, but you can hardly go wrong exploring the birding possibilities anywhere on them.

One of the best parts of the Flint Hills Trail for birding is the 4-mile segment from the Virginia Road Access near the tiny town of **Rantoul** east to the Pressonville Road Access in Miami County. Most of this stretch of the trail follows the south bank of the Marais des Cygnes River through pristine hardwood forest habitat below tall bluffs. The list of about 150 species that have been seen here consists almost entirely of woodland bird species, including 11 flycatcher, all 7 eastern vireo, 26 warbler, and many other species. This stretch is also noted for its diversity of plants, butterflies, and dragonflies. Some birders leave a car at one access and carpool to the opposite one to avoid having to hike out and back.

On the Prairie Spirit Trail, one of the better locations is the access where **NW Mitchell Road** crosses the trail about 3 miles south of Garnett. This is a heavily wooded area adjacent to the South Fork of Pottawatomie Creek. Walk in either direction for a quarter mile or so. At least 110 species of birds have been seen here despite very few checklists having been reported to eBird. These are almost entirely birds of edge habitats and woodlands. Typical breeding species of the deep woods such as Acadian Flycatcher, Wood Thrush, and Kentucky Warbler are present in summer.

The Southwind Trail for the most part does not traverse productive birding habitat, but the southernmost half mile of the trail from the trailhead in Humboldt on Hawaii Road north past the woodlands adjacent to the **Mount Hope Cemetery** is usually worth the brief amount of time required. The cemetery dates from the nineteenth century, and some of the cedars there are huge and attract birds, especially during migration and in winter.

90. Hillsdale Reservoir

County: Miami

eBird Hotspot names: Hillsdale Lake, Hillsdale Lake—Dam & Outlet Area, Hillsdale WA—Antioch Marsh (north), Hillsdale WA—Antioch Marsh (south), Hillsdale WA—Brown Wetland, Hillsdale WA—Bull Creek-W. 223rd Access, Hillsdale SP, Hillsdale SP—Point Area

Hotspot type: Half day

Hillsdale Reservoir is the closest major Kansas reservoir to the Kansas City metro area. Consequently, it receives a lot of recreational use, including birding. The reservoir was completed in 1982 and has 4,500 surface acres and a maximum depth of 57 feet. Hillsdale has two major branches. The western arm of the lake lies along the channel of Big Bull Creek and the eastern arm along the channel of Little Bull Creek. Hillsdale WA and Hillsdale State Park consist of 12,000 acres of land. At the upper ends of the lake are several wetland areas that attract shorebirds, herons, and other species.

To reach **Hillsdale Dam** and the **State Park** areas from Olathe, travel south on US 169 for 14 miles. Turn right (west) onto W 255th Street and go 2.5 miles, passing through the small town of Hillsdale. Just after the curve are the offices for the state park and the Corps of Engineers as well as the trailhead for the Hidden Spring Nature Trail. Cross the dam. The main entrance to **Hillsdale State Park** will be on your right. Follow the entrance road all the way to the end of the point, where there are multiple campsite loops. This point offers the best unobstructed view of the deep water near the dam. Over 230 species of birds have been seen at or from this state park area. Toe Road is a secondary road parallel to and below the dam road. Short-eared Owl is seen in most years during the winter months, foraging above the grasslands along Toe Road as well as in Hillsdale State Park near Lagoon Road.

To access the east side of the reservoir, return to the east end of the dam and go north 1 mile on Harmony Road, which curves into Tontzville Road (check the lake from the parking lot on the south end of Tontzville Causeway or from the Marysville Boat Ramp in the morning light) for a mile, then curves west into South Gardner

Road. After a half mile, turn south onto Lookout Road to reach the **Point Area** of the state park. The mature woodlands at the Point Area have produced at least 22 species of warblers and many other migratory songbirds in spring and fall, and this is probably the best single location for them at Hillsdale.

The **Antioch Marsh** offers the best opportunity for shorebirds and wading birds at Hillsdale. To reach it from the Gardner exit on I-35, go 5 miles straight south on Gardner Road to W 231st Street, then west for 2 miles. At the second curve, there is a small parking area that is the access point for the Antioch Marsh just to the south, where over 220 species of birds have been seen. Depending on the overall water levels in the lake, good mudflat and wetland habitat might be near the parking area and kayak launch, or you may need to walk south to reach the water's edge. Displaying American Woodcock has been heard at dusk from the parking area. From the parking area, you can also hike north for a half mile along an unimproved maintenance road to view the **North Antioch Marsh**, which is birded far less often but has produced over 170 bird species. From the parking area, continue west for another mile to Spoon Creek Road, then north 1 mile to W 223rd Street, then back east for about a half mile to where the road ends at the **Bull Creek W 223rd Access**, another wetland where about 130 species of birds have been found.

Another productive wetland area at Hillsdale is the **Brown Marsh**, located 3 miles south and 2 miles west of Spring Hill, where W 231st dead-ends near the lakeshore. The wetland is just north and west of the parking area and is sometimes completely dry. When it is holding water, the birding can be quite good. In general, there is more of a grassy/weedy habitat at Brown than at Antioch Marsh. From the parking area, it is only a short walk through some woodlands to the shore of the reservoir, where you can scan parts of the lake that are otherwise difficult to access. Over 210 species of birds have been recorded at the Brown Marsh. Good finds have included Cinnamon Teal, LeConte's Sparrow, and Nelson's Sparrow.

As at other eastern reservoirs, over 250 species of birds have been seen at Hillsdale. Waterfowl include Trumpeter Swan, Black and Surf Scoters, and Long-tailed Duck. Five grebe species have included at least one record of Clark's Grebe. Shorebird species have included multiple rare or uncommon species. Antioch Marsh alone has produced 32 shorebird species, including sought-after species such as Hudsonian and Marbled Godwits, Buff-breasted Sandpiper, Red-necked Phalarope, and Short-billed Dowitcher. Ten gull species have included Iceland and Great Black-backed Gulls. While Hillsdale is not ranked among the truly great destinations for migratory songbirds, good numbers of flycatchers, thrushes, sparrows, vireos, warblers, orioles, and other neotropical migrants are always present during spring and fall migrations.

91. Melvern Reservoir

County: Osage

eBird Hotspot names: Eisenhower SP, Melvern Lake, Melvern Lake—Coeur d'Alene Park, Melvern Lake—Dam & Outlet Park, Melvern Lake—S Hoch Rd Causeway, Melvern WA, Melvern WA—Sundance Marsh, Melvern Lake—Sundance Park, Melvern WA—Willow Marsh

Hotspot type: Half day

Melvern Reservoir is one of the larger Kansas lakes, with a total surface area of 7,000 acres, a maximum depth of 60 feet, and 18,000 acres of adjacent public land. The total length of the reservoir from the dam to the west end of the wildlife area is 16 miles. The reservoir is in the Osage Cuestas physiographic region near the eastern edge of the Flint Hills. In addition to the lake itself, Melvern is noteworthy for the thousands of acres of tallgrass prairies it preserves. There are several developed wetlands and significant areas of mature riparian woodland. Over 250 species of birds have been seen at the hotspots described below. Because of its geographical location, Audubon groups from Burroughs (Kansas City), Topeka, and Wichita all conduct field trips to Melvern.

A typical birding trip to Melvern begins at the dam and Outlet Park, located near the intersection of US 75 and K-31 3 miles west of the town of Melvern or 6 miles north of the US 75 exit on I-35. From the south end of the dam, take the road down to the **Outlet Park** area, where there is a 90-acre lake. If you have only limited time for a visit to Melvern, the Outlet Park is the recommended stop. There are often waterfowl, other swimming birds, gulls, and terns on the outlet lake, and for years it has produced sightings of rare waterfowl and gulls. At least 235 species of birds have been found in the Outlet Park area alone. Follow the park road around the north side of the lake to access the River Bottom hiking trail through a woodland tract where over 20 species of warblers have been seen. The paved road on the south side of the outlet channel ends at another, smaller, shallower lake with a swimming beach. This lake can be a good location for shorebirds, depending on water levels and the time of year.

Return to the south end of **Melvern Dam** and drive north across it as slowly as traffic conditions allow. Diving ducks, grebes, loons, and gulls can all be seen from the dam, and the 30 waterfowl species have included all three scoters; 10 gull species, including Iceland Gull; and a variety of others. Continue north on the dam road to the intersection with K-278 and turn left (west). After 2 miles, turn left (south) onto S Fairlawn Road to enter **Eisenhower State Park**, which offers good views of the lake from the end of the point.

Returning to the south end of the dam, the **Coeur d'Alene Park and Picnic Area**, operated by the Corps of Engineers, offers access to over a half mile of the south shore that lies closest to the dam as well as several areas of grassland and woodland habitat. Nearly 200 species of birds have been seen in this park, many of them waterfowl, gulls, and terns seen from shore, including most of those mentioned above for the dam and outlet areas. The 2-mile Tallgrass Heritage hiking trail begins at Coeur d'Alene and goes west for 2 miles through high-quality prairie habitat to the Arrow Rock Park area.

Farther west on the lake are several productive hotspots north and west of the town of Lebo. From the Lebo exit on I-35, go north for 3.5 miles on Hoch Road to the entrance for **Sundance Park** (operated by the Corps) on the east side of the road. In addition to shoreline access, the park consists largely of brushy and weedy habitat where a variety of sparrows can often be found. Take the park road that goes south out of the Sundance area for a half mile. Where the road curves and crosses Coal Creek, there is some good riparian woodland. Beyond the creek is the **Sun Dance Marsh**, which is worth at least a brief look. Return to Hoch Road and go north across the **Hoch Road Causeway**, from which you can see a long stretch of the lake in both directions and a lot of birds. Parking on the causeway is not permitted, so you will have to settle for scoping from either end. At the north end of the causeway, turn left (west) on 313 Street. After about three-quarters of a mile, turn left (south) on a two-track lane and proceed until it dead-ends near the shore. This area can provide a bonanza of shorebirds during spring migration, when mudflats often extend for a good distance. 313 Street continues west toward Willow Marsh, but in recent years, the road has deteriorated and is now nearly impassable. Return eastward on 313 Street to Hoch Road and turn left (north). Go north another 2 miles on Hoch Road to K-170 and turn left (west). Go 2 miles to Docking Road and turn left again (south). After 1.5 miles, the road ends at the **Willow Marsh**, where 166 species of birds have been found, including good numbers of Virginia Rail and Sora, at least 20 species of shorebirds, Sedge and Marsh Wrens, LeConte's Sparrow, and other wetland species.

The **Melvern WA** covers over 15,000 acres of land, and within it are many miles of roads worth exploring. Use eBird or other map resources to explore them as your time allows. In the large tracts of tallgrass prairie, it is still possible to find Greater Prairie-Chicken as well as other prairie specialists such as Short-eared Owl, Upland Sandpiper, Sedge Wren, and Henslow's Sparrow.

92. John Redmond Reservoir and Flint Hills National Wildlife Refuge

Counties: Coffey and Lyon

eBird Hotspot names: Flint Hills NWR, Flint Hills NWR—Burgess Marsh Trail, Flint Hills NWR—Dove Roost Trail, John Redmond Res., John Redmond Res.—Dam & Outlet Park, John Redmond Res.—Redmond Cove Recreation Area, John Redmond Res.—Riverside Park, New Strawn WTP

Hotspot type: Half day

John Redmond Reservoir is a large federal reservoir in Coffey County that is conjoined with the Flint Hills National Wildlife Refuge. The reservoir covers 9,400 surface acres and has a maximum depth of 12 feet, making it by far the shallowest of all major reservoirs in Kansas. It was created by dam construction on the Neosho River that was completed in 1964. All three of the developed parks at Redmond are operated by the Corps of Engineers. The Flint Hills NWR covers 18,463 acres of land immediately northwest of the reservoir that is separated from the Corps-managed area by a straight line drawn across the lake.

Partly because of the shallow depth of the lake and partly because much of the catchment basin is heavily cultivated, silt inflow into the lake had by 2014 caused the reservoir to lose over 40 percent of its capacity, and a costly dredging effort began that is still ongoing. In the initial phase, three million cubic yards of sediment were removed from the area just behind the dam.

John Redmond is a popular birding destination. Birders and birding groups often combine trips to Melvern with a visit to John Redmond and make a full day of it. The entrance for the **Dam and Outlet Park** is in the town of New Strawn, 12 miles south of I-35 on US 75. Turn right (west) at the south edge of town onto 16th Road, then bear left onto Embankment Road. Near the dam, you can turn into the **Redmond Cove Recreation Area**, which is a good place to scope the lake for birds. Continue across the dam (crossing the spillway), and turn left at the sign for the Outlet Park and Riverside West Campground. The outlet channel should be checked. There is almost always some outflow from the lake; the spillway is exceptionally wide; and there are large concrete pylons in the channel that gulls, terns, and cormorants like to perch on. During migration and in winter, the swirling flocks of gulls are impressive. For many years, John Redmond has had a deserved reputation for attracting rare gulls, and the outlet is the best place to look for them. Check the cormorants carefully. Most are Double-crested, but several Neotropics are usually present, even during the winter months. This is probably the most reliable spot for Neotropic Cormorant in Kansas. Adjacent to the outlet channel is the wooded **Riverside West Campground**, which is sometimes closed to vehicles in the off-season but can always be entered on foot. Over 200 species of birds have been seen here.

One of the preferred stops within the national wildlife refuge is the **Dove Roost Trail**, located on the north side of the lake near the intersection of Garner Road and 18th Road NW. This wooded trail is only a half mile long. Over 150 bird species have been seen from the trail, including a variety of warblers and other woodland birds. There are many miles of roads within the refuge that are worth exploring.

About 3 miles north of Hartford on K-130 are the parking lot and trailhead for the **Burgess Nature Trail**, which is a productive birding hotspot. Water conditions fluctuate greatly at this location. Waterfowl, shorebirds, and wading birds can be abundant when conditions are favorable. The quarter-mile trail goes through a woodland and then opens into a broad area of grasslands, wetlands, and deeper pools. There is a boardwalk over the wetland near the end of the trail. Because of the mix of habitats, over 181 species of birds have been seen at this small hotspot, including numerous sparrow and warbler species. Be sure to check the thick brush along the road that leads to the parking lot.

93. Marais des Cygnes Wildlife Area and Marais des Cygnes National Wildlife Refuge

County: Linn

eBird Hotspot names: Marais des Cygnes NWR, Marais des Cygnes WA, Marais des Cygnes WA—1700th Road, Marais des Cygnes WA—1800th Road Marais des Cygnes WA—Unit B, Marais des Cygnes WA—Unit F, Marais des Cygnes WA—Unit G

Hotspot type: Major destination

Located north of Pleasanton, the 7,650-acre **Marais des Cygnes WA** (managed by the KDWP), and the 7,500-acre **Marais des Cygnes NWR** adjoin each other, preserving a beautiful area of prairie, mature deciduous woodland, wetlands, and transitional habitats. No other single site in Kansas is comparable in terms of the overall habitat diversity that exists at Marais des Cygnes, which includes the largest bottomland hardwood forest in Kansas, extensive grasslands, and numerous managed wetland pools. This is one of the best birding hotspots in Kansas. Regardless of the time of year, a visit here will produce a diverse list of grassland, wetland, and woodland species.

Because it has the greatest diversity of habitat and because there is easier public access, the wildlife area is birded more frequently than the national wildlife refuge. The majority of the wildlife area lies west of US 69, and the national wildlife refuge is generally east of the highway. Begin at the modern rest area at the intersection of US 69 and K-52. At the rest area, be sure to check out the exhibits about this wetland complex. You can pick up a brochure and map from one of the headquarters. The Marais des Cygnes NWR headquarters is located a half mile east of the rest area, and the Marais des Cygnes WA headquarters is located a quarter mile west of the rest area.

One of the most popular areas in the wildlife area is reached by going north on Vale Road a half mile and then west on **E 1700 Road**. You will pass **Unit B** on the southwest side of the road; it is a refuge for waterfowl in winter months but generally requires a spotting scope and patience to sort through distant birds. In 2 miles, just before crossing the railroad tracks, turn north onto Queens Road, which runs below a steep wooded slope parallel to the river. This is a good area to spend some time on foot. For the next mile, it leads through excellent hardwood forest, emerging at another large pool of the wildlife area, **Unit G**. Unit G is encircled by a perimeter road. When it is holding adequate water, it is generally considered to be the best wetland location at Marais des Cygnes. Including the surrounding hardwood forest, Unit G has produced well over 250 species of birds.

Another road frequently taken by birders is reached by going north from the KDWP WA headquarters on Vail Road for 1.5 miles to the intersection with E 1800 Road, where there is a small cemetery. Turn west at this corner and drive for about a mile to a prominent gas pipeline control structure on the south side of the road. The following stretch of **E 1800 Road**, between this pipeline and Trego Road, is another good place to look and listen for birds on foot. As you continue west into the refuge, the forest becomes progressively more mature. When you reach the intersection with Trego Road, you may want to walk farther west on foot or drive south and follow this road as it zig-zags south and west to E 1700 Road.

Hotspot 93: No other single site in Kansas is comparable in terms of overall diversity and quality of wetlands, prairie, and woodlands to Marais des Cygnes Wildlife Area. The mature woodlands along E 1800 Road are ideal for a dawn walk searching for neotropical migrants.

Another good area to explore continues south of E 1700 Road as soon as you cross to the west side of the Marais des Cygnes River and turn left onto 1650 Road. The road passes between the river and Unit A until it dead-ends at US 69. Along this stretch are mature riparian woodlands and several great views of the river.

Much of the national wildlife refuge east of US 69 is difficult to access or closed to the public, but there is some good road birding along the edges of the refuge. The tiny town of Trading Post is reached by a road going south from the rest area. From Trading Post, go east on E 1550 Road. This road is not marked in Trading Post, but turn east at the first opportunity and drive 1.5 miles to the intersection with Yardley Road. During the colder months, Short-eared Owl can be found on this stretch. Go south on Yardley for a half mile to E 1500 Road. There is a trailhead at this corner and an opportunity to explore along a service road on foot. Drive east again for another 1.5 miles to State Line Road. There are forests of varying growth stages on the south side of E 1500 Road, and the creek crossing is a favored stop. At State Line Road, you can turn north or south. There is good birding in both directions. The habitat along these roads changes from year to year because of rotational burning but is often favorable for grassland sparrows such as Henslow's, LeConte's, and Nelson's (rare), depending on the time of year. In summer, the scrubby prairie habitat attracts numerous Bell's Vireo and Yellow-breasted Chat.

More than 300 species of birds have been identified at Marais des Cygnes, and more than 100 of them have been documented as nesting species. During early spring and late fall, concentrations of waterfowl often number in the tens of thousands.

Many of these remain through the winter months, which attracts many Bald Eagles and other raptors. Shorebirds are abundant when one or more of the pools is drawn down, creating extensive mudflat habitat. Thirty-four shorebird species have been found. The hardwood forests at Marais des Cygnes are renowned for attracting many species of migrating flycatchers, vireos, thrushes, warblers, tanagers, and sparrows. Over 30 species of warblers have been found there, including many rarities. Among the nesting species are several eastern warblers, including Yellow-throated, Kentucky, Black-and-white, and American Redstart. Other summer species include American Woodcock, Whip-poor-will, Chuck-will's-widow, White-eyed and Yellow-throated Vireos, and Summer Tanager.

About 15 miles southwest of Pleasanton, the **Dingus Natural Area**, located near Mound City, preserves a superb tract of old hardwood forest. It is owned by the KOS, and birders are welcome to wander the 167-acre area freely, although there are no trails or other amenities. To reach the Dingus Natural Area from Mound City, proceed south on K-52 for 1 mile and turn right and then west onto W 750 Road. Go west 1 mile to the T intersection and proceed left up a steep hill on Lane Road to Lewis Road. Turn right on Lewis Road and follow it as it curves left to become W 750 Road again. Continue for 1 mile. After crossing under a high-voltage line, you will see a sign marking the area on your right. Another mile west is Mound City Lake.

94. Lehigh Portland State Park

County: Allen

eBird Hotspot names: Lake Bassola, Lehigh Portland Trails, Lehigh Portland Trails—Western Trailhead

Hotspot type: Worth a stop

Lehigh Portland State Park is a unique birding hotspot located at the southern edge of Iola in Allen County. The property includes a large quarry lake. It was the site of the Lehigh Cement Plant, which closed in 1970. The public Lehigh Portland Trails property was created in 2014 by local volunteers on land owned by Iola Industries through an easement granted to the Sunflower Rail-Trails Conservancy. In early 2023, legislation was passed to make it the newest Kansas state park, which will be formally known as **Lehigh Portland State Park**. At the time of this book's publication, the long-term KDWP plan for this new park had not been developed. It is anticipated that the existing amenities, access, and trails will be improved and enhanced. The site description given here is current as of 2023 and will change as the transition to a state park is implemented. Current information will be available at ksoutdoors.com as the park is developed over time.

As of 2023, there were over 13 miles of hiking trails, including the secondary single-track trails. The current primary trailhead at the east edge of the property is reached from the US 169/US 54 interchange by going east for a half mile to 1800 Street, south a half mile to Nevada Road, west a half mile to 1700 Street, south a half mile to Nebraska Road, and then west a half mile to the parking lot. Multiple trails begin at the parking lot. The Backbone Trail is 2.4 miles in length. It generally follows Elm Creek along the north side of the property through good riparian woodland and ends at the Western Trailhead on State Street near Lake Bassola. The South Loop Trail features unobstructed views of the Quarry Lake from atop 30-foot bluffs. As of 2023, consult the website at lehightrails.com for a complete map of the extensive trail system and other information.

Nearby Lake Bassola is adjacent to the western edge of the public land and can be viewed from public roads. From the Western Trailhead, go south on State Street to Bassett Street and turn left (east) to follow the road around the lake. The lake is privately owned. Bird the lake from the roadside. This is a smaller and shallower quarry lake where about 100 bird species have been reported.

Despite Lehigh Portland Trails having been open to the public only since 2014, 182 species of birds have been reported there. The 25 waterfowl species have included less common species such as Trumpeter Swan, White-winged Scoter, and Long-tailed Duck. Canada Goose, Bufflehead, and Common Goldeneye are common in winter. However, geese and ducks are typically scarce on Bassola and Quarry Lakes. There is no shorebird habitat; consequently, virtually no shorebirds have ever been reported. Gulls and terns are also decidedly scarce or absent. The primary species of interest at this hotspot are woodland and edge habitat species. Expect all of the usual eastern woodland species according to the season. Seven vireo and 26 warbler species have been found here, primarily in May and September.

95. Elm Creek Lake

County: Bourbon

eBird Hotspot names: Elm Creek Lake, Hollister WA

Hotspot type: Worth a stop

Elm Creek Lake is one of the lakes constructed during the 1930s by the WPA and the CCC. The lake covers 100 surface acres when full. Elm Creek is not heavily used, and you often have the place mostly to yourself.

From the junction of K-7 and K-39 in southern Bourbon County, go west for 2 miles to 140th Street, then north for 1.5 miles. From the corner of 140th and Fern Road, you can either go north to access the east side of the lake or west for a half mile to the west entrance on 150th. Both roads connect with Grand Road on the north edge of the property. Park roads allow easy viewing access to the entire lake, and Grand Road crosses the top of the dam. The habitat around the lake is a mix of mowed park areas and native grass. There are modest-sized woodlands below the dam and at the upper end of the lake. Water levels fluctuate at Elm Creek, and on occasion, there are shallows and mudflats that attract shorebirds and other wetland species.

At least 185 species of birds have been found at Elm Creek. The 20 waterfowl species have included Greater Scaup and Black Scoter. Seventeen shorebird species have included American Golden-Plover and Hudsonian Godwit. Other noteworthy aquatic species have included Caspian and Least Terns, Neotropic Cormorant, and White-faced Ibis. In addition to the expected prairie species, the native grasslands around the lake attract Henslow's Sparrow in summer. Henslow's have attempted to breed at the lake several times, but unfortunately timed mowing by the county has sometimes destroyed the nesting habitat. Listen for their quiet vocalizations. Smith's Longspur has been found in late winter. During fall migration, be alert for grassland specialists such as Sedge Wren and LeConte's Sparrow. Despite the relative scarcity of habitat, about a dozen species of warblers have been seen.

Hollister WA covers 2,400 acres and is located immediately east of Elm Creek Lake. The mile of 140th between Fern and Grand is on the boundary between these two public lands. Hollister is 3 miles long from north to south and over 2 miles from east to west. The habitat consists of grassland interspersed with some managed cropland. There are mature riparian woodlands along Elm and Pawnee Creeks. All access to the area is on foot. There are parking areas on Fern, Grand, and Hackberry Roads. Downloading the Hollister map from ksoutdoors.com is suggested. Hackberry Road bisects the area from east to west and allows you to sample the birdlife of the area without expending too much time. There are no bodies of water at Hollister, so the expected species are entirely land birds. The area is seldom visited by birders but certainly has a lot of birding potential.

96. Neosho Wildlife Area

County: Neosho

eBird Hotspot names: Neosho WA, Neosho WA—Udall Rd. Access

Hotspot type: Half day

Covering 3,246-acres, **Neosho WA** is located just east of the town of St. Paul in Neosho County. Developed in 1960, the area includes over 1,700 acres of man-made wetlands that are replenished by pumping and are separated by levees into management pools. The area is managed to attract waterfowl, and there is considerable seasonal and long-term habitat manipulation to achieve this goal. There is a substantial amount of hunting activity during waterfowl season. The extensive water lotus and partially submerged dead trees combine to create the general appearance of a southern swampland.

To reach Neosho WA, proceed east from the town of St. Paul on K-57 for 1 mile. The entrance is on the south side of the highway. The 5-mile perimeter road begins there and encircles the wetland and lake area, eventually returning to K-47. On the west and south sides of the loop, the road is flanked by riparian woodland along Flat Rock Creek and the Neosho River. You can use the trails on top of the levees to access wetland areas within the perimeter road on foot. Pool #3 covers most of the eastern portion of the wetland area and is designated as a refuge area closed to all activities beyond the perimeter road from September 1 to March 31.

An impressive total of nearly 300 species of birds has been recorded at this birding hotspot. Like other Kansas wetlands, Neosho harbors large mixed flocks of waterfowl in early spring and late fall, and many remain for the winter. Twenty-nine waterfowl species have been reported. Trumpeter Swan is seen in most winters, sometimes accompanied by a few Tundra Swans. The extent of shorebird habitat is somewhat variable, but in most years, there are some shallows and mudflats. Thirty-four shore-

Hotspot 96: Neosho Wildlife Area is a man-made wetland that is managed to attract migratory waterfowl. It attracts large concentrations of ducks and geese, including this flock of Snow Geese.

bird species have been found at Neosho, a diversity equal to that of the better-known central Kansas wetlands. Noteworthy shorebirds have included Piping Plover, Snowy Plover, Whimbrel, Long-billed Curlew, Ruddy Turnstone, Dunlin, Buff-breasted Sandpiper, and many others. The refuge attracts herons and ibis, with 14 species having been observed, including American and Least Bitterns and Glossy Ibis. In late summer, large numbers of postbreeding egrets are present. Look through the numerous Double-crested Cormorants that roost in the dead trees for rare but regular Neotropic Cormorant.

The north end of the refuge has large areas of wetland grasses, including prairie cordgrass. In late fall and winter, this is a good place to look for Sedge and Marsh Wrens as well as wetland sparrows such as Swamp and LeConte's among the more common sparrow species. A large colony of nesting Tree Swallow inhabits the dead trees in the southern pool in summer.

The woodlands on the west and south sides of the perimeter road harbor Red-shouldered Hawk, Barred Owl, Pileated Woodpecker, Carolina Wren, Tufted Titmouse, and other typical species of the eastern woodlands year-round. Summer residents include Acadian Flycatcher, Yellow-throated Vireo, Prothonotary Warbler, Northern Parula, and Summer Tanager. There are also a few summer records for Scarlet Tanager. Rusty Blackbird is increasingly rare in Kansas, but Neosho still has a wintering population from November through March in most years. Other winter residents include Hermit Thrush, 12 sparrow species, and Purple Finch in some years.

While woodland species are often seen in the woods along the perimeter road, the best part of the wildlife area for them is reached via Udall Road near the St. Paul city limits. Take this road 1 mile south to the river, where is a slight jog in the road. Park here at the **Udall Road Access**, where there is a parking area and signage. and follow the foot trail eastward along the levee for Pool 6. This is an excellent site for all the woodland species listed above. A total of 29 species of warblers have been found at Neosho, many of them in this part of the wildlife area.

97. Bone Creek Lake

County: Crawford

eBird Hotspot names: Bone Creek Lake

Hotspot type: Worth a stop

Two reservoirs with surrounding public lands in northeastern Crawford County are worthwhile birding stops. These are both located at the edge of the Cherokee Lowlands physiographic region where it meets the Osage Cuestas physiographic region.

Bone Creek Lake is one of the newest artificial lakes in Kansas, having been completed in 2000. The lake covers 540 surface acres and provides drinking water for much of Crawford and Cherokee Counties. To reach the lake from Arma, go 5 miles north on US 69 and then 2 miles west on E 700th Avenue to N 200th Street. The dam is immediately south of this intersection. There is an access road on the north end of the dam where you can view the lake. Follow 200th Street south past the dam until you reach a T intersection. Go north a short distance to where there is a boat ramp, another good place for viewing the lake. Several side roads lead to the lake in the vicinity of the boat ramp, most notably E 690th Avenue, which goes west for 2 miles before ending at the lakeshore. There is no other easy access to the lake, and birding this hotspot requires only an hour or two.

Nearly 200 species of birds have been found at Bone Creek. The primary birds of interest are waterfowl and other swimming bird species, but the number of grassland and woodland species that have been found here is respectable and includes numerous migratory songbirds. Red-breasted Merganser, Horned Grebe, and Common Loon are found in most years. Notable finds at Bone Creek have included Trumpeter Swan, Surf Scoter, Long-tailed Duck, Caspian Tern, Black Vulture, and 14 species of warblers.

98. Crawford State Park

County: Crawford

eBird Hotspot name: Crawford SP

Hotspot type: Worth a stop

In contrast to recently built Bone Creek Lake, **Crawford State Park** is among the oldest lakes in Kansas, having been built by the CCC in the 1930s. This state park is only 5 miles west and 2 miles north of the dam at Bone Creek. The park entrance is via K-277, 1 mile east of K-7. The park covers 530 acres, including the 150-acre lake. Most of the park property consists of mature oak/hickory woodland. Redbud trees are widespread in the park and put on a show in early April when in bloom. The lake is encircled by park roads and campgrounds, allowing easy viewing of the entire lake. Located below the dam, the Drywood Creek hiking trail is a loop through the largest area of undeveloped woodland in the park and usually offers some of the best birding. There are several other stands of undeveloped woodland on both sides of the lake that are all worth checking as your time allows.

While this park has a respectable list of waterfowl and other aquatic species seen on the lake, the emphasis at Crawford State Park is decidedly on woodland birds. Of the 200 species that have been reported at the park, there are 11 flycatcher, 7 vireo, 6 thrush, 17 sparrow, and 25 warbler species. Migratory warbler sightings have included sought-after species such as Golden-winged, Mourning, Blackburnian, Chestnut-sided, and Canada Warblers. Warbler species that remain to nest are typical of eastern Kansas: Louisiana Waterthrush, Northern Parula, Prothonotary Warbler, and Kentucky Warbler. Other summer species include Ruby-throated Hummingbird, Acadian Flycatcher, Yellow-throated Vireo, Summer Tanager, and abundant Indigo Bunting.

99. Elk City Lake

County: Montgomery

eBird Hotspot names: Elk City Lake, Elk City Lake—Overlook Park, Elk City Lake—Outlet Park, Elk City SP, Elk City WA, Elk City WA—Card Creek PUA, Elk City WA—Squaw Creek South PUA, Elk City WA—Wigeon Marsh, Montgomery SFL

Hotspot type: Half day

One of the largest reservoirs in southeastern Kansas, **Elk City Lake** covers 4,500 surface acres. It was created by the construction of a dam on the Elk River in 1966. The lake is relatively shallow, with a maximum depth of 24 feet. The 850-acre Elk City State Park is located on the eastern shore of the lake near the dam. The Corps of Engineers operates three additional public use areas. The 12,240-acre Elk City WA is managed by the KDWP, and an additional 1,600 acres are managed for wildlife by the Corps. The wildlife area extends over 12 miles westward from the dam on the Elk River and includes substantial wetland areas west of Elk City. Elk City Lake has prominent geological features and extensive oak/hickory forests, making it one of most visually appealing public lands in southeastern Kansas. It is also renowned for its nearly 20 miles of hiking trails, two of which have National Trails designation.

The lake is less than 2 miles west of the Independence city limits. Take County Road 4600 west from town to enter the **Squaw Creek South Public Use Area**, which is at the shallow southeastern corner of the reservoir. From there, you can follow the park road north along the shore to **Elk City State Park** and the **Squaw Creek North Public Use Area**. In addition to providing views of the lake, there is a lot of good birding habitat in the Squaw Creek public use area and the state park. The trailhead for the Table Mound Hiking Trail is at the north end of the state park. Exit the park via the north entrance and turn left (north) onto County Road 3300. In about 2 miles, you will reach **Overlook Park**, the trailhead for the Post Oak Trail, and the dam. The Memorial Overlook provides a good view of the lake and is surrounded by hardwood forest. Cross the dam on County Road 5050 to the access road for the

Outlet Park, where Black Vulture is sometimes found roosting along the channel in the warm months.

Another popular Elk City Lake birding destination is the **Card Creek Public Use Area** on the south side of the reservoir. To reach it from Independence, take US 160 west from the city limits to the sign for Card Creek at County Road 2500 and turn north. This road passes through a lot of brushy edge habitat before reaching the wooded campground area. Bell's and White-eyed Vireos, Yellow-breasted Chat, Eastern Towhee, Blue Grosbeak, and Painted Bunting are all possible in summer on this stretch.

Return to US 160 and continue west and north for 12 miles to the small town of Elk City. Turn left (west) at the gas station onto Hickory Street and drive west 2 miles to the intersection with 1500 Road. Turn right, and after another mile, you will reach the parking area for the **Wigeon Marsh**. This extensive wetland can be explored on foot using the paths on the levees. If water levels are favorable, it is worth taking the time to do so. Over 20 species of shorebirds have been seen at the marsh. Good numbers of dabbling ducks, rails, herons, and other wetland species are often found as well. LeConte's and Nelson's Sparrows have been reported in October. Short-eared Owl is possible in winter at the marsh.

Overall, 259 species have been reported from Montgomery County, and most of these have been found in or near the reservoir. Dabbling ducks are often abundant at Wigeon Marsh. On the main body of the lake, numbers of waterfowl often seem lower than at most other major Kansas reservoirs. Eastern woodland species are common at all seasons. Elk City is one of the most likely places to find Black Vulture in Kansas, especially near the dam. Several late-spring and summer sightings of Broad-winged Hawk suggest that a few could remain to breed in the area. Expected summer species include Chuck-will's-widow, Whip-poor-will, White-eyed and Yellow-throated Vireos, Wood Thrush, Kentucky and Prothonotary Warblers, Summer Tanager, and many others. A few western species, including Cinnamon Teal and Western Grebe, have been seen at Elk City. Twenty-five species of warblers have been seen during migration, especially in May and September.

Located 5 miles south of Independence, **Montgomery SFL** is a popular nearby birding hotspot where at least 170 species of birds have been found. The lake covers 105 acres, and there are 300 acres of public land, almost all of which is oak/hickory forest. This location is easy to explore and usually offers good birding potential.

100. Cherokee County Area

County: Cherokee

eBird Hotspot names: Cherokee Lowlands WA—Chestnutt Property, Empire Lake (Riverton), Cherokee Lowlands WA—Michelson Property, Schermerhorn Park & Southeast Kansas Nature Center, Spring River WA

Hotspot type: Full day

Cherokee County in the southeastern corner of Kansas offers an exceptional variety of habitats and birding opportunities. A full day can easily be spent birding the hotspots described here. The heavily forested Ozark Plateau physiographic region barely grazes the southeastern corner of Cherokee County. The remainder of the county lies in the Cherokee Lowlands physiographic region, which covers about 1,000 square miles of Kansas in portions of Bourbon, Cherokee, Crawford, and Labette Counties.

Schermerhorn Park, located along Shoal Creek near the town of Galena, is a longtime favorite birding location in the Ozark Plateau where almost 200 species of birds have been seen. During April and May, the park is one of the best places in Kansas to observe migratory eastern songbirds such as flycatchers, vireos, thrushes, and warblers. During spring migration, over 20 species of warbler have been seen in a single day, and 30 species of warblers have been seen at this small park. Summer residents include four vireo and at least eight warbler species. This is a reliable location for Yellow-throated Warbler, which nests here annually. Listen for the nasal caws of Fish Crow along Shoal Creek. Black Vulture has become an established resident, including a few that remain throughout the winter. Winter birds are typical of eastern Kansas woodlands, often including Purple Finch. Mississippi Kite is uncommon and local in southeastern Kansas but is seen regularly in summer soaring above the park or in nearby Galena.

To reach the park, take K-26 south out of Galena for 1.5 miles. The park entrance is on the east side of the highway, just north of the bridge over Shoal Creek. Park and continue on foot to explore both the area along the creek and the upland forests on the bluffs above. The **Southeast Kansas Nature Center** in the park features exhibits of animals and plants native to the area. Environmental education classes and workshops are offered each month. A large observation window offers views of water features designed to attract birds.

Located 2 miles west of Galena, **Empire Lake** near Riverton is a good place to look for waterfowl, grebes, gulls, and other waterbirds as well as terrestrial species in the adjacent woodlands. The area near the Boater's Clubhouse on the east shore offers a good vantage point of the lake, and there are often birds in the outlet channel below the dam.

The 424-acre **Spring River WA** is located 7 miles north and 3 miles east of Riverton. It is the largest publicly owned tract of preserved natural habitat within the Ozark Plateau physiographic region in Kansas and is dominated by a sandstone bluff overlooking the Spring River. There is abundant oak/hickory forest as well as a large prairie area. The area is seldom visited, and there are few developed trails, but there is high potential for quality birding, especially for woodland species. Downloading a map from ks.outdoors.com is suggested, as there is minimal signage.

Covering 2,110 acres in western Cherokee County, the **Cherokee Lowlands WA** is managed by the KDWP. The wildlife area consists of multiple properties that were purchased as part of the settlement for environmental damage caused by lead and zinc mining in Cherokee County. The wildlife area consists of several disjunct tracts of public land that include extensive wetlands, grasslands, and woodlands. While some decent roadside birding is possible, the most productive birding is on foot. Among the nearly 200 species of birds that have been observed are many species of waterfowl, herons, shorebirds, rails. wrens, and wetland sparrow species, including Nelson's Spar-

Hotspot 100: In the corner of southeast Kansas along Shoal Creek, Schermerhorn Park is a longtime favorite birding location in the Ozark Plateau. At the center of the park, the Southeast Kansas Nature Center features exhibits on local plants and animals.

row. In the winter months, be alert for Short-eared Owl at dawn and dusk. The 720-acre **Chestnutt property** is located 3 miles east and 1 mile north of Chetopa along SW 107th Terrace. The 400-acre **Michelson tract** is 1 mile farther north near the intersection of SW Blackjack Road and SW 110th Street. Two miles to the north and west, The Nature Conservancy in Kansas owns and the KDWP manages the 690-acre **Bogner Tract** located south of SW Quaker Road between SW 110th Street and SW 120th Street. Bogner consists mostly of land enrolled in the Wetland Preserve Program. Because of its berms and other water management features, the Bogner tract offers the best habitat and birding possibilities. Impressive numbers of shorebirds can be found during migration and have included uncommon species such as Hudsonian Godwit and Dunlin.

Birding Resources

Recommended Books

If this book has whetted your appetite to learn more about birds in Kansas, *Birds of Kansas* (2011) by Thompson et al. is the most authoritative book covering all the scientific and distributional information for every bird species (473 when published) documented in Kansas. If you become immersed in Kansas birding, you will want a copy of this book.

Another important book is *The Kansas Breeding Bird Atlas* (2001) by William Busby and John Zimmerman. The atlas was a massive six-year effort to map the distribution and populations of breeding birds throughout the state. The results of the project are summarized in this book, which has separate accounts and range maps for each species.

Even though it was published over 35 years ago, *A Guide to Bird Finding in Kansas and Western Missouri* (1988) by Sebastian Patti and John Zimmerman has aged quite well. It has an in-depth discussion of Kansas physiographic regions and biomes. The seasonal occurrence bar graphs are useful and are not duplicated in any other book devoted to birds in Kansas. The birding location accounts in the book (some discussed in this book as well) are somewhat dated but remain useful.

Less well known but certainly the most consciousness-raising book devoted to Kansas ornithology is *The Birds of Konza: The Avian Ecology of the Tallgrass Prairie* (1993) by the late John Zimmerman. Dr. Zimmerman had a keen intellect and devoted much of his academic career to bird studies at Konza. This fascinating book is the culminating summary of the insights he garnered during his decades of research and study.

For additional information on many of the birding hotspots discussed in this book, *Watching Kansas Wildlife* (1993) by Bob Gress and George Potts is another book you will find useful when exploring natural areas of Kansas. While not geared strictly toward birders, it has useful information pertaining to exploring wildlife areas throughout the state. This is another older book that retains its usefulness.

Printed maps are becoming obsolete, but many birders still carry a copy of the *Kansas Atlas & Gazetteer* published by DeLorme. This eighty-page atlas has highly detailed maps of the state that are invaluable when exploring the back roads of Kansas. Additionally, the KDWP publishes hunting and fishing atlases each year showing public hunting and fishing areas. These have detailed maps of the entire state that are useful navigational aids.

Birding Apps

Almost everyone has a smartphone, iPad, or similar device used to compute, communicate, photograph, navigate, record, or play music. It comes as no surprise that this technology provides many opportunities for birders. There are app versions of field guides as well as apps useful for learning birdsongs and calls. There are apps for recording birds and finding birds, and some are used to identify photos and sounds of birds. Below are some of the most popular ones.

eBird

This free birding app is used by most birders. See the introduction to the Hotspots section of this book for a more detailed discussion of eBird. Created by the Cornell Lab of Ornithology, it has revolutionized birding. As you enter your bird data, it keeps track of your species lists, annual lists, and locations and enables you to include your bird photos and sound recordings. It can organize and keep track of your data in many ways. With the companion phone app, you can record your sightings in real time and upload them instantly. Another eBird tool is the Rare Bird Alert that you can set for your state, your county, or even your favorite hotspot. The Needs Alert notifies you when a bird you have not seen before is reported in your designated area. eBird is the world's largest citizen science project and combines your birding data with the data submitted by millions of other users to create a massive database of information. If you become serious about birding, you will most likely become an eBird user.

Merlin Bird ID

Do you need assistance identifying what bird you are hearing or what bird you have photographed? Merlin Bird ID by Cornell Lab of Ornithology accesses the massive eBird database to help you identify what you are seeing or hearing. Using Merlin's Sound ID tool, your cell phone will listen to singing birds and make identification suggestions for the birds you are hearing. You can also upload photos for identification, and it is amazingly accurate. Merlin also provides information on identification, range maps, species photos, and the calls and songs of most species. Merlin is not perfect and does make identification mistakes, but its accuracy is continually improving.

iBird Pro Guide to Birds

This app is available only for iPhones and is a self-contained app that functions as a field guide to the birds of North America. It features illustrations, photos, field marks, maps, and over four thousand songs. Be aware that this app uses a substantial amount of memory storage.

Sibley Birds, second edition

This is the app version of *The Sibley Guide to Birds*, second edition. It includes all the artwork from David Sibley's book as well as descriptions and distribution maps. In addition, it has recordings of over 2,700 bird calls and songs covering most species. A useful feature allows you to compare similar bird species side by side on the same screen.

Audubon Bird Guide

This is another free app. It covers over 800 species and includes over 3,000 photos and audio clips of songs and calls. It includes range maps and text by bird expert Kenn Kaufman.

The Warbler Guide

This app is devoted entirely to warblers and features illustrations, sounds, and a unique 3D model that enables you to rotate bird images in any direction to study an angle that might replicate your field view. It also shows comparisons to similar species.

Raptor ID

Raptor ID, created by Hawkwatch International, is devoted to the thirty-four species of raptors in North America. It helps with identification of raptors in flight and covers subspecies and color morphs to help sort through the sometimes confusing diversity of birds of prey.

BirdsEye Bird Finding Guide

This app interfaces with eBird and uses your phone's GPS to track your location and report bird sightings in your area over a range of 1–50 miles. It is designed to help you find birds and to target specific birds you are seeking. You can use it to create a map showing where your target birds have recently been seen. If you use eBird, you can import your eBird sightings and customize your search for birds.

Birding Organizations

This is a summary of birding clubs and organizations relevant to Kansas birding. For each, we have provided a mailing address and/or internet URL, if available. All the contact information below is current as of 2024 and is subject to change.

American Birding Association (ABA)

The ABA is a national nonprofit organization of birders that has many members in Kansas. It represents the North American birding community and supports birders through publications, workshops, tours, partnerships, and networks. The ABA publishes three periodicals invaluable to birders. It has an ever-expanding menu of internet resources that is too diverse to detail here.

Website: www.aba.org
Email: info@aba.org
Address: American Birding Association, PO Box 3070, Colorado Springs, CO 80934
Telephone: 800-850-2473

Kansas Ornithological Society (KOS)

The KOS is a statewide organization devoted exclusively to birds. Its members range from backyard and amateur birdwatchers to professional ornithologists. It informally serves as an umbrella organization for Kansas birders and birding groups as well as professionally for students and academics. A newsletter and a scientific, peer-reviewed journal are published quarterly. A weekend spring field trip is usually held in early May at some location in the state. The annual fall meeting is usually held in early October, frequently at a college or university. At the fall meeting, academicians and members present papers on current research primarily on Kansas birds and take half-day field trips. The Kansas Bird Records Committee receives reports on rare birds and maintains and publishes the checklist of Kansas birds. It coordinates and publishes results of Christmas Bird Counts held across the state. The KOS provides research funds for projects conducting research on Kansas birds. A book royalty fund is maintained to help support publishing books on Kansas birds. The society owns and maintains the Dingus Natural Area in Linn County. The website, https://ksbirds.org, contains a great deal of information on Kansas birds and the history of the KOS.

Website: https://ksbirds.org
Facebook: www.facebook.com/KSOrnithologicalSociety
Email: Changes annually. See website for current address.

Audubon of Kansas (AOK)

Audubon of Kansas is an independent grassroots nonprofit organization that defends wildlife and wild areas through advocacy, conservation, and education. The AOK's emphasis is on conservation in Kansas and the central Great Plains. Members and supporters are from across Kansas and beyond. The AOK is governed by a board of trustees consisting of conservation leaders, scientists, farmers, ranchers, bird watchers, and others. It is neither funded nor administered by the National Audubon Society but works in partnership with local Audubon chapters and many other organizations.

The AOK's advocacy gives wildlife a voice when legislation and policy are being proposed at county, state, and federal levels. Its conservation goal is to show how people and wildlife can coexist through its sanctuary program. The AOK's education program includes monthly newsletters, a yearly magazine called *Prairie Wings*, public seminars on conservation topics, and sponsoring two annual birding festivals. The Celebration of Cranes is held at Quivira NWR during the first weekend of November. The Kansas Lek Treks Prairie-Chicken Festival is held in mid-April on the prairies of north-central and western Kansas.

Website: www.audubonofkansas.org
Facebook: www.facebook.com/audubonofkansas
Email: aok@audubonofkansas.org
Address: Audubon of Kansas, PO Box 1106, Manhattan, KS 66505-1106
Phone: 785-537-4385

National Audubon Society Chapters

Kansas chapters of the National Audubon Society offer informative meetings and field trips throughout the year. Meetings and trips are open to the public, do not require membership, and are usually free. The following contact information is subject to change.

Burroughs Audubon Society (Kansas City), www.burroughs.org
Lawrence Bird Alliance (Lawrence), www.lawrencebirdalliance.org
Northern Flint Hills Audubon Society (Manhattan), www.nfhas.org
Smoky Hills Audubon Society (Salina), www.smokyhillsaudubon.com
Southeast Kansas Audubon Society (Parsons)
Sperry-Galligar Audubon Society (Pittsburg), www.sperry-galligar.com
Sunflower, A Chapter of Audubon (Hays)
Topeka Audubon Society (Topeka), www.topekaaudubonsociety.org
Wichita Audubon Society (Wichita), www.wichitaaudubon.org

KSBIRD-L Listserv and Archives

While the rise of multiple social media platforms has led to the general decline of email listservs, many Kansas birders still subscribe to this free listserv devoted to Kansas birds. KSBIRD is a useful tool. Use it to become acquainted with birders across the state and to learn more about birds and birding. The web address for the KSBIRD Listserv archives is http://listserv.ksu.edu/archives/ksbird-l.html, from which you can access the subscription web page. Use the archives by typing the name of a bird species, birding location, or related topic that you would like to learn more about. You will be able to access prior discussions on the topic that in some cases date back to the early 1990s. As of 2023, the listserv had about seven hundred subscribers.

Social Media
Facebook

In the ever-changing landscape of social media, there are numerous available platforms, many of which offer bird-related content. These include Instagram, WhatsApp, Twitter, Twitch, TikTok, YouTube, and more. It is impossible to project what the future holds for social media, and the information that we are including here is likely to be obsolete if you are reading this book even just a few years after it has been published. With that said, there are several active groups and pages specifically devoted to Kansas birding topics. In terms of membership, the Kansas Birding group has the most members and most posts per day. As of mid-2023, it had nearly 10,000 members and averaged over twenty posts per day. This is the unofficial umbrella Facebook group for Kansas birders. It has a broad spectrum of members in terms of birding experience and interests. The Kansas Bird Photography group has over 2,500 members and averages eight posts per day. The Kansas Rare Birds and Notable Sightings group has nearly one thousand members and focuses on bird sightings that are noteworthy in some aspect, averaging less than one post per day. The Kansas County Listers group is much smaller, with fewer than 150 members who enjoy traveling the state and adding bird species to their individual county life lists. The ABA, KOS, AOK, and several of the local Audubon chapters, including the Burroughs, Topeka, and Wichita chapters, all have their own Facebook groups as well.

Other Internet Resources
Kansas Birding Trail

The well-conceived and -designed Kansas Birding Trail was launched in 2021 by the KDWP. If you have found the hotspot section of this book useful, this website is for you! The home page for the project is www.ksbirdingtrail.com. There are individual web pages for twelve birding trails located throughout Kansas. In many instances, these trails include birding hotspots discussed in this book. The trails all meet the following criteria: they 1) can be completed in one day, 2) are anchored in cities with standard amenities, 3) include up to ten specific birding locations, 4) can be entered or exited at any point, and 5) use all paved roads. From a map of the state, you can click on the icon for any of the twelve trails. There you will find a map of the overall route for that trail and a menu of the birding locations on it. There is a link for each birding location on the trail that includes a detailed discussion of the location and the birds you can expect to find there. At the end of each location discussion are links to websites with additional information. These always include the eBird list for the location, links to site-specific websites such as the KDWP and/or Corps of Engineers, and other site-specific web pages. Because it is entirely web-based, the embedded links and other content will remain current over time.

Kansas Department of Wildlife and Parks

https://ksoutdoors.com
In addition to the Birding Trails resource, the KDWP provides a great deal of additional useful information for all fishing lakes, state parks, and wildlife areas that it manages. From the three pages listed below, you can navigate to specific locations and download PDF files of the area information and maps for each of these locations statewide.

State fishing lakes information: www.ksoutdoors.com/State-Fishing-Lakes
State parks information: www.ksoutdoors.com/State-Parks
Wildlife area information:
www.ksoutdoors.com/KDWP-Info/Locations/Wildlife-Areas

Kansas County Checklist Project

This is a page linked to the KOS website: www.ksbirds.org/checklist/checklist_index
There is a huge amount of information available from this page that is too extensive to fully discuss here. Much of the distributional information used in this book, as well as the range maps used in the *Birds of Kansas* book (2011), came directly from the data available on this page. It is updated twice a year. For each species, there is a county map of the state with dots on those counties where a species has occurred and different dot symbols for counties with confirmed breeding records of that species. There are links to checklists for each of the 105 counties, listing every species of bird ever observed in each county and which of those species have had confirmed breeding records. Try clicking on your home county. You will be surprised at how many species of birds have been seen there! The distributional information on this page is more comprehensive than the information available on eBird because it is drawn from records going back as far as the nineteenth century, while eBird records are in general less than twenty years old, and not all birders submit their sightings to eBird.

Natural Kansas

http://www.naturalkansas.org
This is a great omnibus website with an exceptional number of useful features and links. It has a wealth of information on all aspects of the natural world in Kansas, including outstanding wildlife viewing sites, hiking trails, driving tours, and upcoming events related to wildlife viewing. Bookmark this one.

Playa Lakes Joint Venture

https://pljv.org

The Playa Lakes Joint Venture (PLJV) is a nonprofit organization that is "dedicated to conserving the playas, prairies and landscapes of the western Great Plains to benefit birds, other wildlife, and people." The PLJV is an umbrella group that coordinates the efforts of multiple organizations and agencies. At this website, you can learn a great deal about playas and the work of the PLJV. The website is well designed, with many interesting links. Of most interest to birders are downloadable county maps in PDF format that show all potential playa lakes. Many ardent birders traveling to a given county print that county's playa map. Most or all of these playas are on private land, but many can be viewed from public roads. These are classified only as likely locations for playas. In years with minimal rainfall, they often have no water, but when there has been recent local rain and conditions are right, surprising numbers of shorebirds and other water-dependent bird species can occur. The playa maps are a great navigational tool for exploring these unique wetlands.

BirdsInFocus Bird Photography

www.BirdsInFocus.com

If you love birds, then check out BirdsInFocus! This book draws heavily from the BirdsInFocus photo collection, but there is much more to see. Bob Gress, Judd Patterson, and David Seibel routinely explore the planet in search of new birds, and their site currently shares more than 16,000 photos of over 2,600 species from 38 countries.

Seasonal Distribution of Kansas Warblers

Key located at bottom of table

Species	Status	Expected Location	January	February	March	April	May	June	July	August	September	October	November	December
Ovenbird	C trans	Statewide						* * * * *	* * * * *	* * * *				
Worm-eating Warbler	R trans	E						* * * * *	* * * * *	* * * * *	* * * * *	* *		
Louisiana Waterthrush	U trans, Su res	E ½												
Northern Waterthrush	U trans	Statewide												
Golden-winged Warbler	R trans	E ¼					* * * *			* *	* * * *			
Blue-winged Warbler	R trans	E ⅛				*	* * * * *			* *	* * * * *	*		
Black-and-white Warbler	C trans, Su res	Statewide												
Prothonotary Warbler	U trans, Su res	E ½												
Swainson's Warbler	Vagrant	9 records					* * * * *	*						
Tennessee Warbler	C trans	Statewide												
Orange-crowned Warbler	C trans, R Wi res	Statewide	* * * *	* * * * *	* * * * *	*						*	* * * * *	* * * * *
Nashville Warbler	C trans	Statewide												
Virginia's Warbler	R trans	southwest					* *	* * * * *		*	* * * * *			
Connecticut Warbler	R trans	E ⅓					* *				* *			
MacGillivray's Warbler	U trans	W ¼						*						
Mourning Warbler	C trans	Statewide												
Kentucky Warbler	C trans, Su res	E ⅓												
Common Yellowthroat	C trans, C Su res	Statewide	* * * *					*	* * * * *	* *			* * * * *	* * * * *
Hooded Warbler	R trans	E ¼					*	*	* *					
American Redstart	C trans, Su res	Statewide								*				
Cape May Warbler	R trans	E ¼								*	* * * *			
Cerulean Warbler	R trans, Su res	E ¼						* * * * *	* * * * *	* * * * *	* *			

490

Species	Expected Location	Status
Northern Parula	C, E	C trans, Su res
Magnolia Warbler	Statewide	U trans
Bay-breasted Warbler	E ½	R trans
Blackburnian Warbler	E ⅓	U trans
Yellow Warbler	Statewide	C trans, Su res
Chestnut-sided Warbler	E ¾	U trans
Blackpoll Warbler	Statewide	U trans
Black-throated Blue Warbler	Statewide	R trans
Palm Warbler	E ⅔	U trans
Pine Warbler	E ½	U trans
Yellow-rumped Warbler	Statewide	C trans, U wi res
Yellow-throated Warbler	E ¼	U trans, Su res
Prairie Warbler	Extreme E	R trans, R Su res
Black-throated Gray Warbler	Extreme W	R trans
Townsend's Warbler	W ¼	U trans
Hermit Warbler	1 record *	Accidental
Black-throated Green Warbler	E ⅔	U trans
Canada Warbler	E ½	U trans
Wilson's Warbler	Statewide	C trans
Painted Redstart	7 records	Accidental

Status: C = common, U = uncommon, R = rare, Su = summer, trans = transient (migrant), res = resident (Su res = likely or probable breeder), Wi res = winter resident

Expected Location: N = north, S = south, E = east, W = west, C = central, and standard combinations thereof

Seasonal Distribution: The darker the shading the greater the abundance. * = accidental, vagrant, or rare species and when they've been recorded—not expected every year in this season

For more complete status, please refer to *Birds of Kansas*, Thompson, et al., 2011.

Compiled by Chuck Otte, February 2023 © Chuck Otte

Seasonal Distribution of Kansas Sparrows

Key located at bottom of table

Species	Status	Expected Location	January	February	March	April	May	June	July	August	September	October	November	December
Cassin's Sparrow	U Su res	W									░			
Bachman's Sparrow	Vagrant	(1 specimen)				* *								
Grasshopper Sparrow	C Su res	Statewide	* * * *	* * * * *	* * * *	* ▓	▓	▓	▓	▓	▓	* * ▓	* *	* *
Black-throated Sparrow	Vagrant	S			* *	* * * * *	* *			* *	* * * * *	* *	* * * *	* * * *
Lark Sparrow	C Su res	Statewide				▓	▓	▓	▓	▓	▓	▓	▓	
Lark Bunting	U Su res	W	░	░	░	░							░	░
Chipping Sparrow	C Su res	Statewide	* * * *	* * * * *	* *	▓	▓	▓	▓	▓	▓	▓	* *	* *
Clay-colored Sparrow	C trans	Statewide				░	░				░	░		
Field Sparrow	U Su res	Statewide			▓	▓	▓	▓	▓	▓	▓	▓		
Brewer's Sparrow	R trans	W					░			░	░			
Fox Sparrow	U trans & Wi res	Statewide	░	░	░	░						░	░	░
American Tree Sparrow	A Wi res	Statewide	▓	▓	▓								▓	▓
Dark-eyed Junco	A trans & Wi res	Statewide	▓	▓	▓	▓							▓	▓
Yellow-eyed Junco	Accidental	(1 record)											* * * *	
White-crowned Sparrow	A trans & Wi res	Statewide	▓			▓	▓					▓	▓	▓
Golden-crowned Sparrow	Vagrant	Statewide	* * * *	* * * * *	* * * * *	* * * * *	*					* *	* * * *	* * * *
Harris's Sparrow	A trans & Wi res	Statewide	▓	▓	▓	▓	▓					▓	▓	▓

Species	Status	Expected Location
White-throated Sparrow	C trans & Wi res	E 2/3
Sagebrush Sparrow	Vagrant	SW
Vesper Sparrow	A trans	Statewide
Le Conte's Sparrow	U trans	E 2/3
Nelson's Sparrow	R trans	E 2/3
Baird's Sparrow	R trans	Statewide
Henslow's Sparrow	U Su res	E
Savannah Sparrow	A trans U Wi res	Statewide
Song Sparrow	C trans & Wi res	Statewide
Lincoln's Sparrow	C trans	Statewide
Swamp Sparrow	U trans & Wi res	Statewide
Canyon Towhee	R trans	Morton Co.
Rufous-crowned Sparrow	Casual	SW
Green-tailed Towhee	U trans	W
Spotted Towhee	U Wi res	Statewide
Eastern Towhee	U Su res	E

Status: R = rare, U = uncommon, C = common, A = abundant, res = resident, Casual = sporadic and limited occurance, trans = transient/migrant, Wi = winter, Su = summer

Expected Location: N = north, S = south, E = east, W = west, C = central, and standard combinations thereof

Seasonal Distribution: The darker the shading the greater the abundance. * = accidental, vagrant, or rare species and when they've been recorded—not expected every year in this season

For more complete status, please refer to *Birds of Kansas*, Thompson, et al., 2011.

Compiled by Chuck Otte, February 2023 © Chuck Otte

Complete Species List for Kansas (as of September 1, 2023)

WHISTLING-DUCKS
_____ Black-bellied Whistling-Duck
_____ Fulvous Whistling-Duck

GEESE, SWANS
_____ Snow Goose
_____ Ross's Goose
_____ Greater White-fronted Goose
_____ Brant
_____ Cackling Goose
_____ Canada Goose
_____ Trumpeter Swan
_____ Tundra Swan

DUCKS
_____ Wood Duck
_____ Garganey†
_____ Blue-winged Teal
_____ Cinnamon Teal
_____ Northern Shoveler
_____ Gadwall
_____ Eurasian Wigeon
_____ American Wigeon
_____ Mallard
_____ Mexican Duck†
_____ American Black Duck
_____ Mottled Duck
_____ Northern Pintail
_____ Green-winged Teal
_____ Canvasback
_____ Redhead
_____ Ring-necked Duck
_____ [Tufted Duck]†
_____ Greater Scaup
_____ Lesser Scaup
_____ King Eider†
_____ Common Eider†
_____ Harlequin Duck†
_____ Surf Scoter
_____ White-winged Scoter
_____ Black Scoter
_____ Long-tailed Duck

_____ Bufflehead
_____ Common Goldeneye
_____ Barrow's Goldeneye
_____ Hooded Merganser
_____ Common Merganser
_____ Red-breasted Merganser
_____ Ruddy Duck

QUAILS
_____ Northern Bobwhite
_____ Scaled Quail

TURKEYS, GROUSE, PHEASANTS
_____ Wild Turkey
_____ Ruffed Grouse
_____ Sharp-tailed Grouse
_____ Greater Prairie-Chicken
_____ Lesser Prairie-Chicken
_____ Ring-necked Pheasant

FLAMINGOS
_____ American Flamingo†

GREBES
_____ Pied-billed Grebe
_____ Horned Grebe
_____ Red-necked Grebe
_____ Eared Grebe
_____ Western Grebe
_____ Clark's Grebe

PIGEONS, DOVES
_____ Rock Pigeon
_____ Band-tailed Pigeon†
_____ Eurasian Collared-Dove
_____ Inca Dove
_____ Common Ground Dove
_____ White-winged Dove
_____ Mourning Dove

CUCKOOS
_____ Groove-billed Ani

_____ Greater Roadrunner
_____ Yellow-billed Cuckoo
_____ Black-billed Cuckoo

NIGHTJARS
_____ [Lesser Nighthawk]†
_____ Common Nighthawk
_____ Common Poorwill
_____ Chuck-will's-widow
_____ Eastern Whip-poor-will
_____ [Mexican Whip-poor-will]†

SWIFTS
_____ Chimney Swift
_____ White-throated Swift†

HUMMINGBIRDS
_____ Mexican Violetear†
_____ Rivoli's Hummingbird†
_____ Ruby-throated Hummingbird
_____ Black-chinned Hummingbird
_____ Anna's Hummingbird
_____ Costa's Hummingbird†
_____ Calliope Hummingbird
_____ Rufous Hummingbird
_____ Allen's Hummingbird†
_____ Broad-tailed Hummingbird
_____ Broad-billed Hummingbird†

RAILS, GALLINULES
_____ King Rail
_____ Virginia Rail
_____ Sora
_____ Common Gallinule
_____ American Coot
_____ Purple Gallinule
_____ Yellow Rail
_____ Black Rail

LIMPKINS
_____ Limpkin†

CRANES
_____ Sandhill Crane
_____ Common Crane†
_____ Whooping Crane

STILTS, AVOCETS
_____ Black-necked Stilt
_____ American Avocet

PLOVERS
_____ Black-bellied Plover
_____ American Golden-Plover
_____ Killdeer
_____ Semipalmated Plover
_____ Piping Plover
_____ [Wilson's Plover]†
_____ Mountain Plover
_____ Snowy Plover

SANDPIPERS
_____ Upland Sandpiper
_____ Whimbrel
_____ Eskimo Curlew
_____ Long-billed Curlew
_____ Hudsonian Godwit
_____ Marbled Godwit
_____ Ruddy Turnstone
_____ Red Knot
_____ Ruff
_____ Stilt Sandpiper
_____ Curlew Sandpiper†
_____ Red-necked Stint†
_____ Sanderling
_____ Dunlin
_____ Purple Sandpiper†
_____ Baird's Sandpiper
_____ Little Stint†
_____ Least Sandpiper
_____ White-rumped Sandpiper
_____ Buff-breasted Sandpiper
_____ Pectoral Sandpiper
_____ Semipalmated Sandpiper

_____ Western Sandpiper
_____ Short-billed Dowitcher
_____ Long-billed Dowitcher
_____ American Woodcock
_____ Wilson's Snipe
_____ Spotted Sandpiper
_____ Solitary Sandpiper
_____ Lesser Yellowlegs
_____ Willet
_____ Spotted Redshank†
_____ Greater Yellowlegs
_____ Wilson's Phalarope
_____ Red-necked Phalarope
_____ Red Phalarope

JAEGERS
_____ Pomarine Jaeger†
_____ Parasitic Jaeger
_____ Long-tailed Jaeger†

ALCIDS
_____ Long-billed Murrelet†

GULLS
_____ Black-legged Kittiwake
_____ Sabine's Gull
_____ Bonaparte's Gull
_____ Black-headed Gull
_____ Little Gull
_____ Ross's Gull†
_____ Laughing Gull
_____ Franklin's Gull
_____ Short-billed Gull
_____ Ring-billed Gull
_____ California Gull
_____ Herring Gull
_____ Iceland Gull
_____ Lesser Black-backed Gull
_____ [Glaucous-winged Gull]†
_____ Glaucous Gull
_____ Great Black-backed Gull

TERNS
_____ Least Tern
_____ Gull-billed Tern†
_____ Caspian Tern
_____ Black Tern
_____ Common Tern
_____ [Arctic Tern]†
_____ Forster's Tern
_____ Royal Tern †

SKIMMERS
_____ Black Skimmer†

LOONS
_____ Red-throated Loon
_____ Pacific Loon
_____ Common Loon
_____ Yellow-billed Loon†

STORKS
_____ Wood Stork

FRIGATEBIRDS
_____ Magnificent Frigatebird†

BOOBIES
_____ Brown Booby†

DARTERS
_____ Anhinga

CORMORANTS
_____ Double-crested Cormorant
_____ Neotropic Cormorant

PELICANS
_____ American White Pelican
_____ Brown Pelican

HERONS
_____ American Bittern
_____ Least Bittern
_____ Great Blue Heron

____ Great Egret
____ Snowy Egret
____ Little Blue Heron
____ Tricolored Heron
____ Reddish Egret†
____ Western Cattle Egret
____ Green Heron
____ Black-crowned Night Heron
____ Yellow-crowned Night Heron

IBISES
____ White Ibis
____ Glossy Ibis
____ White-faced Ibis
____ Roseate Spoonbill

VULTURES
____ Black Vulture
____ Turkey Vulture

OSPREYS, HAWKS, KITES, EAGLES
____ Osprey
____ White-tailed Kite†
____ Swallow-tailed Kite
____ Golden Eagle
____ Northern Harrier
____ Sharp-shinned Hawk
____ Cooper's Hawk
____ American Goshawk
____ Bald Eagle
____ Mississippi Kite
____ Harris's Hawk†
____ Gray Hawk†
____ Red-shouldered Hawk
____ Broad-winged Hawk
____ Swainson's Hawk
____ Red-tailed Hawk
____ Rough-legged Hawk
____ Ferruginous Hawk

BARN OWLS
____ American Barn Owl

OWLS
____ [Flammulated Owl]†
____ Western Screech-Owl†
____ Eastern Screech-Owl
____ Great Horned Owl
____ Snowy Owl
____ Burrowing Owl
____ Barred Owl
____ Long-eared Owl
____ Short-eared Owl
____ Northern Saw-whet Owl

KINGFISHERS
____ Belted Kingfisher

WOODPECKERS
____ Lewis's Woodpecker
____ Red-headed Woodpecker
____ Red-bellied Woodpecker
____ Williamson's Sapsucker†
____ Yellow-bellied Sapsucker
____ Red-naped Sapsucker
____ American Three-toed Woodpecker†
____ Downy Woodpecker
____ Ladder-backed Woodpecker
____ Hairy Woodpecker
____ Northern Flicker
____ Pileated Woodpecker

FALCONS
____ Crested Caracara†
____ American Kestrel
____ Merlin
____ Gyrfalcon†
____ Peregrine Falcon
____ Prairie Falcon

FLYCATCHERS
____ Ash-throated Flycatcher
____ Great Crested Flycatcher
____ Great Kiskadee†
____ Piratic Flycatcher†
____ Cassin's Kingbird

_____ Western Kingbird
_____ Eastern Kingbird
_____ Scissor-tailed Flycatcher
_____ [Fork-tailed Flycatcher]†
_____ Olive-sided Flycatcher
_____ Western Wood-Pewee
_____ Eastern Wood-Pewee
_____ Yellow-bellied Flycatcher
_____ Acadian Flycatcher
_____ Alder Flycatcher
_____ Willow Flycatcher
_____ Least Flycatcher
_____ Hammond's Flycatcher
_____ Gray Flycatcher
_____ Dusky Flycatcher
_____ Western Flycatcher
_____ Black Phoebe†
_____ Eastern Phoebe
_____ Say's Phoebe
_____ Vermilion Flycatcher

VIREOS
_____ Black-capped Vireo†
_____ White-eyed Vireo
_____ Bell's Vireo
_____ Gray Vireo†
_____ Yellow-throated Vireo
_____ Cassin's Vireo
_____ Blue-headed Vireo
_____ Plumbeous Vireo
_____ Philadelphia Vireo
_____ Warbling Vireo
_____ Red-eyed Vireo

SHRIKES
_____ Loggerhead Shrike
_____ Northern Shrike

JAYS, MAGPIES, CROWS
_____ Pinyon Jay
_____ Steller's Jay
_____ Blue Jay
_____ Woodhouse's Scrub-Jay

_____ Mexican Jay†
_____ Clark's Nutcracker
_____ Black-billed Magpie
_____ American Crow
_____ Fish Crow
_____ Chihuahuan Raven
_____ Common Raven

CHICKADEES, TITMICE
_____ Carolina Chickadee
_____ Black-capped Chickadee
_____ Mountain Chickadee
_____ [Juniper Titmouse]†
_____ Tufted Titmouse

LARKS
_____ Horned Lark

SWALLOWS
_____ Bank Swallow
_____ Tree Swallow
_____ Violet-green Swallow
_____ Northern Rough-winged Swallow
_____ Purple Martin
_____ Barn Swallow
_____ Cliff Swallow
_____ Cave Swallow

BUSHTITS
_____ Bushtit

KINGLETS
_____ Ruby-crowned Kinglet
_____ Golden-crowned Kinglet

WAXWLNGS
_____ Bohemian Waxwing
_____ Cedar Waxwing

SILKY-FLYCATCHERS
_____ Phainopepla†

498

NUTHATCHES
____ Red-breasted Nuthatch
____ White-breasted Nuthatch
____ Pygmy Nuthatch
____ Brown-headed Nuthatch†

CREEPERS
____ Brown Creeper

GNATCATCHERS
____ Blue-gray Gnatcatcher

WRENS
____ Rock Wren
____ Canyon Wren†
____ Bewick's Wren
____ Carolina Wren
____ Northern House Wren
____ Pacific Wren†
____ Winter Wren
____ Sedge Wren
____ Marsh Wren

THRASHERS
____ Gray Catbird
____ Curve-billed Thrasher
____ Brown Thrasher
____ Sage Thrasher
____ Northern Mockingbird

STARLINGS
____ European Starling

THRUSHES
____ Eastern Bluebird
____ Western Bluebird†
____ Mountain Bluebird
____ Townsend's Solitaire
____ Veery
____ Gray-cheeked Thrush
____ Swainson's Thrush
____ Hermit Thrush
____ Wood Thrush

____ American Robin
____ Varied Thrush

OLD WORLD FLYCATCHERS
____ [Northern Wheatear]†

OLD WORLD SPARROWS
____ House Sparrow

PIPITS
____ American Pipit
____ Sprague's Pipit

FINCHES
____ Brambling†
____ Evening Grosbeak
____ Pine Grosbeak
____ Gray-crowned Rosy-Finch†
____ House Finch
____ Purple Finch
____ Cassin's Finch
____ Common Redpoll
____ Red Crossbill
____ White-winged Crossbill
____ Pine Siskin
____ Lesser Goldfinch
____ American Goldfinch

LONGSPURS
____ Lapland Longspur
____ Chestnut-collared Longspur
____ Smith's Longspur
____ Thick-billed Longspur
____ Snow Bunting

SPARROWS
____ Cassin's Sparrow
____ Bachman's Sparrow†
____ Grasshopper Sparrow
____ Black-throated Sparrow
____ Lark Sparrow
____ Lark Bunting
____ Chipping Sparrow

_____ Clay-colored Sparrow

_____ Field Sparrow

_____ Brewer's Sparrow

_____ Fox Sparrow

_____ American Tree Sparrow

_____ Dark-eyed Junco

_____ Yellow-eyed Junco†

_____ White-crowned Sparrow

_____ Golden-crowned Sparrow

_____ Harris's Sparrow

_____ White-throated Sparrow

_____ Sagebrush Sparrow†

_____ Vesper Sparrow

_____ LeConte's Sparrow

_____ Nelson's Sparrow

_____ Baird's Sparrow

_____ Henslow's Sparrow

_____ Savannah Sparrow

_____ Song Sparrow

_____ Lincoln's Sparrow

_____ Swamp Sparrow

_____ Canyon Towhee†

_____ Rufous-crowned Sparrow

_____ Green-tailed Towhee

_____ Spotted Towhee

_____ Eastern Towhee

CHATS

_____ Yellow-breasted Chat

BLACKBIRDS, ORIOLES

_____ Yellow-headed Blackbird

_____ Bobolink

_____ Eastern Meadowlark

_____ Western Meadowlark

_____ Orchard Oriole

_____ Hooded Oriole†

_____ Bullock's Oriole

_____ Baltimore Oriole

_____ Scott's Oriole†

_____ Red-winged Blackbird

_____ Bronzed Cowbird†

_____ Brown-headed Cowbird

_____ Rusty Blackbird

_____ Brewer's Blackbird

_____ Common Grackle

_____ Great-tailed Grackle

WARBLERS

_____ Ovenbird

_____ Worm-eating Warbler

_____ Louisiana Waterthrush

_____ Northern Waterthrush

_____ Golden-winged Warbler

_____ Blue-winged Warbler

_____ Black-and-white Warbler

_____ Prothonotary Warbler

_____ Swainson's Warbler†

_____ Tennessee Warbler

_____ Orange-crowned Warbler

_____ Nashville Warbler

_____ Virginia's Warbler

_____ Connecticut Warbler

_____ MacGillivray's Warbler

_____ Mourning Warbler

_____ Kentucky Warbler

_____ Common Yellowthroat

_____ Hooded Warbler

_____ American Redstart

_____ Cape May Warbler

_____ Cerulean Warbler

_____ Northern Parula

_____ Magnolia Warbler

_____ Bay-breasted Warbler

_____ Blackburnian Warbler

_____ Yellow Warbler

_____ Chestnut-sided Warbler

_____ Blackpoll Warbler

_____ Black-throated Blue Warbler

_____ Palm Warbler

_____ Pine Warbler

_____ Yellow-rumped Warbler

_____ Yellow-throated Warbler

_____ Prairie Warbler

_____ Black-throated Gray Warbler

_____ Townsend's Warbler

_____ Hermit Warbler†
_____ Black-throated Green Warbler
_____ Canada Warbler
_____ Wilson's Warbler
_____ Painted Redstart†

GROSBEAKS, BUNTINGS
_____ Hepatic Tanager†
_____ Summer Tanager
_____ Scarlet Tanager

_____ Western Tanager
_____ Northern Cardinal
_____ Pyrrhuloxia†
_____ Rose-breasted Grosbeak
_____ Black-headed Grosbeak
_____ Blue Grosbeak
_____ Lazuli Bunting
_____ Indigo Bunting
_____ Painted Bunting
_____ Dickcissel

† Fewer than ten Kansas records
[ ] Hypothetical species

This list was compiled from records of the Kansas Ornithological Society, Kansas Breeding Bird Atlas Project, and Kansas Biological Survey.

483 species

Of the 486 species documented to have occurred in Kansas, three are not included on this list because they no longer occur. The Passenger Pigeon and Carolina Parakeet are extinct, and the Gunnison Sage-Grouse has been extirpated from the state.

The taxonomic sequence and nomenclature used in this list follow the *Check-list of North American Birds*, seventh edition, American Ornithologists Union, 1998, updated through the 63rd Supplement, 2022, *The Auk* 139, no. 3, July 7, 2022, ukac020, https://doiorg/10.1093 /ornithology/ukac020.

Acknowledgments

First and foremost, we want to extend deepest thanks to Audubon of Kansas and the Kansas Ornithological Society for generously providing the financial support needed to publish this second edition, without which it would not have been possible.

Chuck Otte provided advice, graphs, a wealth of data, and indispensable background information. Dallas Hewett created all the maps and tolerated many changes and revisions that the authors asked for during the process. David Rintoul provided text commentary in addition to several outstanding photographs.

Many others provided assistance in a variety of ways, including input and advice on the Kansas hotspot accounts in both this and the original edition of the book (2008). These were Henry Armknecht, Jackie Augustine, Andrew Burnett, Bill Busby, Jeff Calhoun, Kathy Carroll, Henry Castro-Miller, Donna Chance, Kelli Egbert, Tom Ewert, Kat Farres, Chad Gardner, Matt Gearheart, Dan Gish, Malcolm Gold, Kevin Groeneweg, Dave Henness, Thomas Jones, Mark Land, Robert Mangile, Jim Mason, Don Merz, Carol Morgan, Bethany Mowry, Tom Nagel, Chuck Otte, Judd Patterson, Sebastian Patti, Linda Phipps, Jenn Rader, Mike Rader, Ed Raynor, David Seibel, Tom and Sara Shane, John Shuckman, Bill Watson, Lisa Weeks, Dave Williams, and Tracy Wohl.

Special thanks to contributing photographers Roni and LaVern Allen, James Arterburn, Vic Berardi, Lillis Boyer, Alvan Buckley, David Butel, Bruce Hogle, Joseph Miller, Tom Parker, Judd Patterson, Wayne Rhodus, David Rintoul, Steve Rottenborn, and David Seibel for the use of their outstanding photos!

Index to Place Names

Page numbers in boldface denote the principal reference for featured "hotspot" birding locations discussed in the text.

Index to Bird Names

For those species with multiple index entries, page numbers in boldface refer to the primary discussion of the species within the text.